MW00760716

# Mathematical Engineering

**Series editors**

Jörg Schröder, Essen, Germany
Bernhard Weigand, Stuttgart, Germany

Today, the development of high-tech systems is unthinkable without mathematical modeling and analysis of system behavior. As such, many fields in the modern engineering sciences (e.g. control engineering, communications engineering, mechanical engineering, and robotics) call for sophisticated mathematical methods in order to solve the tasks at hand.

The series Mathematical Engineering presents new or heretofore little-known methods to support engineers in finding suitable answers to their questions, presenting those methods in such manner as to make them ideally comprehensible and applicable in practice.

Therefore, the primary focus is—without neglecting mathematical accuracy—on comprehensibility and real-world applicability.

To submit a proposal or request further information, please use the PDF Proposal Form or contact directly: *Dr. Jan-Philip Schmidt, Publishing Editor (jan-philip.schmidt@springer.com).*

More information about this series at http://www.springer.com/series/8445

Livija Cveticanin · Miodrag Zukovic
Jose Manoel Balthazar

# Dynamics of Mechanical Systems with Non-Ideal Excitation

Livija Cveticanin
Donát Bánki Faculty of Mechanical
and Safety Engineering
Óbuda University
Budapest
Hungary

Jose Manoel Balthazar
Mechanical-Aeronautics Division
Aeronautics Technological Institute
São José dos Campos, São Paulo
Brazil

Miodrag Zukovic
Faculty of Technical Sciences
University of Novi Sad
Novi Sad
Serbia

ISSN 2192-4732 ISSN 2192-4740 (electronic)
Mathematical Engineering
ISBN 978-3-319-54168-6 ISBN 978-3-319-54169-3 (eBook)
DOI 10.1007/978-3-319-54169-3

Library of Congress Control Number: 2017931575

© Springer International Publishing AG 2018
This work is subject to copyright. All rights are reserved by the Publisher, whether the whole or part of the material is concerned, specifically the rights of translation, reprinting, reuse of illustrations, recitation, broadcasting, reproduction on microfilms or in any other physical way, and transmission or information storage and retrieval, electronic adaptation, computer software, or by similar or dissimilar methodology now known or hereafter developed.
The use of general descriptive names, registered names, trademarks, service marks, etc. in this publication does not imply, even in the absence of a specific statement, that such names are exempt from the relevant protective laws and regulations and therefore free for general use.
The publisher, the authors and the editors are safe to assume that the advice and information in this book are believed to be true and accurate at the date of publication. Neither the publisher nor the authors or the editors give a warranty, express or implied, with respect to the material contained herein or for any errors or omissions that may have been made. The publisher remains neutral with regard to jurisdictional claims in published maps and institutional affiliations.

Printed on acid-free paper

This Springer imprint is published by Springer Nature
The registered company is Springer International Publishing AG
The registered company address is: Gewerbestrasse 11, 6330 Cham, Switzerland

# Preface

Forced vibrations of the system are, usually, theoretically considered under influence of a periodical force whose frequency and amplitude are constant. The force is assumed to be absolutely independent of the motion of the system. The source of this force is called 'ideal', and the vibrating system is 'ideal'. However, in the real system, the action of the excitation force, which is produced by a particular motor energy source (for example), may depend on the motion of the oscillating system. For that case, the excitation force is the function of the parameters of oscillatory motion. The force and energy source is called 'nonlinear' while the oscillator–non-ideal source system is named 'non-ideal'. The study of non-ideal vibrational systems, where the excitation source is influenced by the dynamics of the driven nonlinear system behavior, has been considered a great challenge in the theoretical and practical research in engineering science. It is worth to be said that the model of non-ideal vibrating system is closer to real situations encountered in practice. Only such a non-ideal model can eliminate the non-correspondence between theoretical predictions of motion with observed properties of the oscillating system. Generally, in the system the power supply is limited and it causes the behavior of the vibrating systems to depart from the case of ideal power supply. For this kind of non-ideal dynamic system, an equation that describes the interaction of the power supply with the driven system must be added. Thus, the non-ideal vibrating systems have one degree of freedom more than the corresponding ideal system.

In this book, dynamics of the non-ideal oscillatory system in which the excitation is influenced by the response of oscillator is considered. Various types of non-ideal systems are investigated: linear and nonlinear oscillators with one or more degrees of freedom interacted with one or more energy sources. For example: oscillating system excited by an elastic connection which is deformed by a crank driven by a rotating motor, system excited by an unbalanced rotating mass which is driven by a motor, system of parametrically excited oscillator and an energy source, frictionally self-excited oscillator and an energy source, energy harvesting system, portal frame–non-ideal source system, non-ideal rotor system, planar mechanism–non-ideal source interaction. For the systems the regular motion and irregular motion are tested. The effect of self-synchronization is discussed. Chaos and methods for suppressing chaos in non-ideal systems are considered. In the book, various types of motion control are suggested. The most important property of the non-ideal system connected with the jump-like transition from a resonant state to a non-resonant one is discussed. The so-called Sommerfeld effect, when the resonant state becomes unstable and the system jumps into a new stable state of motion, which is above the resonant region, is deeply explained. Mathematical model of the system is solved analytically and numerically. Approximate analytical solving procedures are developed. The obtained results are numerically proved. Besides, the simulation of the motion of the non-ideal system is presented. The obtained results are compared with those for the ideal case. Significant difference is evident.

This book aims to present the already known results and to increase the literature in non-ideal vibrating systems. Besides, the intention of the book is to give a prediction of the effects for a system where the interaction between an oscillator and the energy source exists. This book is recommended for engineers and technicians dealing with the problem of source–machine system, and also for Ph.D. students and researchers interested in nonlinear and non-ideal problems.

Budapest, Hungary — Livija Cveticanin
Novi Sad, Serbia — Miodrag Zukovic
São José dos Campos, Brazil — Jose Manoel Balthazar

# Contents


**1 Introduction** .......... 1
References .......... 6

**2 Linear Oscillator and a Non-ideal Energy Source** .......... 9
2.1 Simple Degree of Freedom Oscillator Coupled with a Non-ideal Energy Source .......... 10
2.1.1 Analytical Solving Procedure .......... 12
2.1.2 Steady-State Solution and Sommerfeld Effect .......... 14
2.1.3 Model Analogy and Numerical Simulation .......... 18
2.1.4 Stability Analysis .......... 21
2.2 Oscillator with Variable Mass Excited with Non-ideal Source .......... 22
2.2.1 Model of the System with Variable Mass .......... 23
2.2.2 Model of the System with Constant Mass .......... 25
2.2.3 Comparison of the Systems with Constant and Variable Mass .......... 27
2.3 Oscillator with Clearance Coupled with a Non-ideal Source .......... 30
2.3.1 Model of the System .......... 31
2.3.2 Transient Motion of the System .......... 33
2.3.3 Steady-State Motion of the System .......... 37
2.3.4 Chaotic Motion .......... 42
2.3.5 Chaos Control .......... 44
2.4 Conclusion .......... 45
References .......... 46

**3 Nonlinear Oscillator and a Non-ideal Energy Source** .......... 49
3.1 Nonlinear Oscillator Coupled with a Non-ideal Motor with Nonlinear Torque .......... 50
3.1.1 Nonlinear Motor Torque Property .......... 51
3.1.2 Solution Procedure in General .......... 53
3.1.3 Steady-State Motion and Its Stability .......... 57

3.1.4 Characteristic Points on the Steady State Curves ........ 58
3.1.5 Suppression of the Sommerfeld Effect ................ 59
3.1.6 Conclusion ........................................ 60
3.2 Pure Nonlinear Oscillator and the Motor with Nonlinear Torque ............................................ 60
3.2.1 Approximate Solution Procedure ..................... 63
3.2.2 Steady-State Motion and Its Properties ............. 64
3.2.3 Characteristic Points .............................. 66
3.2.4 Suppression of the Sommerfeld Effect ............... 67
3.2.5 Numerical Examples ................................. 68
3.3 Pure Strong Nonlinear Oscillator and a Non-ideal Energy Source ............................................ 71
3.3.1 Model of the System ................................ 73
3.3.2 Analytical Solving Procedure ....................... 74
3.3.3 Resonant Case and the Averaging Solution Procedure .... 76
3.3.4 Suppression of the Sommerfeld Effect ............... 81
3.3.5 Numerical Examples of Non-ideal Driven Pure Nonlinear Oscillators ............................. 82
3.3.6 Conclusion ......................................... 90
3.4 Stable Duffing Oscillator and a Non-ideal Energy Source ....... 91
3.4.1 Asymptotic Solving Method .......................... 93
3.4.2 Stability of the Steady State Solution and Sommerfeld Effect ............................................ 95
3.4.3 Numerical Simulation and Chaotic Behavior ........... 100
3.4.4 Chaos Control ...................................... 103
3.4.5 Conclusion ......................................... 105
3.5 Bistable Duffing Oscillator Coupled with a Non-ideal Source .... 105
3.5.1 Semi-trivial Solutions and Quenching of Amplitude...... 109
3.5.2 Non-trivial Solutions and Their Stability ............ 110
3.5.3 Conclusion ......................................... 112
Appendix: Ateb Functions ................................... 113
References ................................................. 116

**4 Two Degree-of-Freedom Oscillator Coupled to a Non-ideal Source** ................................................ 121
4.1 Model of the System .................................... 122
4.2 Analytical Solution .................................... 124
4.2.1 Steady-State Motion ................................ 127
4.2.2 Stability Analysis ................................. 129
4.3 Special Cases .......................................... 130
4.3.1 Resonance Frequencies in Orthogonal Directions Are Equal ......................................... 130
4.3.2 Resonance Frequency in One Direction Is Half of the Resonance frequency in Other Direction ......... 134

4.4 Numerical Simulation . . . . . . . . . . 137
4.5 Conclusions . . . . . . . . . . 139
References . . . . . . . . . . 140

**5 Dynamics of Polymer Sheets Cutting Mechanism** . . . . . . . . . . 141
5.1 Structural Synthesis of the Cutting Mechanism . . . . . . . . . . 143
5.1.1 Comparison of the Simple, Eccentric and Two Slider-Crank mechanisms . . . . . . . . . . 145
5.2 Kinematics of the Cutting Mechanism . . . . . . . . . . 146
5.3 Dynamic Analysis of the Mechanism with Rigid Support . . . . . . . . . . 147
5.3.1 Mathematical Model of the Mechanism . . . . . . . . . . 147
5.3.2 Numerical Simulation . . . . . . . . . . 152
5.3.3 Analytical Consideration . . . . . . . . . . 153
5.3.4 Comparison of Analytical and Numerical Results . . . . . . . . . . 155
5.4 Dynamics of the Cutting Mechanism with Flexible Support and Non-ideal Forcing . . . . . . . . . . 155
5.4.1 Mathematical Model of Motion of the Cutting Mechanism . . . . . . . . . . 156
5.4.2 Ideal Forcing Conditions . . . . . . . . . . 161
5.4.3 Non-ideal Forcing Conditions . . . . . . . . . . 163
5.4.4 Non-stationary Motion . . . . . . . . . . 168
5.5 Conclusion . . . . . . . . . . 170
References . . . . . . . . . . 171

**6 Non-ideal Energy Harvester with Piezoelectric Coupling** . . . . . . . . . . 173
6.1 Constitutive Equation of the Piezoceramic Material . . . . . . . . . . 175
6.2 Harvesting System with Ideal Excitation . . . . . . . . . . 176
6.2.1 Analytical Procedure . . . . . . . . . . 179
6.2.2 Harvester with Linear Piezoelectricity . . . . . . . . . . 182
6.2.3 Harvester with Nonlinear Piezoelectricity . . . . . . . . . . 185
6.3 Harvesting System with Non-ideal Excitation . . . . . . . . . . 187
6.3.1 Model of the Non-ideal Mechanical System with Harvesting Device . . . . . . . . . . 187
6.3.2 Analytical Solving Procedure . . . . . . . . . . 192
6.3.3 Steady-State Motion . . . . . . . . . . 193
6.3.4 Harvested Energy . . . . . . . . . . 196
6.3.5 Comparison of the Analytical and Numerical Solutions . . . 197
6.3.6 Linear Energy Harvester . . . . . . . . . . 199
6.3.7 Nonlinear Energy Harvesting . . . . . . . . . . 200
6.3.8 Conclusion . . . . . . . . . . 200
6.4 Harvester with Exponential Type Non-ideal Energy Source . . . . . . . . . . 201
6.4.1 Numerical Simulation Results . . . . . . . . . . 203
6.4.2 Linear Energy Harvesting . . . . . . . . . . 204
6.4.3 Nonlinear Energy Harvesting . . . . . . . . . . 205

6.4.4 Chaos in the System . . . . . . . . 206
6.4.5 Control of the System . . . . . . . . 207
6.4.6 Conclusion . . . . . . . . 209
6.5 Non-ideal Portal Frame Energy Harvester Controlled with a Pendulum . . . . . . . . 211
6.5.1 Numerical Simulation . . . . . . . . 214
6.5.2 Conclusion . . . . . . . . 218
References . . . . . . . . 218

**7 Instead Conclusions: Emergent Problems in Nowadays and Future Investigation** . . . . . . . . 221

**Index** . . . . . . . . 227

# Chapter 1
# Introduction

In the real coupled system, which contains an energy source and a system which is supplied with power from that source, the energy source acts on the system but the system supplied with power from the source also affects the motion of the source. Such systems in which there is the interaction in motion between the energy source and the working element, for example an oscillator, is named 'non-ideal system' (Nayfeh and Mook 1979). The source which is influenced by the response of the system to which it supplies power is the 'non-ideal source'. Examples of non-ideal sources are brushed DC electric motors, induction motors, drives with dissipative couplings or any kind of load dependent slip, etc. Most often, we find that drives are assumed to be ideal. Such ideal drive assumption is influenced by two factors: a motive to simplify the problem or the assumption that the operating range of the system is so limited that there is sufficient power available from the drive and thus, the drive remains uninfluenced (or marginally influenced) by the system's dynamics. Nevertheless, most forced vibrating systems are non-ideal. The dynamic behavior of a non-ideal system may significantly differ from the corresponding idealized case when the power supply to the system becomes limited. Our aim is to consider the non-ideal vibrating systems.

The study of non-ideal vibrational systems, where the excitation source is influenced by the dynamics of the driven nonlinear system behavior, has been considered a great challenge in the theoretical and practical research in engineering science. When the excitation is not influenced by this behavior, such systems are known as ideal systems or systems with ideal power supply. The behavior of ideal systems is very well known in recent literature, but there are fewer results devoted to non-ideal systems. A revision of non-ideal problems has been published in Balthazar et al. (2003, 2004a), Balthazar and Pontes (2005), Cveticanin (2010). For this kind of non-ideal dynamic system an equation that describes the interaction of the power supply with the driven system must be added. In comparison, therefore, a non-ideal system has one more degree of freedom. Moreover, the differential equations of the non-ideal system become strongly non-linear. Although we find the non-ideal

© Springer International Publishing AG 2018

L. Cveticanin et al., *Dynamics of Mechanical Systems with Non-Ideal Excitation*, Mathematical Engineering, DOI 10.1007/978-3-319-54169-3_1

supposition to be a more realistic viewpoint, there is not so a large number of studies on non-ideal systems (Sommerfeld 1902; Kononenko 1964; Nayfeh and Mook 1979; Alifov and Frolov 1990). Most of those deal with the dynamics of a system composed of an unbalanced electric motor placed on a flexible support or its different variations (Sommerfeld 1902; Kononenko 1964; Blekhman 2000).

The study of interaction between motors and structure is not new. Probably the first notice about dynamics of motor - structure system was given by A Sommerfeld in Sommerfeld (1902). He proposed an experiment of a motor mounted on a flexible wooden table and observed that the energy supplied to the motor was converted in the form of table vibration instead of being converted to increase angular velocity of the motor. This observation was used to explain a class of motors called non-ideal energy sources. Later, Laval was the first to perform an experiment with a steam turbine to observe that quick passage through critical speed would reduce significantly the levels of vibration when compared to steady state excitation. It has been shown that under certain conditions, the structural vibration of the system, which is excited by a non-ideal drive, may act like an energy sink, i.e. instead of the drive energy being spent to increase the drive speed, a major part of that energy is diverted to vibrate the structure. This is formally known as the Sommerfeld effect. Usually, the Sommerfeld effect concerns the dynamics of a system, which is composed of an unbalanced electric motor placed on a flexible support. As the energy supply to the motor in this system is gradually increased, the rotational speed of the motor increases until it approaches the frequency of foundation and thereafter it remains stuck there for some more increase in power input. Upon exceeding a critical limit of power input, the rotational speed of the motor suddenly jumps to a much higher value and the amplitude of structural vibration jumps to a much lower value. Similar phenomena are observed when the power is gradually reduced. However, when the power is gradually reduced, the transition points are not the same as those observed during gradual increase of the power input. Therefore, the form of the resonance curve depends on the direction of the gradual variation of the frequency of the excitation and it is impossible to realize certain motor speeds near the resonance frequency. Detailed studies on these characteristic jump-phenomena are given by Balthazar et al. (2002), Dantas and Balthazar (2003), Bolla et al. (2007) and Felix and Balthazar (2009). Sommerfeld effect is also observed in rotor dynamics. This is because of the dependence of the flexural vibrations (due to the unbalance forces or circulatory forces) on the motion of the energy source. This dependence is manifested as a coupling between the differential equations of whirling motion of the rotor–shaft system and the spin rate of rotor–shaft connected to the source (Balthazar et al. 2004a). In such a paradigm, under certain conditions, the energy supplied by the source to the flexible spinning shaft is spent to excite the bending modes rather than to increase the drive speed (Vernigor and Igumnov 2003). The unbalance force and the circulatory force (the force produced due to rotating internal damping) depend on the speed of the motor and thus both these forces load the drive, i.e., they draw power from the source. The power supply to a non-ideal system becomes limited when the motor speed approaches the resonance frequency or the stability threshold. While it is impossible to pass through the stability threshold, the driving power of motor decides whether it is possible to pass through the resonance. Sometimes the passage through resonance requires more input power than that can be provided by the source. As a consequence, the vibrating system cannot pass the resonance and remains stuck in the resonance conditions. If the motor power is insufficient then as the motor speed approaches resonance during the coast up operation, the power transmission to the structural vibrations increases and the motor cannot accelerate sufficiently which is why the system gets stuck in resonance. On the other hand, if there is sufficient motor power to cross the resonance then a jump phenomenon occurs from near resonance speed to a much higher motor speed (Balthazar et al. 2004b). A detailed study on such non-ideal problems concerning passage through resonance was presented also in Timoshenko (1961) and further results have been reported by Balthazar et al. (2001, 2003, 2004a). Depending upon the motor power, one may get almost smooth passage, passage with a temporary slowdown followed by acceleration, and complete capture at the resonance (Dimentberg et al. 1997). Various other known systems exhibit similar phenomena (see Hübner 1965; Frolov and Krasnopolskaya 1987; Rand 1992; Ryzhik et al. 2001; Zukovic and Cveticanin 2007). However, the motor speed cannot exceed the stability threshold. This is another kind of Sommerfeld effect where it is possible that a strong interaction results in fluctuating motor speed and large vibration amplitudes (Mukherjee et al. 1999). At the stability threshold, any extra power supplied to the motor is spent in exciting the structure, i.e., increasing the amplitude of the structural vibrations.

Finally, the jump phenomena in the vibration amplitude and the increase of the power required by the source to operate next to the system resonance are both manifestations of a non-ideal problem. This phenomenon suggests that the vibratory response of the non-ideal system emulates an 'energy sink' in the regions next to the system resonance, by transferring the power from the source to vibrates of the support structure, instead of the speeding up the driver machine. In other words, one of the problems confronted by mechanical engineers is how to drive a system through a system resonance and avoid this 'disappearance of energy'. Considering a DC motor, usually the angular velocity increases according to the power supplied to the source. However, near the resonance, with additional energy the mean angular velocity of the DC motor remains unchanged until it suddenly jumps to a much higher value upon exceeding a critical input power. Simultaneously, the amplitude of oscillations of the excited system jumps to a much lower value. Before the jump, the non-ideal oscillating system can not pass through the resonance frequency of the system or requires an intensive interaction between the vibrating system and the energy source to be able to do so (Goncalves et al. 2014).

The Sommerfeld effect is studied also in the two degrees-of-freedom systems next to the 1:1 resonance by Tsuchida et al. (2003). Using the linear and nonlinear curves of a DC torque motor, they have observed an increase of the vibration amplitude of a two degrees-of-freedom mechanical structure excited by an eccentric mass in the motor axis. The same phenomenon was also observed near the 1:2 subharmonic resonance (Tsuchida et al. 2005). In these works the Sommerfeld effect occurs due to the nonlinearity of the non-ideal coupling between the structure and the driver motor, resulting in large vibration of the structure supporting the motor. Zniber and

Quinn (2006) investigated the dynamic behavior of a two-degree-of-freedom oscillator connected with a non-ideal source for 1:1 resonance not only analytically but also experimentally. An investigation on the influence of the two orthogonal resonance frequencies is presented. Considering the previous analysis of such systems, Palacios et al. (2002) applied the Bogoliubov averaging method to study of the vibrations of an elastic foundation, forced by a non-ideal energy source. Goncalves et al. (2016) studied the dynamics of complex structure consisting of a portal frame coupled to a non-ideal unbalanced motor. The planar portal frame supporting a direct current motor with limited power was with quadratic nonlinearity and the case of internal resonance 1:2 was analyzed.

Quinn (1997) presented a complete study of the conditions of resonance capture in a three degree-of-freedom system modeling the dynamics of an unbalance rotor subject to a small constant torque supported by orthogonal, linearly elastic supports which is constrained to move in the plane. In the physical system the resonance exists between translational motions of the frame and the angular velocity of the unbalanced rotor.

In the literature the non-ideal excitation in continual systems is also treated. Thus, in terms of continuous systems with coupled motors, Krasnopolskaya (2006) studied an infinite plate immersed in an acoustic medium. The plate was subject to a point excitation by an electric motor of limited power supply and it was shown that chaos might occur in the system due to the feedback influence of waves in the infinite hydro-elastic subsystem in the regime of motor shaft rotation.

The highly nonlinear interactions make it difficult to perform analytical studies on transient behavior due to non-ideal vibrations. Moreover, most of studies on dynamics of non-ideal systems are based on approximate solutions and numerical studies. Numerical studies show that in these nonlinear system chaotic motion may appear. Belato et al. (2001) obtained this phenomena for certain conditions in a vertical pendulum, whose base is actuated horizontally through a slider crank mechanism, where the crank is driven through a DC motor. Investigations on the properties of the transient response of this nonlinear and non-ideal problem showed that near the fundamental resonance region, the system may exhibit multi-periodic, quasi-periodic, and chaotic motion. It was later shown that the loss of stability occurs by a sequence of events, which include intermittence and crisis, when the system reaches a chaotic attractor. A characteristic boundary-crisis feature in the bifurcation diagram of this system has been attributed to the non-ideal supposition made in the mathematical model (Belato et al. 2005).

A group of researchers treated the problem of suppressing chaotic motion (Souza et al. 2005) and has been also focused on control of the passage through resonance with a non-ideal source (Eckert 1996; Wauer and Suherman 1997). Thus, Samantaray et al. (2010) show that direct or external damping and the rotating internal damping significantly influence the dynamics of non-ideal systems. To mitigate the effects of chaos and to damp the nonlinear vibrations, several authors have proposed the use of passive and active control methods. In Felix et al. (2005) a nonlinear control method is studied based on the phenomenon of mode saturation which is applied to a portal frame support and unbalanced motor with limited power. A passive control scheme is proposed. Using a tuned liquid column damper, a special type of damper relying on the internal motion of a column of liquid in a U-tube container is forced to counteract the forces acting on the structure. These authors obtained good results, controlling the structural vibrations in the resonant regions and showing that the successes of the damper applications depends on the variation of fluid resistance and inertness. Active control methods include the use of a linear and nonlinear electromechanical vibration absorber (Felix and Balthazar 2009) as controlled damping devices. The application of nonlinear damp control has been effective to reduce the vibration amplitude of the non-ideal system, showing a considerable reduction of the jump and resonance capture phenomena. Castao et al. (2010) use a semi-active method to mitigate the Sommerfeld effect in non-ideal systems by nonlinear controlled damping with magnetorheological fluids. The non-ideal system consists of a mass connected to a support by a nonlinear controlled spring and a magnetic rheological damper. The block mass supports a DC motor that is powered by a continuous current source with an eccentric mass in its axis. The non-ideal behavior of the system is due to the interaction between the block and the DC motor.

The book has six chapters. After the introduction, in the Chap. 2 the dynamics of a linear oscillator driven with a nonlinear energy source is considered. The oscillators with constant and variable mass are investigated. The system is mathematically modeled. An analytical procedure for solving of the equations is developed. The conditions of steady-state motion are determined and the stability of motion is discussed. The Sommerfeld effect is also explained. In this chapter the numerical simulation to the problem is done. Analytical and numerical solutions are compared. As a special problem the dynamics of the oscillator with clearance excited with a non-ideal source is investigated. The transient motion is clarified. The attention is given to deterministic chaos in the system and to chaos control.

In the Chap. 3 the dynamics of the nonlinear oscillator - non-ideal energy source system is investigated. Beside linear, the influence of the nonlinear torque on the motion of the system is considered. The Sommerfeld effect and its suppression are explained. As a special type of nonlinear oscillators, the Duffing oscillator with cubic nonlinearity is assumed. For the Duffing oscillator - non-ideal energy system the Sommerfeld phenomena, chaos and chaos control are presented. Besides, the properties of the bistable Duffing oscillator coupled with non-ideal source is considered. Semi-trivial solutions and quenching of amplitude is discussed. Non-trivial solutions are obtained and their stability is analyzed. In this chapter the motion of the pure nonlinear oscillator excited with torque which is a linear or nonlinear function of the angular velocity is presented. The resonant motion is considered. Using the analytical averaging solving procedure the characteristic values for Sommerfeld effect is obtained. Procedures for suppressing chaos are shown. Some numerical examples are calculated. The chapter ends with an appendix about the Ateb functions (inverse Beta functions) which are the mathematical solutions of the pure nonlinear equations.

The considerations given for the one degree-of-freedom oscillator are extended to two degree-of-freedom oscillator in the Chap. 4. Analytical and numerical solving procedures are developed. Steady state solutions are obtained and their stability is discussed. We considered two special cases: (a) resonant frequencies in orthogonal directions are equal and (b) resonance frequency in one direction is half of the resonance frequency in other direction.

As it is well known, the effect of non-ideal source is evident in real machines and mechanism. It is the reason that in Chap. 5 we investigate the dynamics of a polymer sheets cutting mechanism. Various structures of mechanism are compared: the simple, eccentric and the slider-crank mechanism. Kinematics and dynamics analysis of the ideally and non-ideally forced systems supported with rigid and flexible supports is investigated. Advantages and disadvantages of systems are presented.

In Chap. 6, the non-ideal energy harvester with piezoelectric coupling is investigated. The mathematical model of the non-ideal energy harvester is given. The harvester system with ideal and non-ideal excitation is considered. Electric power is harvested from the mechanical component. The harvester is with linear or nonlinear piezoelectricity. Harvester is of beam type or in the form of a portal frame. Analytical and numerical consideration of the system is introduced. Both, linear and nonlinear harvesting systems are evident. Besides the steady state motion, chaotic motion is realized. The motion of the system has to be controlled. The non-ideal portal frame energy harvester is suggested to be controlled with a pendulum.

All of the chapters end with the reference list.

## References

Alifov, A. A., & Frolov, K. V. (1990). *Interaction of Non-linear Oscillatory Systems with Energy Sources*. London: Taylor & Francis.

Balthazar, J. M., Cheshankov, B. I., Rushev, D. T., Barbanti, L., & Weber, H. I. (2001). Remarks on the passage through resonance of a vibrating system, with two degree of freedom. *Journal of Sound and Vibration*, *239*(5), 1075–1085.

Balthazar, J. M., Mook, D. T., Brasil, R. M. L. R. F., Fenili, A., Belato, D., Felix, J. L. P., et al. (2002). Recent results on vibrating problems with limited power supply. *Meccanica*, *330*(7), 1–9.

Balthazar, J. M., Mook, D. T., Weber, H. I., Brasil, R. M. I. R. F., Fenili, A., Beltano, D., et al. (2003). An overview on non-ideal vibrations. *Meccanica*, *38*(6), 613–621.

Balthazar, J. M., Brasil, R. M. L. R. F., Weber, H. I., Fenili, A., Belato, D., Felix, J. L. P., & Garzeri, F. J. (2004a). A review of new vibrating issues due to non-ideal energy sources. In: F. Udwadia, H. I. Weber, & G. Leitmann (Eds.), Dynamics Systems and Control, Stability and Control Theory. Methods and Applications, vol. 22 (pp. 237–258). London: Chapman & Hall.

Balthazar, J. M., Brasil, R. M. L. R. F., & Garzeri, F. J. (2004b). On non-ideal simple portal frame structural model: Experimental results under a non-ideal excitation. *Applied Mechanics and Materials*, *1–2*, 51–58.

Balthazar, J. M., & Pontes, B. R, Jr. (2005). On friction induced nonlinear vibrations: a source of fatigue. In D. Inman (Ed.), *Damage Prognosis - Fro Aerospace, Civil and Mechanical System*. New York: Wiley.

Belato, D., Weber, H. I., Balthazar, J. M., & Mook, D. T. (2001). Chaotic vibrations of a nonideal electro-mechanical system. *International Journal of Solids Structstructure*, *38*, 1699–1706.

Belato, D., Weber, H. I., & Balthazar, J. M. (2005). Using transient and steady state considerations to investigate the mechanism of loss of stability of a dynamical system. *Applied Mathematics and Computation*, *164*, 605–613.

Blekhman, I. I. (2000). *Vibrational Mechanics: Nonlinear Dynamic Effects, General Approach, Applications*. Singapore: World Scientific.

Bolla, M. R., Balthazar, J. M., Felix, J. L. P., & Mook, D. T. (2007). On an approximate analytical solution to a nonlinear vibrating problem, excited by a nonideal motor. *Nonlinear Dynamics, 50*(4), 841–847.

Castao, K. A. L., Goes, C. S., & Balthazar, J. M. (2010). A note on the attenuation of the Sommerfeld effect of a non-ideal system taking into account a MR damper and the complete model of a DC motor. *Journal of Vibration and Control, 17*(7), 112–1118.

Cveticanin, L. (2010). Dynamic of the non-ideal mechanical systems: A review. *Journal of the Serbian Society for Computational Mechanics, 4*(2), 75–86.

Dantas, M. J. H., & Balthazar, J. M. (2003). On the appearance of a Hopf bifurcation in a non-ideal mechanical problem. *Mechanics Research Communications, 30*(5), 493–503.

De Souza, S. L. T., Caldas, I. L., Viana, R. L., Balthazar, J. M., & Brasil, R. M. L. R. F. (2005). Impact dampers for controlling chaos in systems with limited power supply. *Journal of Sound and Vibration, 279*(3–5), 955–967.

Dimentberg, M. F., McGovern, L., Norton, R. L., Chapdelaine, J., & Harrison, R. (1997). Dynamics of an unbalanced shaft interacting with a limited power supply. *Nonlinear Dynamics, 13*, 171–187.

Eckert, M. (1996). The Sommerfeld effect: theory and history of a remarkable resonance phenomenon. *European Journal of Physics, 17*(5), 285–289.

Felix, J. L. P., Balthazar, J. M., & Brasil, R. M. (2005). On saturation control of a non-ideal vibrating portal frame foundation type shear-building. *Journal of Vibration and Control, 11*(1), 121–136.

Felix, J. L. P., & Balthazar, J. M. (2009). Comments on a nonlinear and nonideal electromechanical damping vibration absorber, Sommerfeld effect and energy transfer. *Nonlinear Dynamics, 55*(1–2), 1–11.

Frolov, K. V., & Krasnopolskaya, T. S. (1987). Sommerfeld effect in systems without internal damping. *Prikladnaya Mekhanika, 23*(12), 19–24.

Goncalves, P. J. P., Silveira, M., Pontes, B. R, Jr., & Balthazar, J. M. (2014). The dynamic behavior of a cantilever beam coupled to a non-ideal unbalanced motor through numerical and experimental analysis. *Journal of Sound and Vibration, 333*(2), 5115–5129.

Goncalves, P. J. P., Silveira, M., Petrocino, E. A., & Balthazar, J. M. (2016). Double resonance capture of a two-degree-of-freedom oscillator coupled to a non-ideal motor. *Meccanica, 51*(9), 2203–2214.

Hübner, W. (1965). Die Wechsetwirkung zwischen Schwinger und Antrieb bei Schwingungen. *Ingenieur-Archive, 34*, 411–422.

Kononenko, V. O. (1964). *Vibrating Systems with Limited Excitation*. Moscow (in Russian): Nauka.

Krasnopolskaya, T. (2006). Chaos in acoustic subspace raised by the Sommerfeld-Kononenko effect. *Meccanica, 41*(3), 299–310.

Mukherjee, A., Karmakar, R., & Samantaray, A. K. (1999). Modelling of basic induction motors and source loading in rotor-motor systems with regenerative force field. *Simulation Practice and Theory, 7*, 563–576.

Nayfeh, A. H., & Mook, D. T. (1979). *Nonlinear Oscilltions*. New York: Wiley-Interscience.

Palacios, J., Balthazar, J., & Brasil, R. (2002). On non-ideal and non-linear portal frame dynamics analysis using Bogoliubov averaging method. *Journal of Brazilian Society of Mechanical Science, 24*(4), 257–265.

Quinn, D. D. (1997). Resonance capture in a three degree-of freedom mechanical system. *Nonlinear Dynamics, 14*(4), 309–333.

Rand, R. H., Kinsey, R. J., & Mingori, D. L. (1992). Dynamics of spinup through resonance. *International Journal of Non-Linear Mechanincs, 27*(3), 489–502.

Samantaray, A. K., Dasgupta, S. S., & Bhattacharyya, R. (2010). Sommerfeld effect in rotationally symmetric planar dynamical systems. *International Journal of Engineering Science, 48*, 21–36.

Ryzhik, A., Amer, T., Duckstein, H., & Sperling, L. (2001). Zum Sommerfeldeffect beim selbsttätigen Auswuchten in einer Ebene. *Technische Mechanik, 21*(4), 297–312.

Sommerfeld, A. (1902). Beiträge zum dynamischen Ausbau der Festigkeitslehe. *Physikal Zeitschrift, 3*, 266–286.

Timoshenko, S. (1961). *Vibration Problems in Engineering*. New York: Van Nostrand, Princeton.

Tsuchida, M., Guilherme, K. L., Balthazar, J. M., Silva, G. N., & Chechancov, B. I. (2003). On regular and irregular vibrations of a non-ideal system with two degrees of freedom. 1:1 resonance. *Journal of Sound and Vibration*, *260*, 949–960.

Tsuchida, M., Guilherme, K. L., & Balthazar, J. M. (2005). On chaotic vibrations of a non-ideal system with two degree of freedom. 1:2 resonance and Sommerfeld effect. *Journal of Sound and Vibration*, *282*, 1201–1207.

Vernigor, V. N., & Igumnov, I. N. (2003). Sommerfeld effect study on the basis of rotating rotor mechanical model. *Problemy Mashinostraeniya i Nadezhnos'ti Mashin*, *3*, 3–8.

Wauer, J., & Suherman, S. (1997). Vibration suppression of rotating shafts passing through resonances by switching shaft stiffness. *Journal of Vibration and Acoustics*, *120*, 170–180.

Zniber, A., & Quinn, D. D. (2006). Resonance capture in a damped three-degree-of-freedom system: Experimental and analytical comparison. *International Journal of Non-Linear Mechanics*, *41*(10), 1128–1142.

Zukovic, M., & Cveticanin, L. (2007). Chaotic responses in a stable duffing system of non-ideal type. *Journal of Vibration and Control*, *13*(6), 751–767.

# Chapter 2
# Linear Oscillator and a Non-ideal Energy Source

In the non-ideal oscillator-motor system there is an interaction between the motions of the oscillator and of the motor: the motor has an influence on the oscillator and vice versa the motion of the oscillator affects the motion of the motor. It is in contrary to the ideal system, where only the motor has an influence on the oscillator motion and the influence of the oscillator on the motor is negligible (Kononenko 1969, 1980; Nayfeh and Mook 1979). In this chapter the linear oscillator coupled to a non-ideal energy source is considered. A significant number of researches in dynamics of linear oscillator coupled with the non-ideal energy source is already done (see overviews Balthazar et al. 2003; Cveticanin 2010; References given in these papers and, recently published papers Souza et al. 2005a, b; Dantas and Balthazar 2007; Felix et al. 2009, 2011; Samantaray 2010; Kovriguine 2012; Tusset et al. 2012a, b etc). In the non-ideal system with linear oscillator the connection of the system with the fixed element is with an elastic element with linear property. Usually, a motor is supported on a cantilever beam which has linear properties or a motor is settled on the linear foundation (Dimentberg et al. 1997; Warminski et al. 2001). In the literature the vibration of the system is determined analytically and the result is compared with numerically obtained value. Discussing the results the special attention is given to the phenomenon called 'Sommerfeld effect' which is a property of the non-ideal systems.

The chapter is divided into three sections where three types of oscillator-motor systems are considered: the one degree-of-freedom linear oscillator connected with a non-ideal energy source, oscillator with variable mass excited with anon-ideal source and the oscillator with clearance.

In the Sect. 2.1, a motor supported on a cantilever beam with linear elastic properties is considered. The system is modelled and an analytical solving procedure is developed for obtaining of the approximate solutions. The steady-state motion in the resonant working regime is given and the Sommerfeld effect is explained. An analog mechanical model is introduced for better explanation of the problem

© Springer International Publishing AG 2018

L. Cveticanin et al., *Dynamics of Mechanical Systems with Non-Ideal Excitation*, Mathematical Engineering, DOI 10.1007/978-3-319-54169-3_2

(Goncalves et al. 2014). Conditions for motion stability of the non-ideal system with linear oscillator are determined.

In the Sect. 2.2, the oscillator with variable mass with non-ideal excitation is investigated. The system is described with two coupled equations with time variable parameters. The analytical and numerical solution of the problem is considered. The influence of the parameter variation on the behavior of the system is discussed.

Section 2.3 deals with an oscillator with clearance forced with a non-ideal source. The elastic force in the spring is discontinual and it causes some additional disturbances in the motion of the system. Both, the transient and the steady-state motion of the system are investigated. In this system beside regular motion, chaotic motion is evident (Lin and Ewins 1993). Conditions for chaos are obtained and a method for chaos control is developed (Zukovic and Cveticanin 2009).

## 2.1 Simple Degree of Freedom Oscillator Coupled with a Non-ideal Energy Source

Let us consider a motor settled on a table where it is supposed that the motor is a non-ideal energy source while the support represents an oscillator. The model of a motor - support system is usually modelled as a cantilever beam with a concentrated mass positioned at its free end (see Fig. 2.1).

The beam is made of steel and its properties are defined by Young's modulus $E$, density $\rho$, the length $L$, the cross-sectional area $S$ and the second moment of area $I$. The beam bending stiffness is $k = 3EI/L^3$, and its mass is $m_b = \rho SL$. If $M_c$ represents the mass of the electric motor, the total concentrated mass at the end of the beam is $m_1 = M_c + 0.23m_b$ (Goncalves et al. 2014). The motor has moment of inertia $J$. It is known that the rotor of the motor suffers from unbalance. Rotor is never perfectly balanced and the unbalance mass is $m_2 = m_b$ that rotates at a distance $d$ from the motor shaft center. For low amplitudes of oscillation when the higher order modes are neglected, it is possible to write an expression for the first bending natural frequency $\omega_0$ of the beam with concentrated mass as

$$\omega_0 = \sqrt{\frac{3EI}{L^3(m_1 + m_2)}}. \tag{2.1}$$

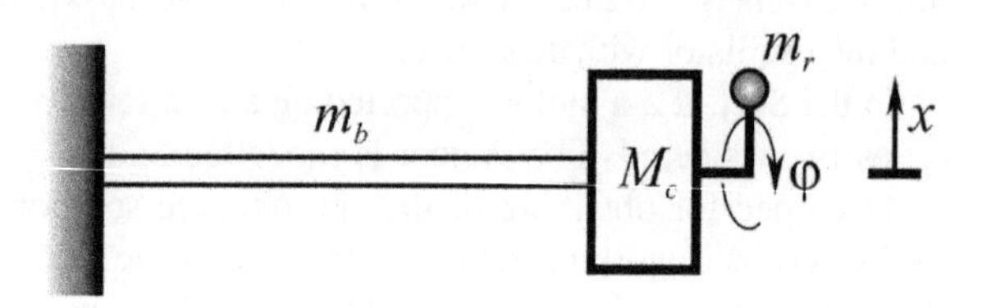


**Fig. 2.1** Cantilever beam with a concentrated mass and an unbalanced motor

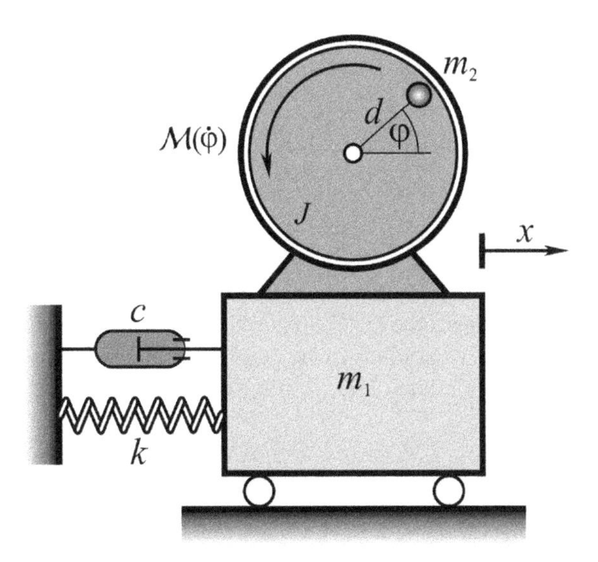


**Fig. 2.2** Model of a spring-mass-damper oscillator and a motor with unbalanced mass

Due to rigidity and damping properties of the beam, it can be modified as a spring-mass-damper oscillator. Then, the cantilever beam - motor system is represented as a cart with mass $m_1$, connected with a spring and damper to the fixed plane, and coupled with unbalanced mass (Fig. 2.2).

The system has two degrees of freedom. The cart displacement is defined by $x$ and the motor angular position is represented by $\varphi$. The kinetic energy of the system $T$ is

$$T = \frac{m_1}{2}\dot{x}^2 + \frac{J}{2}\dot{\varphi}^2 + \frac{m_2}{2}v_2^2, \tag{2.2}$$

where $v_2$ is the velocity of the unbalance mass. For position coordinates of the motor unbalanced mass $m_2$: $x_2 = x + d\cos\varphi$ and $y_2 = d\sin\varphi$, the velocity follows as

$$v_2 = \sqrt{\dot{x}_2^2 + \dot{y}_2^2} = \sqrt{\dot{x}^2 + d^2\dot{\varphi}^2 - 2d\dot{x}\dot{\varphi}\sin\varphi}. \tag{2.3}$$

Substituting (2.3) into (2.2) we obtain

$$T = \frac{1}{2}(m_1 + m_2)\dot{x}^2 + \frac{1}{2}(J + m_2 d^2)\dot{\varphi}^2 - m_2 d\dot{x}\dot{\varphi}\sin\varphi. \tag{2.4}$$

If the gravity potential energy is neglected, the systems potential energy is

$$U = \frac{1}{2}kx^2. \tag{2.5}$$

Equations of motion of the system are obtained by using the Lagrange's differential equations of motion

$$\frac{d}{dt}\frac{\partial T}{\partial \dot{x}} - \frac{\partial T}{\partial x} + \frac{\partial U}{\partial x} = Q_x,$$
$$\frac{d}{dt}\frac{\partial T}{\partial \dot{\varphi}} - \frac{\partial T}{\partial \varphi} + \frac{\partial U}{\partial \varphi} = Q_\varphi, \tag{2.6}$$

where $Q_x$ and $Q_\varphi$ are generalized forces. The non-conservative force in $x$ direction is the damping force $Q_x = -c\dot{x}$ with the damping coefficient $c$, while the generalized force $Q_\varphi$ corresponds to the torque $\mathcal{M}(\dot{\varphi})$ applied to motor.

Using relations (2.4)–(2.6) and the generalized force, the cart equation of motion is

$$(m_1 + m_2)\,\ddot{x} + c\dot{x} + kx = m_2 d\left(\ddot{\varphi}\sin\varphi + \dot{\varphi}^2\cos\varphi\right), \tag{2.7}$$

and the motion of the motor shaft is given by

$$\left(J + m_2 d^2\right)\ddot{\varphi} = m_2 d\ddot{x}\sin\varphi + \mathcal{M}\left(\dot{\varphi}\right). \tag{2.8}$$

Equations (2.7) and (2.8) are autonomous and nonlinear.

### 2.1.1 Analytical Solving Procedure

Let us introduce the dimensionless length and time variables

$$y = \frac{x}{d}, \tag{2.9}$$

$$\tau = \omega t, \tag{2.10}$$

and parameters

$$\varepsilon\zeta_1 = \frac{c}{\sqrt{k\left(m_1 + m_2\right)}}, \quad \varepsilon = \frac{m_2 d}{X\left(m_1 + m_2\right)},$$
$$\varepsilon\eta_1 = \frac{m_2 d X}{J + m_2 d^2}, \quad \mathcal{M}\left(\varphi'\right) = \frac{\mathcal{M}\left(\dot{\varphi}\right)}{\omega^2\left(J + m_2 d^2\right)}. \tag{2.11}$$

Dimensionless differential equations of motion of the oscillatory system follow as

$$y'' + y = -2\varepsilon\zeta_1 y' + \varepsilon\left(\varphi''\sin\varphi + \varphi'^2\cos\varphi\right),$$
$$\varphi'' = \varepsilon\eta_1 y''\sin\varphi + \varepsilon\,\mathcal{M}\left(\varphi'\right), \tag{2.12}$$

where

$$\omega = \sqrt{\frac{k}{m_1 + m_2}}, \tag{2.13}$$

is the frequency of the system, $(') \equiv d/d\tau$ and $X$ is the length characteristic of the amplitude of the motion of the motor Nayfeh and Mook (1979). Assuming that the parameter $\varepsilon$ is small, Eq. (2.12) are with small nonlinear terms. For computational reasons it is convenient to rewrite (2.12) into a system of first order equations. Accordingly, we let

$$y = a\cos(\varphi + \psi), \tag{2.14}$$

where $a$, $\varphi$ and $\psi$ are functions of $\tau$. Generally, it cannot be expected that the frequency of the rectilinear motion $(\varphi' + \psi')$ to be the same as the angular speed of the rotor $\varphi'$. Hence, $\psi$ is included in the argument.

We are considering the motion near resonance and it is convenient to introduce a detuning parameter $\Delta$ as follows

$$\varphi' = 1 + \Delta. \tag{2.15}$$

Hence $\psi$ is used to distinguish between the speed of the rotor and the actual frequency of the rectilinear motion, while $\Delta$ is used to distinguish between the speed of the rotor and the natural frequency of the rectilinear motion.

Using the method of variation of parameters, we put

$$a'\cos(\varphi + \psi) - a(\Delta + \psi')\sin(\varphi + \psi) = 0, \tag{2.16}$$

so that

$$y' = -a\sin(\varphi + \psi), \tag{2.17}$$

and

$$y'' = -a'\sin(\varphi + \psi) - a(1 + \Delta + \psi')\cos(\varphi + \psi). \tag{2.18}$$

Substituting (2.15), (2.17) and (2.18) into (2.12) leads to

$$\begin{aligned} &-a'\sin(\varphi + \psi) - a(\Delta + \psi')\cos(\varphi + \psi) \\ &= \varepsilon\Delta'\sin\varphi + \varepsilon(1 + \Delta)^2\cos\varphi + 2\varepsilon\zeta_1 a\sin(\varphi + \psi), \end{aligned} \tag{2.19}$$

and

$$\Delta' = -\varepsilon\eta_1\left(a'\sin(\varphi + \psi) + \varepsilon a(1 + \Delta + \psi')\cos(\varphi + \psi)\right)\sin\varphi + \varepsilon\mathcal{M} \tag{2.20}$$

where $\mathcal{M} = \mathcal{M}(\varphi')$. Solving (2.16) and (2.19) for $a'$ and $\psi'$ produces

$$a' = \varepsilon\Delta' \sin\varphi - \varepsilon(1+\Delta)^2 \cos\varphi - 2\varepsilon\zeta_1 a \sin(\varphi+\psi)\sin(\varphi+\psi), \tag{2.21}$$

$$\psi' = -\Delta - \frac{\varepsilon}{a}\Delta' \sin\varphi - \frac{\varepsilon}{a}(1+\Delta)^2 \cos\varphi - 2\varepsilon\zeta_1 \sin(\varphi+\psi)\cos(\varphi+\psi). \tag{2.22}$$

Equations (2.20)–(2.22) are equivalent to the system (2.12). No approximations have been made yet. These equations show that $\Delta'$ and $a'$ are $O(\varepsilon)$. We restrict our attention to a narrow band of frequencies around the natural frequency

$$\Delta = \varepsilon\sigma. \tag{2.23}$$

$\Delta$ and $\Delta'$ are $O(\varepsilon)$, and it follows from (2.22) that $\psi'$ is also $O(\varepsilon)$. As a first simplification we neglect all terms $O(\varepsilon^2)$ appearing in (2.20)–(2.22) and we obtain

$$\Delta' = \varepsilon\left(\mathcal{M} - \eta_1 a \cos(\varphi+\psi)\sin\varphi\right), \tag{2.24}$$

$$a' = -\varepsilon\left(\cos\varphi + 2\zeta_1 a \sin(\varphi+\psi)\right)\sin(\varphi+\psi), \tag{2.25}$$

$$\psi' = -\Delta - \frac{\varepsilon}{a}\left(\cos\varphi + 2\zeta_1 \sin(\varphi+\psi)\right)\cos(\varphi+\psi). \tag{2.26}$$

To solve the Eqs. (2.24)–(2.26) in exact analytical form is not an easy task. The approximate solution is obtained by applying of the averaging procedure. We consider $a$, $\sigma$ and $\psi$ to be constant over one cycle and average the equations over one cycle. The result is

$$\Delta' = \varepsilon\left(\mathcal{M} + \frac{1}{2}\eta_1 a \sin\psi\right), \tag{2.27}$$

$$a' = -\varepsilon\left(\frac{1}{2}\sin\psi + a\zeta_1\right), \tag{2.28}$$

$$\psi' = -\varepsilon\left(\sigma + \frac{1}{2a}\cos\psi\right). \tag{2.29}$$

At this point, let us mention the difference between ideal and non-ideal systems. For the ideal system relation (2.27) is not a governing equation as $\sigma$ is specified and (2.28) and (2.29) are solved for $a$ and $\psi$. For the non-ideal system the Eqs. (2.27)–(2.29) are solved for $a$, $\Delta$ and $\psi$ with $\mathcal{M}$ as a control setting.

### 2.1.2 Steady-State Solution and Sommerfeld Effect

For steady-state responses, $a'$, $\Delta'$ and $\psi'$ are zero, i.e., Eqs. (2.27)–(2.29) transform into

$$\mathcal{M} + \frac{1}{2}\eta_1 a \sin\psi = 0, \tag{2.30}$$

$$\frac{1}{2}\sin\psi + a\zeta_1 = 0, \tag{2.31}$$

$$\sigma + \frac{1}{2a}\cos\psi = 0. \tag{2.32}$$

Combining (2.31) and (2.32) yields

$$a = \frac{1}{2\sqrt{\zeta_1^2 + \sigma^2}}, \tag{2.33}$$

while combining (2.30) and (2.31) gives

$$\mathcal{M} = \eta_1 \zeta_1 a^2. \tag{2.34}$$

The phase is given by

$$\psi = \cos^{-1}(-2a\sigma) = \sin^{-1}(-2a\zeta_1). \tag{2.35}$$

Recalling the definitions of the dimensionless variables, we rewrite these results in terms of the original physical variables as

$$Xa = \frac{\omega m_2 d}{\sqrt{c^2 + 4(\omega - \dot{\varphi})^2 (m_1 + m_2)^2}}, \tag{2.36}$$

$$\psi = \tan^{-1}\left(-\frac{2(m_1 + m_2)(\omega - \dot{\varphi})}{c}\right), \tag{2.37}$$

and

$$\mathcal{M}(\dot{\varphi}) = \frac{c\omega k m_2 d X}{2\left(c^2 + 4(\omega - \dot{\varphi})^2 (m_1 + m_2)^2\right)}, \tag{2.38}$$

where $(Xa)$ is the physical amplitude of the motion and according to (2.35), $-\pi/2 < \psi < 0$. The solving procedure is as it follows: first, the (2.38) is solved for $\dot{\varphi}$ and then, the amplitude and phase from (2.36) and (2.37) are obtained.

To solve the Eq. (2.38) it is necessary to know the torque function of the motor. Namely, the torque of the motor contains two terms: the characteristic of the motor $L(\dot{\varphi})$ and the resisting moment $H(\dot{\varphi})$ due primarily to windage of the rotating parts outside the motor

$$\mathcal{M}(\dot{\varphi}) = L(\dot{\varphi}) - H(\dot{\varphi}).$$

Generally, $L(\dot{\varphi})$ and $H(\dot{\varphi})$ are determined experimentally. Various types of mathematical description of the motor property are suggested. One of the most often applied and the simplest one is the linear mode which is a function of the angular velocity $\dot{\varphi}$

$$\mathcal{M}(\dot{\varphi}) = M_0 \left(1 - \frac{\dot{\varphi}}{\Omega_0}\right), \tag{2.39}$$

and depends on two constant parameters $M_0$ and $\Omega_0$ which define the limited source of power as the angular velocity increases. The expression (2.39) defines the characteristic curve of the motor shown in Fig. 2.3, where for angular velocities greater than $\Omega_0$ the torque reduces to zero and when the angular velocity is zero the torque is maximum. Figure 2.3 is an illustrative example in which Eq. (2.39) is valid for positive values of torque and angular velocity.

In this section the calculation is done for the linear torque function (2.39). Solving (2.36)–(2.38) with (2.39) the frequency - response relation is obtained. In Fig. 2.4 the corresponding diagram is plotted. For the ideal linear system the frequency-response diagram is a continual curve presented with a solid line in Fig. 2.4. For the non-ideal system the curves are obtained by allowing the system to achieve a steady-state motion while the control was fixed. Then the amplitude of the steady-state response was calculated. The control was then changed very slightly and held in the new position until a new steady state was achieved. For increasing of angular frequency the amplitude increases up to T and jumps to H and moves in right-hand side direction along the amplitude curve. For the decreasing frequency the amplitude increases up to R and suddenly jumps to P and decreases continually along the solid line curve. We note that there are gaps where no steady-state response exists. The gaps are not the same in the two directions but there is some overlap. The arrows indicate the change brought about by slowly increasing or decreasing the control setting in a

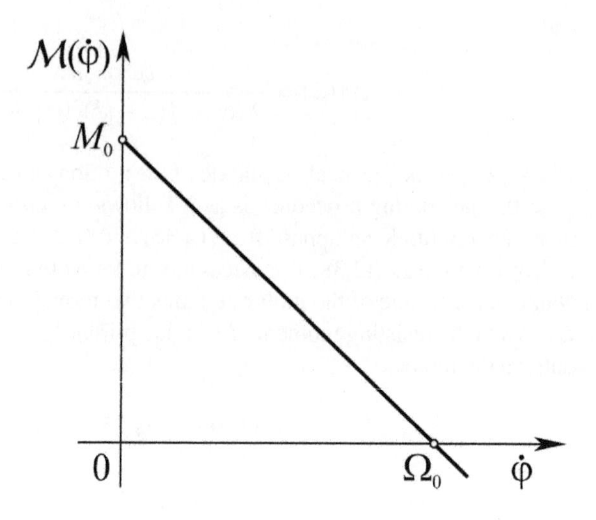


**Fig. 2.3** Motor torque characteristic curve

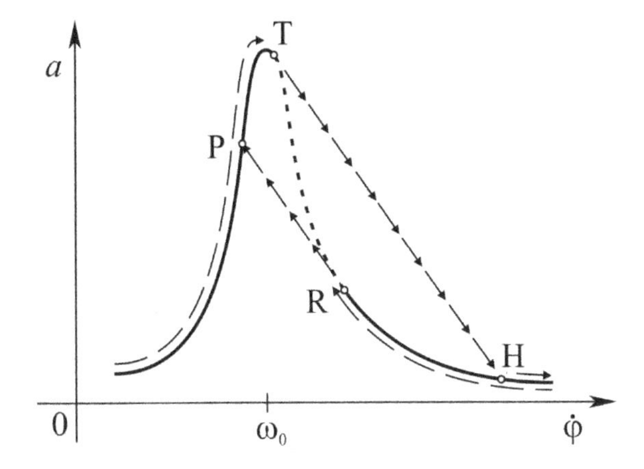


**Fig. 2.4** Frequency-response curves for ideal system (----) and non-ideal system (→→) (Nayfeh and Mook 1979)

non-ideal system. We note that the non-ideal system cannot be made to respond at a frequency between T and H by simply increasing the control setting form a low value. In contrast, the ideal system can respond also at frequencies between T and H. When the control setting is continually decreased, the system cannot be made to respond between R and P. In other words, the right side of the resonance spike between T and R cannot be reached by either continually increasing or continually decreasing the control setting. Though the system is linear, the non-ideal source causes a jump phenomenon to occur.

At the left-hand side of the frequency-response curve the input power is relatively low. As the input power increases, the amplitude of the response increases noticeably while the frequency changes only slightly, especially along the portion of the curve between P and T. Here a relatively large increase in power causes a relatively large increase in the amplitude and practically no change in the frequency.

At T the character of the motion suddenly changes. An increase in the input power causes the amplitude to decrease and the frequency to increase considerably. This phenomenon is called Sommerfeld effect. It was discovered by Arnold Sommerfeld in 1902, commented in a book of Kononenko (1969) and described in the book of Nayfeh and Mook (1976). The jump phenomena in the amplitude-frequency curve for the non-ideal system is remarked during passage through resonance. In this working regime special properties of the non-ideal source are caused. Namely, in the region before resonance as the power supplied to the source increases, the RPM of the energy source (motor) increases accordingly. But, the closed motor speed moves toward the resonance frequency the more power the source requires to increase the motor speed. Near resonance it appears that additional power supplied to the motor will only increase the amplitude of the response with little effect on the RPM of the motor, and the amplitude of vibration increases. In non-ideal vibrating systems the passage through resonance requires more input power than is available. As a consequence the vibrating system cannot pass through resonance or requires an intensive interaction between the vibrating system and the energy source to be able to do it. Strong interaction leads to fluctuating motor speed and fairly large vibration amplitudes appear. The motor may not have enough power to reach higher regimes with low energy consumption as most of its energy is applied to move the structure and not to accelerate the shaft. In fact, the vibrating response provides a certain energy sink.

### *2.1.3 Model Analogy and Numerical Simulation*

To explain the motion in non-ideal system a model analogy is introduced. Let us consider the Eq. (2.38). Substituting (2.39) into (2.8) it is

$$\left(J + m_2 d^2\right)\ddot{\varphi} = m_2 d\ddot{x}\sin\varphi + M_0\left(1 - \frac{\dot{\varphi}}{\Omega_0}\right). \tag{2.40}$$

Discussion of (2.40) follows.

Let us consider a motor mounted on a rigid base. Motion of the cart is eliminated ($\ddot{x} = 0$) and according to (2.40) the mathematical model is

$$\left(J + m_2 d^2\right)\ddot{\varphi} = M_0\left(1 - \frac{\dot{\varphi}}{\Omega_0}\right). \tag{2.41}$$

The physical model which corresponds to (2.41) is a wheel climbing on a ramp. The slope of the ramp is related to the motor inertia defining the rate of the angular velocity.

For a motor with no resistive torque when $\mathcal{M} = M_0$, the angular acceleration is constant and therefore, the angular velocity of the wheel increases by a constant rate (see Fig. 2.5a). The motor torque should be switched off as the desired angular velocity $\Omega_0$ is achieved.

When considering the motor with resistive torque, the angular acceleration is no longer constant and decreases as the angular velocity increases. The system shown in

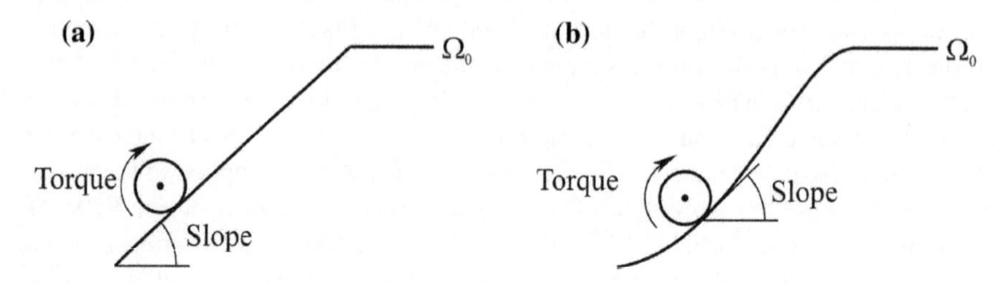


**Fig. 2.5** Analogy by a wheel climbing a ramp: **a** a motor with no resistive torque, **b** a motor with resistive torque (Goncalves et al. 2014)

Fig. 2.5b is used to represent the motor with resistive torque where it is more difficult to reach the energy level $\Omega_0$. The rate of velocity changing is no longer linear.

When the motor is mounted on a flexible base, its motion is described with Eq. (2.40). It is clear that the angular acceleration is also a function of the cart motion $x$. Besides, the motion of the cart is a function of the acceleration and angular velocity of the motor (2.7). In Fig. 2.6 a system which is analog with the motor mounted on a flexible base is represented. Similar to Fig. 2.5a, wheel must climb a ramp to reach the level of energy defined by $\Omega_0$. In this case the ramp path is modified by the cart resonance frequency $\omega_0$.

The resonance frequency is represented by the valley in the ramp path. The deep and the width of the valley in the ramp are related to the amplitude of the motion of the cart and in some cases the wheel can get stuck inside the valley in the ramp path. Numerical simulation is done for frequencies around the cart resonance frequency $\omega_0$. Figure 2.7 shows that when $\Omega_0$ is slightly bigger than $\omega_0$ the angular velocity does not increase. Setting $\Omega_0 = 1.1\omega_0$ the motor does not reach the angular velocity $1.1\omega_0$, instead it will oscillate with angular velocity $\omega_0$. As a consequence, the additional energy increases the amplitude of the displacement of the cart.

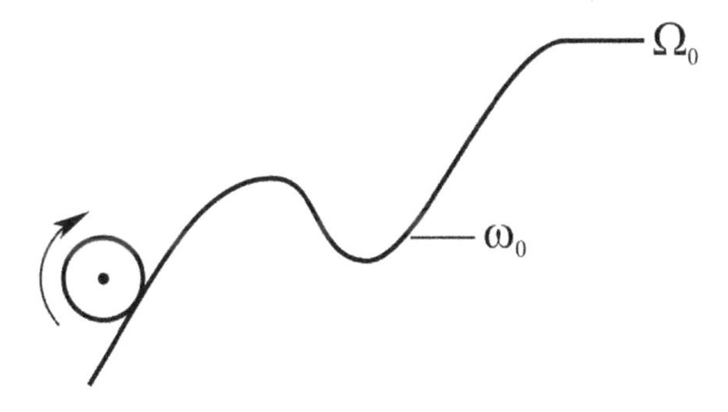


**Fig. 2.6** Analogy of the resonance frequency in the ramp path (Goncalves et al. 2014)

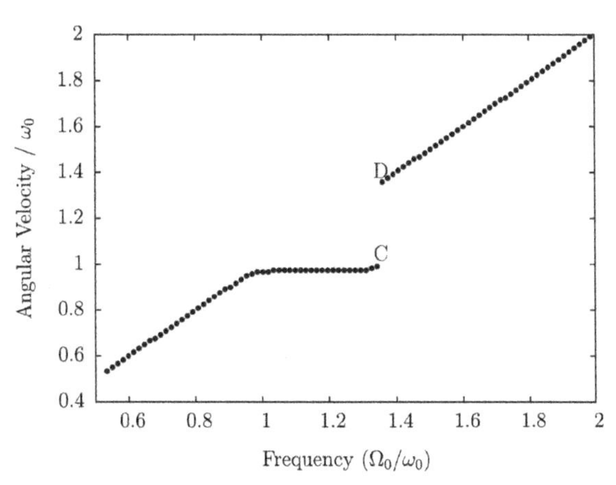


**Fig. 2.7** The angular velocity as a function of motor constant $\Omega_0$ (Goncalves et al. 2014)

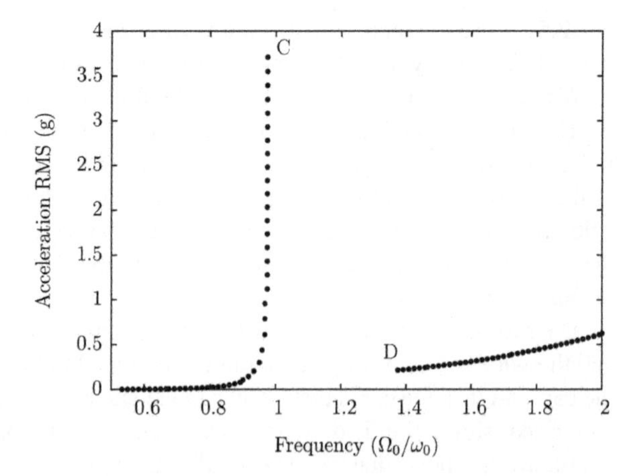


**Fig. 2.8** Root mean square acceleration as a function of the motor constant $\Omega_0$ (Goncalves et al. 2014)

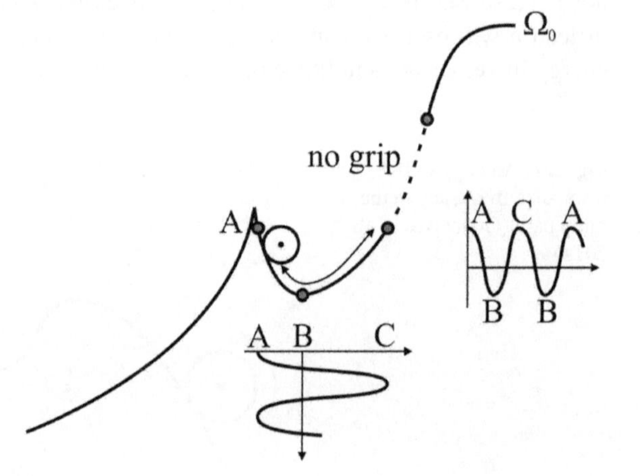


**Fig. 2.9** The analogy of a wheel climbing a ramp with the influence of a resonance frequency (Goncalves et al. 2014)

In Fig. 2.8 the cart root mean square (RMS) magnitude of the acceleration as a function of the oscillation frequency rate $\Omega_0/\omega_0$ is plotted. Based on plots in Figs. 2.7 and 2.8 it is noted that there is a region between C and D without data points. This region corresponds to a jump phenomenon described by Sommerfeld. The wheel climbing a ramp is used to explain the jump using the sketch in Fig. 2.9.

The system passing through resonance frequency is represented by the wheel inside the valley and oscillating between points ABC. There is a region in the ramp path between points C and D where the wheel does not have grip. The system cannot stay in energy levels in the region as it will fall in the valley. The width and the depth of the valley ABC are controlled by the resonance amplitude. The higher the damping, the shallower the valley.

### 2.1.4 Stability Analysis

For a given setting of control there can be one, two or three steady state solutions (see Fig. 2.4). To determine which of these steady-state solutions actually corresponds to a realizable motion, we need to consider the stability of motion. Namely, we determine the stability of the steady-state solutions by determining the nature of the singular points which are the solutions of (2.27)–(2.29). To accomplish this, we let

$$a = a_S + a_1, \quad \psi = \psi_S + \psi_1, \quad \Delta = \Delta_S + \Delta_1. \tag{2.42a}$$

Substituting (2.42a) into (2.27)–(2.29) and neglecting all but the linear terms in $a_1$, $\psi_1$ and $\Delta_1$, we obtain

$$\begin{aligned} \Delta_1' &= \varepsilon\left(\frac{d\mathcal{M}}{d\varphi'}\varphi_S'\Delta_1 + \frac{1}{2}\eta_1 a_1 \sin(\psi_S) + \frac{1}{2}\eta_1 a_S \psi_1 \cos(\psi_S)\right), \\ a_1' &= -\varepsilon\left(\frac{1}{2}\psi_1 \cos(\psi_S) + a_1\alpha_1\right), \\ \psi_1' &= -\varepsilon\left(\sigma_1 + \frac{1}{2}\left(\frac{a_1}{a_S^2}\right)\cos(\psi_S) - \frac{1}{2a_S}\psi_1 \sin(\psi_S)\right), \end{aligned} \tag{2.43}$$

where $\varepsilon\sigma_1 = \Delta_1$. Linear equation (2.43) have a solution in the form

$$(a_1, \psi_1, \Delta_1) = (a_{10}, \psi_{10}, \Delta_{10})\exp(\lambda t),$$

i.e.,

$$a_1 = a_{10}\exp(\lambda t), \quad \psi_1 = \psi_{10}\exp(\lambda t), \quad \Delta_1 = \Delta_{10}\exp(\lambda t), \tag{2.44}$$

where $\lambda$ is the eigenvalue coefficient matrix and $a_{10}$, $\psi_{10}$, $\Delta_{10}$ are constants. Substituting (2.44) into (2.43) the characteristic determinant is obtained as

$$\begin{vmatrix} \varepsilon\varphi_S'(\frac{d\mathcal{M}}{d\varphi'})_{\varphi_S'} - \lambda & \frac{\varepsilon}{2}\eta_1 \sin(\psi_S) & \frac{\varepsilon}{2}\eta_1 a_S \cos(\psi_S) \\ 0 & \lambda + \varepsilon\alpha_1 & \frac{1}{2}\varepsilon\cos(\psi_S) \\ 1 & \frac{\varepsilon}{2a_S^2}\cos(\psi_S) & \lambda - \frac{\varepsilon}{2a_S}\sin(\psi_S) \end{vmatrix} = 0,$$

and the characteristic equation is a cubic one

$$0 = \lambda^3 + \lambda^2\left[\varepsilon\alpha_1 - \frac{\varepsilon}{2a_S}\sin(\psi_S) - \varepsilon\varphi_S'\left(\frac{d\mathcal{M}}{d\varphi'}\right)_{\varphi_S'}\right]$$

$$-\lambda\left[\frac{\varepsilon^2\alpha_1}{2a_S}\sin(\psi_S)+\frac{\varepsilon^2}{4a_S^2}\cos^2(\psi_S)+\left(\varepsilon\alpha_1-\frac{\varepsilon}{2a_S}\sin(\psi_S)\right)\varepsilon\varphi_S'\left(\frac{d\mathcal{M}}{d\varphi'}\right)_{\varphi_S'}\right.$$
$$\left.-\frac{\varepsilon}{2}\eta_1 a_S\cos(\psi_S)\right]$$
$$+\left(\varepsilon\varphi_S'\left(\frac{d\mathcal{M}}{d\varphi'}\right)_{\varphi_S'}\left[\frac{\varepsilon^2\alpha_1}{2a_S}\sin(\psi_S)+\frac{\varepsilon^2}{4a_S^2}\cos^2(\psi_S)\right]\right.$$
$$+\frac{\varepsilon^2}{2}\eta_1\alpha_1 a_S\cos(\psi_S)-\frac{\varepsilon^2}{8}\eta_1\sin(2\psi_S). \tag{2.45}$$

The solutions are stable and hence the corresponding motions realizable, if the real part of each eigenvalue is negative or zero. Without solving the Eq. (2.45) and using the Routh–Hurwitz principle we can determine the conditions for the stable solution up to the small value $O(\varepsilon^2)$

$$\left[\alpha_1-\varphi_S'\left(\frac{d\mathcal{M}}{d\varphi'}\right)_{\varphi_S'}\right]a_S\sin(\psi_S)-\frac{1}{4}\sin(2\psi_S)-\varphi_0'\left(\frac{d\mathcal{M}}{d\varphi'}\right)_{\varphi_S'}\eta_1 a_S^3\cos(\psi_S)>0. \tag{2.46}$$

Analyzing the relation (2.46) it turns out that the solutions between T and R are unstable, while all those outside this region are stable (Fig. 2.4). As $(2.43)_1$ indicates the parameter, that gives the influence of the motor on the stability, is the slope of the characteristic.

## 2.2 Oscillator with Variable Mass Excited with Non-ideal Source

There is a significant number of equipment and machines which can be modelled as one degree-of-freedom oscillators with time variable mass. Let us mention some of them: centrifuges, sieves, pumps, transportation devices, etc. For all of them it is common that their mass is varying slowly during the time. The mass variation is assumed to be continual. The excitation of the motion of the equipment is ideal (the excitation force is a harmonic function and the influence of the oscillator on the source is negligible) or non-ideal, where not only the energy source has an influence on the oscillator, but vice versa. In this chapter the oscillator with non-ideal excitation is considered. First the model of the one-degree-of-freedom oscillator with non-ideal excitation is formed. It is a system of two coupled differential equations with time variable parameters. Vibrations close to the resonant regime are considered. For the case when the mass variation is slow the amplitude and frequency of vibration are determined. The Sommerfeld effect for the system where the parameters depend on the slow time is discussed. Analytical solutions are compared with numerically obtained ones.

### 2.2.1 Model of the System with Variable Mass

In Fig. 2.10 the model of an oscillator with time variable mass $m_1$ connected with a motor which is a non-ideal energy source is plotted. The motor is settled on a cart whose mass $m_1$ is varying in time due to leaking of the contain with velocity $u$. It is supposed that the mass variation is slow in time. The connection of the oscillating cart to the fixed element has the rigidity $k$ and damping $c$.

The motor has the moment of inertia $J$, unbalance $m_2$ and eccentricity $d$. The excitation torque of the motor, $\mathcal{M}(\dot{\varphi})$, is the function of the angular velocity $\dot{\varphi}$

$$\mathcal{M}(\dot{\varphi}) = M_0 \left(1 - \frac{\dot{\varphi}}{\Omega_0}\right), \tag{2.47a}$$

where $\Omega_0$ is the steady-state angular velocity. This mathematical model corresponds to asynchronous AC motor (Dimentberg et al. 1997).

To describe the motion of the system, let us assume the two generalized coordinates: the displacement of the oscillator $x$ and the rotation angle of the motor $\varphi$. Variation of the mass of the oscillator is assumed to be slow and to be the function of the slow time $\tau = \varepsilon t$, where $\varepsilon << 1$ is a small constant parameter. Equations of motion of the system with time variable mass is in general (Cveticanin 2015)

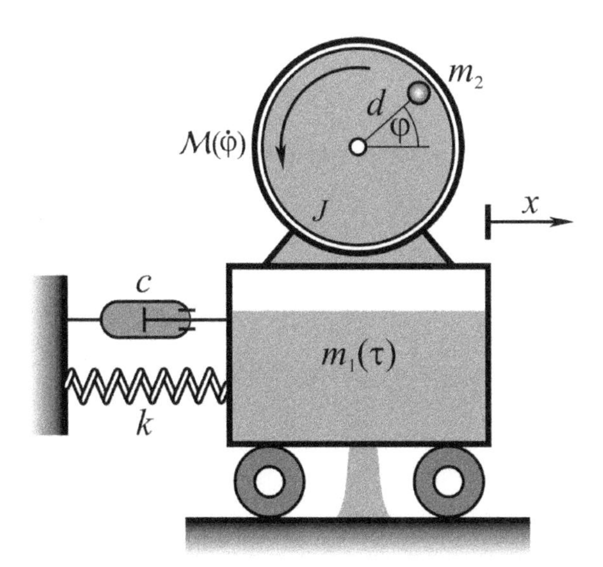


**Fig. 2.10** Model of the non-ideal mass variable system

$$\frac{d}{dt}\frac{\partial T}{\partial \dot{x}} - \frac{\partial T}{\partial x} + \frac{\partial U}{\partial x} = Q_x + Q_R,$$
$$\frac{d}{dt}\frac{\partial T}{\partial \dot{\varphi}} - \frac{\partial T}{\partial \varphi} + \frac{\partial U}{\partial \varphi} = Q_\varphi, \tag{2.48}$$

where $Q_x$ and $Q_\varphi$ are generalized forces and $Q_R$ is the generalized reactive force caused by mass variation. If the mass is added or separated with the absolute velocity $u$ in $x$ direction, the generalized reactive force is the product of the velocity $u$ and mass variation $dm_1/dt$, i.e.

$$Q_R = \frac{dm_1}{dt}u. \tag{2.49}$$

The non-conservative force in $x$ direction is the damping force $Q_x = -c\dot{x}$ with the damping coefficient $c$, while the generalized force $Q_\varphi$ corresponds to the torque $\mathcal{M}(\dot{\varphi})$ applied to motor. The kinetic energy of the system is according to (2.4)

$$T = \frac{1}{2}[m_1(\tau) + m_2]\dot{x}^2 + \frac{1}{2}(J + m_2 d^2)\dot{\varphi}^2 - m_2 d\dot{x}\dot{\varphi}\sin\varphi. \tag{2.50}$$

and the potential energy of the system is according to (2.5)

$$U = \frac{kx^2}{2}. \tag{2.51}$$

Using (2.50) and (2.51) and also (2.49) equations of motion are due to (2.48)

$$[m_1(\tau) + m_2]\ddot{x} + c\dot{x} + kx = \frac{dm_1(\tau)}{dt}(u - \dot{x}) + m_2 d\left(\ddot{\varphi}\sin\varphi + \dot{\varphi}^2\cos\varphi\right), \tag{2.52}$$

$$\left(J + m_2 d^2\right)\ddot{\varphi} = m_2 d\ddot{x}\sin\varphi + \mathcal{M}(\dot{\varphi}). \tag{2.53}$$

Assuming that the velocity $u$ is zero, the Eqs. (2.52) and (2.53) transform into

$$[m_1(\tau) + m_2]\ddot{x} + c\dot{x} + kx = -\frac{dm_1(\tau)}{dt}\dot{x} + m_2 d\left(\ddot{\varphi}\sin\varphi + \dot{\varphi}^2\cos\varphi\right), \tag{2.54}$$

$$\left(J + m_2 d^2\right)\ddot{\varphi} = m_2 d\ddot{x}\sin\varphi + \mathcal{M}(\dot{\varphi}). \tag{2.55}$$

Let us rewrite (2.52) and (2.53) into

$$\ddot{x} + \omega^2(\tau)x = -\varepsilon\zeta(\tau)\dot{x} - \frac{\varepsilon}{m_1(\tau) + m_2}\frac{dm_1(\tau)}{d\tau}\dot{x} \tag{2.56}$$
$$+ \varepsilon\mu(\tau)\left(\ddot{\varphi}\sin\varphi + \dot{\varphi}^2\cos\varphi\right),$$
$$\ddot{\varphi} = \varepsilon\eta\ddot{x}\sin\varphi + \varepsilon\gamma\mathcal{M}(\dot{\varphi}), \tag{2.57}$$

where

$$\omega^2(\tau) = \frac{k}{m_1(\tau)+m_2}, \quad \varepsilon\zeta(\tau) = \frac{c}{m_1(\tau)+m_2}, \quad \varepsilon\gamma = \frac{1}{J+m_2d^2}$$
$$\varepsilon\mu(\tau) = \frac{m_2 d}{m_1(\tau)+m_2}, \quad \varepsilon\eta = \frac{m_2 d}{J+m_2d^2}. \tag{2.58}$$

In the Eqs. (2.56) and (2.57) the right-hand side terms are of the order of small parameter $\varepsilon$. Analyzing the dimensionless parameters (2.58) it is obvious that the dimensionless frequency $\omega$ , damping $\zeta$ and excitation $\gamma$ are functions of slow time. Namely, the mass variation affects these values.

### 2.2.2 Model of the System with Constant Mass

Let us consider the system with constant mass when $m_1 = const$. Then the dimensionless values $\omega$, $\zeta$ and $\gamma$ in (2.58) are also constant. Assuming that the mass of the system is constant and omitting the terms with the second and higher order of the small parameter $\varepsilon$, relations (2.56) and (2.57) simplify into

$$\ddot{x} + \omega^2 x = -\varepsilon\zeta\dot{x} + \varepsilon\mu\dot{\varphi}^2\cos\varphi, \tag{2.59}$$
$$\ddot{\varphi} = \varepsilon\eta\ddot{x}\sin\varphi + \varepsilon\gamma\mathcal{M}(\dot{\varphi}). \tag{2.60}$$

Solution of (2.59) and (2.60) is

$$x = a\cos(\varphi+\psi), \tag{2.61}$$

with time derivatives

$$\dot{x} = -a\omega\sin(\varphi+\psi). \tag{2.62}$$

and

$$\dot{\varphi} = \Omega, \tag{2.63}$$

where $a$ is the amplitude of vibration, $\psi$ is the phase angle and $\Omega$ is the time derivative of the solution $\varphi$. Substituting (2.61)–(2.63) and after some modification the Eqs. (2.59) and (2.60) are rewritten into first order differential equations of motion in new variables $a$, $\psi$ and $\Omega$

$$\dot{a} = -\varepsilon\zeta a\sin^2(\varphi+\psi) - \varepsilon\mu\frac{\Omega^2}{\omega}\sin(\varphi+\psi)\cos\varphi, \tag{2.64}$$

$$\dot{\psi} = (\omega - \Omega) - \varepsilon\mu\frac{\Omega^2}{a\omega}\cos(\varphi+\psi)\cos\varphi - \varepsilon\zeta\frac{\sin 2(\varphi+\psi)}{2}, \tag{2.65}$$

$$\dot{\Omega} = \varepsilon\gamma\mathcal{M}(\Omega) - \varepsilon\eta a\omega^2\cos(\varphi+\psi)\sin\varphi. \tag{2.66}$$

Equations (2.64)–(2.66) are three coupled strong nonlinear equations.

**Averaging Procedure**

For simplicity, let us introduce the averaging over the period of the trigonometric function $\varphi$ for the period $2\pi$. After averaging it is

$$\dot{a} = -\frac{1}{2}\varepsilon\zeta a - \frac{1}{2}\varepsilon\mu\frac{\Omega^2}{\omega}\sin\psi, \tag{2.67}$$

$$\dot{\psi} = (\omega - \Omega) - \frac{\varepsilon\mu\Omega^2}{2a\omega}\cos\psi, \tag{2.68}$$

$$\dot{\Omega} = \varepsilon\gamma\mathcal{M}(\dot{\varphi}) + \frac{1}{2}\varepsilon\eta a\omega^2\sin\psi. \tag{2.69}$$

For the steady state motion, when $\dot{a} = 0$, $\dot{\psi} = 0$ and $\dot{\Omega} = 0$, the Eqs. (2.67)–(2.69) transform into

$$\frac{\varepsilon\mu\Omega^2}{2\omega}\sin\psi = -\frac{1}{2}\varepsilon\zeta a, \tag{2.70}$$

$$\frac{\varepsilon\mu\Omega^2}{2\omega}\cos\psi = (\omega - \Omega)a, \tag{2.71}$$

$$\frac{1}{2}\varepsilon\eta a\omega^2\sin\psi = -\varepsilon\gamma\mathcal{M}(\dot{\varphi}). \tag{2.72}$$

Using relations (2.70) and (2.71) the amplitude -frequency relation is obtained

$$a = \frac{\varepsilon\mu\Omega^2}{\omega\sqrt{(\varepsilon\zeta)^2 + 4(\omega-\Omega)^2}}. \tag{2.73}$$

Eliminating $\psi$ in the Eqs. (2.70) and (2.72), we have

$$\frac{(\varepsilon\eta)(\varepsilon\zeta)\omega^3}{2\varepsilon\mu\Omega^2}a^2 = \varepsilon\gamma\mathcal{M}(\dot{\varphi}) \equiv \varepsilon\gamma M_0\left(1 - \frac{\Omega}{\Omega_0}\right). \tag{2.74}$$

### 2.2.3 Comparison of the Systems with Constant and Variable Mass

Let us compare the properties of the systems with various values of mass. The characteristic points, which represent the intersection of curves (2.73) and (2.74), will be analyzed. In Fig. 2.11 the points of intersection of amplitude-frequency and characteristic curve are presented: in Fig. 2.11a the intersection of an amplitude-frequency curve and various characteristic curves and in Fig. 2.11b for one characteristic curve and amplitude-frequency curves for various values of mass are plotted.

In Fig. 2.11a the intersection of the amplitude-frequency diagram for $m_1 = m_{10}$ and various values of motor torque are plotted. It can be seen that there may be one or three points of intersection: two of them are stable and one is unstable. In Fig. 2.11b only one motor characteristic for $m_1 = m_{10}$ is plotted.

Namely, the influence of the small mass variation on the motor characteristic is negligible. The intersection of this motor characteristic and of amplitude-frequency diagrams obtained for various values of mass $m_1$ is plotted in Fig. 2.11b It is seen that for $m_1 = m_{10}$ there are three intersection points. If the mass is higher than $m_{10}$, i.e., $m_1 = 1.1m_{10}$, the amplitude-frequency diagram is moved to left the number of intersections decreases from three to only one. If the mass is smaller than $m_{10}$, i.e., $m_1 = 0.9m_{10}$, the amplitude-frequency diagram is moved to right in comparison to the previous one. There exists only one steady state position. It can be concluded that the value of mass has an influence on the number and position of characteristic points. Besides, it can be seen that the maximal amplitude depends on the non-dimensional damping coefficient, as it is affected with mass value: for the higher value of mass the maximal amplitude of vibration is smaller than for $m_{10}$. Otherwise, the smaller the mass, the higher the value of the maximal amplitude.

In Fig. 2.12a–c the influence of mass increase on the position of the characteristic point with small amplitude and high frequency is plotted. The amplitude of steady state position decreases from 1 to 3, while the frequency increases.

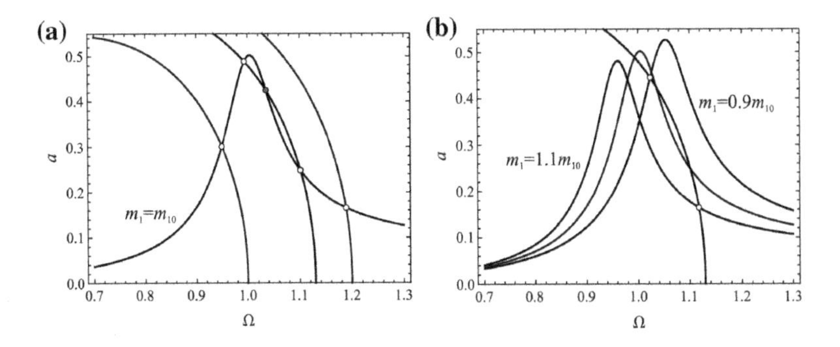


**Fig. 2.11** **a** Intersection of the amplitude-frequency diagram and various values of motor torque; **b** Intersection of the motor torque and amplitude-frequency diagrams for various values of masses

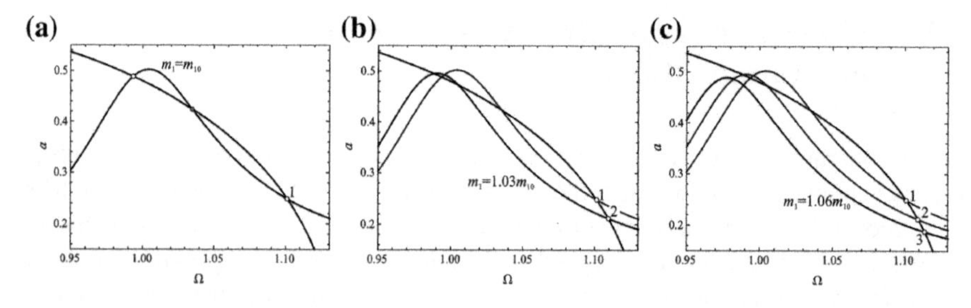


**Fig. 2.12** The motion of the lower intersection point for increasing mass: **a** $m_1 = m_{10}$, **b** $m_1 = 1.03m_{10}$, **c** $m_1 = 1.06m_{10}$

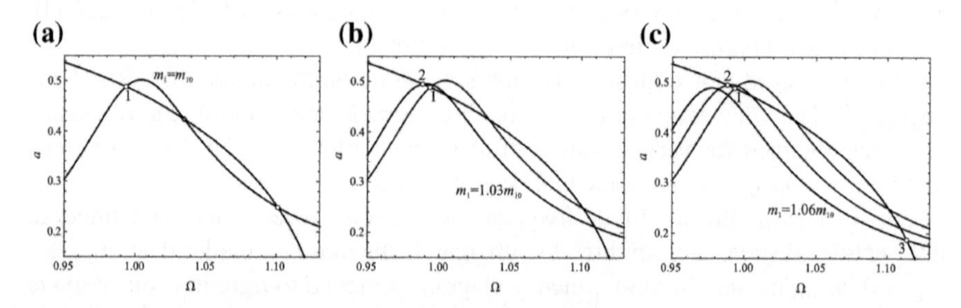


**Fig. 2.13** The motion of the upper intersection point for increasing mass: **a** $m_1 = m_{10}$, **b** $m_1 = 1.03m_{10}$, **c** $m_1 = 1.06m_{10}$

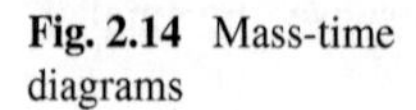
**Fig. 2.14** Mass-time diagrams

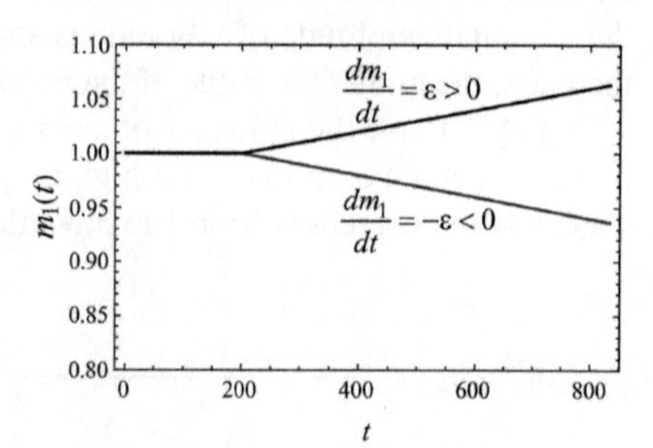


The second characteristic point which corresponds to the steady state motion also moves due to mass increase (see Fig. 2.13a–c). First the intersection point moves toward higher amplitude and smaller frequency (point 2) and then jumps to the position 3 with small amplitude and high frequency.

In Fig. 2.14 the mass-time diagrams are plotted: for $t \in [0, 200]$ the mass is constant, while for $t > 200$ mass is increasing ($\varepsilon > 0$) of decreasing ($\varepsilon < 0$). In Fig. 2.15 the displacement and frequency time history diagrams for mass increase are plotted.

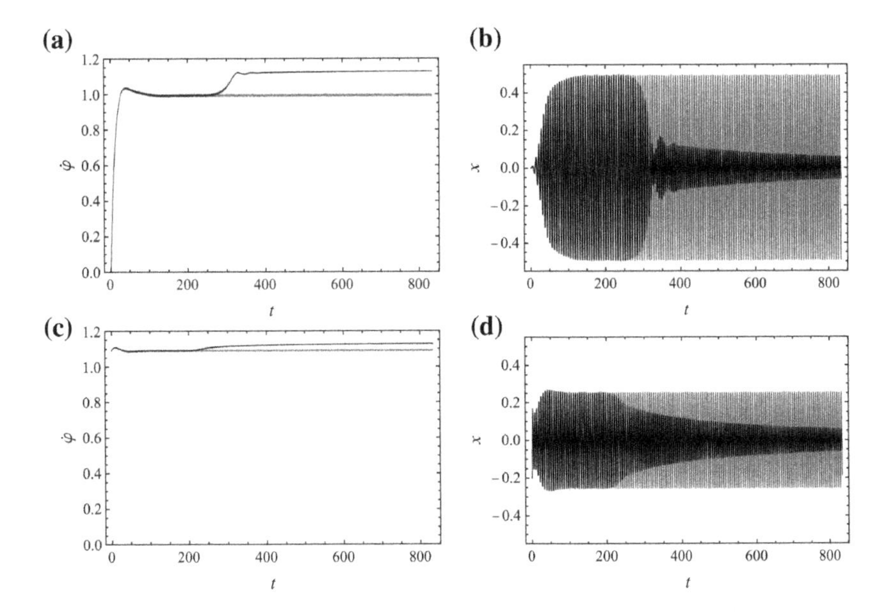


**Fig. 2.15** **a** Frequency-time diagram for upper steady state position, **b** Displacement-time diagram for upper steady state position, **c** Frequency-time diagram for lower steady state position, **d** Displacement-time diagram for lower steady state position. Mass is constant (*gray line*) and mass is increasing (*black mass*)

In Fig. 2.15a the frequency-time diagram for upper steady state position is plotted. For the constant mass after the transient motion the frequency is constant. If the mass is increasing the frequency increase, too. In Fig. 2.15b the displacement-time diagram for upper steady state position and constant mass is the gray line. Increasing the mass the amplitude decreases with mass increase. The same tendency of motion is evident for the lower steady state position (Fig. 2.15c, d). In Fig. 2.16 the case when the mass is decreasing is plotted. For the case when the mass is constant the displacement-time (gray line) and frequency-time diagrams are constant, while for decreasing mass the frequency-time diagrams decrease (Fig. 2.16a, c). Decrease of mass causes the displacement-time diagram (black line) for the upper position to decrease (Fig. 2.16b) while for the lower position to increase (Fig. 2.16d).

Finally, it can be concluded that the mass variation is suitable to be applied as a method for control of motion in non-ideal systems.

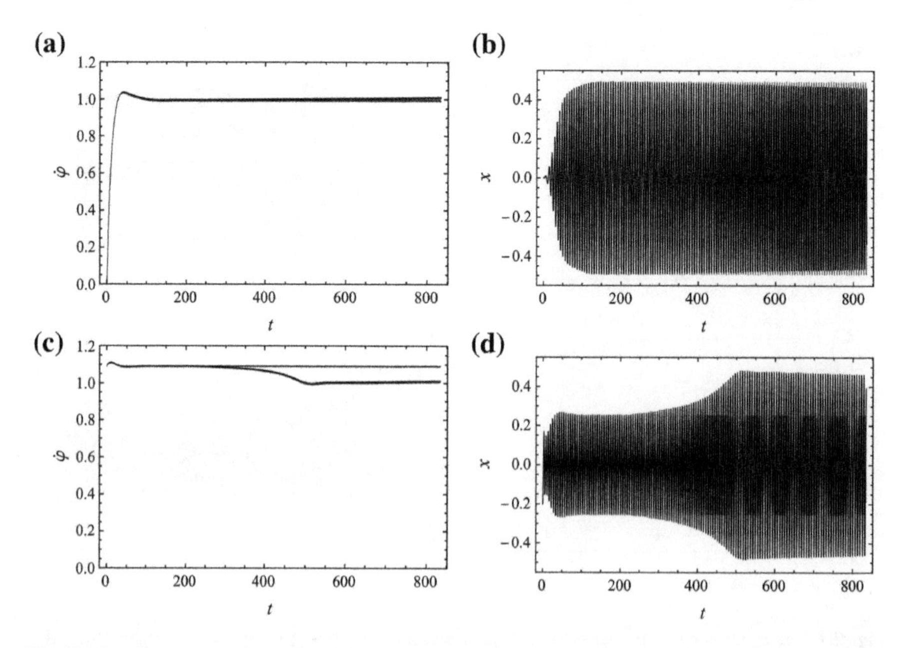


**Fig. 2.16** **a** Frequency-time diagram for upper steady state position, **b** Displacement-time diagram for upper steady state position, **c** Frequency-time diagram for lower steady state position, **d** Displacement-time diagram for lower steady state position. Mass is constant (*black line*) and mass is decreasing (*gray mass*)

## 2.3 Oscillator with Clearance Coupled with a Non-ideal Source

In the previous sections we discussed the cases when the connection between the oscillator and the fixed element is continual. Introducing the clearance in the connection between the oscillator and the fixed element, the discontinual elastic force acts. It has to be mentioned that the elastic property is a linear displacement function, but due to discontinuity the system may be treated as the nonlinear one. The mathematical model of the system is given with two coupled nonlinear differential equations. For the case of small nonlinearity the asymptotic methods are applied for determining of the transient and steady-state motion and their stability. In the system the Sommerfeld effect occurs. Beside the regular, the chaotic motion in non-ideal mechanical systems with clearance exists. It is of interest to obtain conditions for transformation of the chaotic motion into periodic motion (Zukovic and Cveticanin 2009).

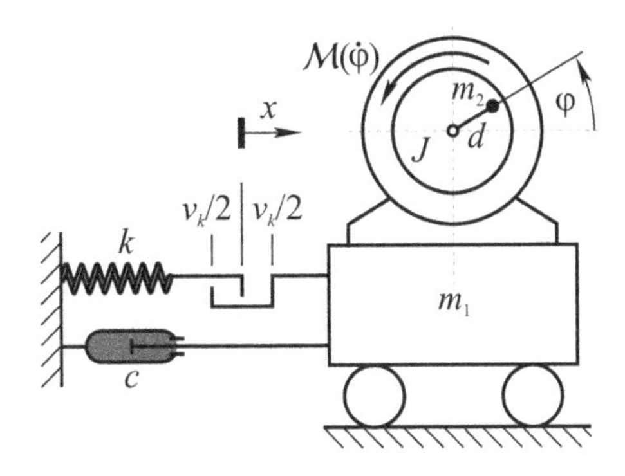


**Fig. 2.17** Model of the non-ideal mechanical system with clearance

## 2.3.1 *Model of the System*

The non-ideal system with clearance is modeled as an oscillatory system with unbalanced motor (Fig. 2.17).

The mechanical model contains the oscillatory mass $m_1$ and the motor with moment of inertia $J$, unbalance $m_2$ and eccentricity $d$. The connection of the oscillator to the fixed element has the rigidity $k$, damping $c$ and clearance $v_k$. The excitation torque of the motor, $\mathcal{M}(\dot{\varphi})$, is the function of the angular velocity $\dot{\varphi}$

$$\mathcal{M}(\dot{\varphi}) = M_0 \left(1 - \frac{\dot{\varphi}}{\Omega_0}\right), \tag{2.75}$$

where $\Omega_0$ is the steady-state angular velocity. This mathematical model corresponds to asynchronous AC motor (Dimentberg et al. 1997).

For the generalized coordinates: the displacement of the oscillator $x$ and the rotation angle of the motor $\varphi$, the motion of the system is described with a system of two coupled non-linear differential equations

$$\begin{aligned}(m_1 + m_2)\,\ddot{x} + c\dot{x} + F_k &= m_2 d\left(\ddot{\varphi}\sin\varphi + \dot{\varphi}^2\cos\varphi\right),\\ \left(J + m_2 d^2\right)\ddot{\varphi} &= m_2 d\ddot{x}\sin\varphi + \mathcal{M}(\dot{\varphi}),\end{aligned} \tag{2.76}$$

where $F_k$ is the elastic force in the spring. The weight of the elements is neglected as the motion of the system is in horizontal plane.

For the clearance $v_k$, the spring has not an influence on the motion of the system as the elastic force $F_k$ is zero (Fig. 2.18a).

For the case of spring extension it is assumed that the elastic force is the linear displacement function

$$F_k(x) = kx + f_k = kx + \begin{cases} -k\frac{v_k}{2} & \text{if} \quad x > \frac{v_k}{2} \\ -kx & \text{if} \quad -\frac{v_k}{2} \le x \le \frac{v_k}{2} \\ k\frac{v_k}{2} & \text{if} \quad x < -\frac{v_k}{2} \end{cases}. \tag{2.77}$$

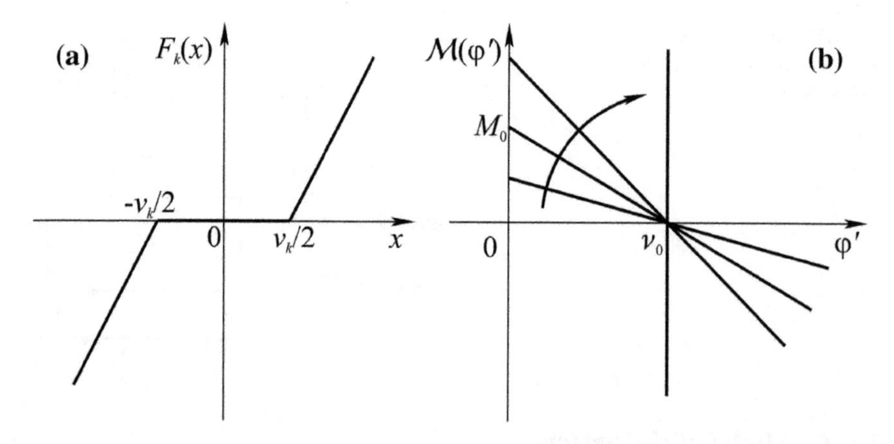


**Fig. 2.18** Properties of the system: **a** elastic force distribution, **b** motor-torque characteristics

where $kx$ is the linear part of the force and $f_k$ is different in the interval in front of and beyond clearance and also in the clearance.

By introducing the dimensionless displacement

$$y = \frac{x}{d}, \tag{2.78}$$

and dimensionless time

$$\tau = \omega t, \tag{2.79}$$

and also (2.77) into (2.76), the dimensionless differential equations of motion of the oscillatory system are obtained

$$\begin{aligned} y'' + y &= -\zeta y' - \kappa f_k + \mu \left(\varphi'' \sin\varphi + \varphi'^2 \cos\varphi\right), \\ \varphi'' &= \eta y'' \sin\varphi + \xi \mathcal{M}\left(\varphi'\right), \end{aligned} \tag{2.80}$$

where $\omega = \sqrt{\frac{k}{m_1+m_2}}$ is the eigenfrequency of the system, $(') \equiv d/d\tau$ and

$$\begin{aligned} \zeta &= \frac{c}{\sqrt{k}\left(m_1+m_2\right)}, \quad \kappa = \frac{1}{dk}, \quad \mu = \frac{m_2}{\left(m_1+m_2\right)}, \\ \eta &= \frac{m_2 d^2}{\left(J+m_2 d^2\right)}, \quad \xi = \frac{1}{\omega^2\left(J+m_2 d^2\right)}. \end{aligned} \tag{2.81}$$

The dimensionless elastic force is

$$\kappa f_k(y) = \begin{cases} -\frac{V_k}{2} & if \quad y > \frac{V_k}{2} \\ -y & if \quad -\frac{V_k}{2} \leq y \leq \frac{V_k}{2} \\ \frac{V_k}{2} & if \quad y < -\frac{V_k}{2} \end{cases}, \tag{2.82}$$

where $V_k = \frac{v_k}{d}$ is the dimensionless clearance. In dimensionless coordinates the torque is the function of the dimensionless angular velocity $\varphi'$

$$\mathcal{M}(\varphi') = M_0 \left(1 - \frac{\varphi'}{\nu_0}\right) \tag{2.83}$$

where $\nu_0 = \frac{\Omega_0}{\omega}$. The parameter $M_0$ has a significant influence on the gradient of the curve (see Fig. 2.18b): for higher values of parameter the gradient is higher and tends to vertical position when the power is unlimited and the system is ideal.

### 2.3.2 *Transient Motion of the System*

The motion of the system, with small nonlinearity and energy supply close to ideal, is considered. Due to real properties of the system it can be concluded that the parameters $\alpha$, $\mu$, $\kappa$, $\eta$ and $\xi$ in (2.80) are small. The parameters are described as

$$\zeta = \varepsilon\zeta_1, \qquad \mu = \varepsilon\mu_1, \qquad \eta = \varepsilon\eta_1, \qquad \xi = \varepsilon\xi_1, \tag{2.84}$$

where $\varepsilon << 1$ is a small parameter.

For (2.84) the differential equations of motion (2.80) are transformed into

$$\begin{aligned} y'' + y &= -\varepsilon\alpha_1 y' + \varepsilon\mu_1 \left(\varphi'' \sin\varphi + \varphi'^2 \cos\varphi\right) - \kappa f_k, \\ \varphi'' &= \varepsilon\eta_1 y'' \sin\varphi + \varepsilon\xi_1 \mathcal{M}(\varphi'), \end{aligned} \tag{2.85}$$

where $(') \equiv (d/d\tau)$ and $('') \equiv (d^2/d\tau^2)$. After some simplification and neglecting the second order small values in (2.85) the following system of differential equations of motion is obtained

$$\begin{aligned} y'' + y &= -\varepsilon\zeta_1 y' + \varepsilon\mu_1 \varphi'^2 \cos\varphi - \kappa f_k, \\ \varphi'' &= -\varepsilon\eta_1 y \sin\varphi + \varepsilon\xi_1 \mathcal{M}(\varphi'). \end{aligned} \tag{2.86}$$

Substituting (2.82) and (2.83) into (2.86), neglecting the damping term and assuming that the clearance is small, i.e., $y \gtrless \frac{\varepsilon V_k}{2}$, it follows

$$y'' + y = \varepsilon\mu_1\varphi'^2\cos\varphi \pm \frac{\varepsilon V_k}{2},$$
$$\varphi'' = -\varepsilon\eta_1 y\sin\varphi + \varepsilon\xi_1 M_0\left(1 - \frac{\varphi'}{\nu_0}\right) \quad for \quad y \gtrless \frac{\varepsilon V_k}{2}. \tag{2.87}$$

Analyzing the first relation in (2.87), it is concluded that for

$$-\frac{\varepsilon V_k}{2} \leq y \leq \frac{\varepsilon V_k}{2},$$

the deflection $y$ is of order $O(\varepsilon)$ and that the first term on the right side of the second equation is a small value of order $O(\varepsilon^2)$. Neglecting the second order small values, it follows

$$y'' = \varepsilon\mu_1\varphi'^2\cos\varphi,$$
$$\varphi'' = \varepsilon\xi_1 M_0\left(1 - \frac{\varphi'}{\nu_0}\right) \quad for \quad -\frac{\varepsilon V_k}{2} \leq y \leq \frac{\varepsilon V_k}{2}. \tag{2.88}$$

Introducing the series expansion

$$y = y_0 + \varepsilon y_1 + \cdots, \qquad \varphi = \varphi_0 + \varepsilon\varphi_1 + \cdots, \tag{2.89}$$

into (2.87), and separating the terms with the same order of small parameter $\varepsilon$ the following system of differential equations is obtained

$$\varepsilon^0 \quad : \quad y_0'' + y_0 = 0, \qquad \varphi_0'' = 0, \tag{2.90}$$
$$\varepsilon^1 \quad : \quad y_1'' + y_1 = \mu_1\left(\varphi_0'\right)^2\cos\varphi_0 \mp \frac{V_k}{2}, \tag{2.91}$$
$$\varphi_1'' \quad = \quad -\eta_1 y_0\sin\varphi_0 + \xi_1 M_0\left(1 - \frac{\varphi_0'}{\nu_0}\right),$$
$$\cdots$$

with initial conditions

$$\varepsilon^0 \quad : \quad y_0(\tau_0) = Y_0, \quad y_0'(\tau_0) = Y_0', \quad \varphi_0(\tau_0) = \Phi_0, \quad \varphi_0'(\tau_0) = \Phi_0', \tag{2.92}$$

$$\varepsilon^1 \quad : \quad y_1(\tau_0) = 0, \quad y_1'(\tau_0) = 0, \quad \varphi_1(\tau_0) = 0, \quad \varphi_1'(\tau_0) = 0, \tag{2.93}$$
$$\cdots$$

The solution of (2.90) is

$$y_0 = A_0\cos(\tau + \alpha_0), \qquad \varphi_0 = B_0\tau + C_0, \tag{2.94}$$

and of (2.91)

$$y_1 = A_1 \cos(\tau + \alpha_1) \mp \frac{V_k}{2} + \frac{\mu_1 B_0^2}{1 - B_0^2} \cos(B_0 \tau + C_0), \tag{2.95}$$

$$\begin{aligned}\varphi_1 =& \frac{(1 - B_0)^2}{2} \eta_1 A_0 \sin[(1 - B_0)\tau + (\alpha_0 - C_0)] \\ &+ \frac{(1 + B_0)^2}{2} \eta_1 A_0 \sin[(1 + B_0)\tau \\ &+ (\alpha_0 + C_0)] + \xi_1 M_0 \left(1 - \frac{B_0}{\nu_0}\right) \frac{\tau^2}{2} + B_1 \tau + C_1,\end{aligned} \tag{2.96}$$

where $A_0$, $B_0$, $C_0$ and $\alpha_0$ are integrating constants which have to be determined according to (2.92), and $A_1$, $B_1$, $C_1$ and $\alpha_1$ according to (2.93). In general, the solution in the first approximation for the case when the elastic force acts is

$$\begin{aligned}y =& A_0 \cos(\tau + \alpha_0) + \varepsilon A_1 \cos(\tau + \alpha_1) \\ &\mp \frac{\varepsilon V_k}{2} + \frac{\varepsilon \mu_1 B_0^2}{1 - B_0^2} \cos(B_0 \tau + C_0),\end{aligned} \tag{2.97}$$

$$\begin{aligned}\varphi =& (B_0 + \varepsilon B_1)\tau + (C_0 + \varepsilon C_1) + \varepsilon \xi_1 M_0 \left(1 - \frac{B_0}{\nu_0}\right) \frac{\tau^2}{2} \\ &+ \frac{\varepsilon \eta_1 A_0}{2(1 - B_0)^2} \sin[(1 - B_0)\tau + (\alpha_0 - C_0)] \\ &+ \frac{\varepsilon \eta_1 A_0}{2(1 + B_0)^2} \sin[(1 + B_0)\tau + (\alpha_0 + C_0)],\end{aligned} \tag{2.98}$$

and

$$y' = -A_0 \sin(\tau + \alpha_0) - \left[\varepsilon A_1 \sin(\tau + \alpha_0) + \frac{\varepsilon \mu_1 B_0^3}{1 - B_0^2} \sin(B_0 \tau + C_0)\right], \tag{2.99}$$

$$\begin{aligned}\varphi' =& B_0 + \varepsilon B_1 + \varepsilon \xi_1 M_0 \left(1 - \frac{B_0}{\nu_0}\right) \tau + \frac{\varepsilon \eta_1 A_0}{2(1 - B_0)} \cos[(1 - B_0)\tau + (\alpha_0 - C_0)] \\ &+ \frac{\varepsilon \eta_1 A_0}{2(1 + B_0)} \cos[(1 + B_0)\tau + (\alpha_0 + C_0)].\end{aligned} \tag{2.100}$$

For the deflection $y = (\mp \varepsilon V_k/2)$ using (2.97) the value of time $\tau_V$ is calculated. Substituting this value of time $\tau_V$ into (2.98)–(2.100) the position and velocities $\varphi(\tau_v)$, $y'(\tau_v)$ and $\varphi'(\tau_v)$ are determined. The values

$$y(\tau_v) = \mp \varepsilon V_k/2, \qquad y'(\tau_v),\ \varphi(\tau_v),\ \varphi'(\tau_v), \tag{2.101}$$

are the initial conditions for the motion of the system without elastic force. According to (2.88)

$$\varphi = K_0 + \nu_0\tau + \frac{\nu_0^2 K_1}{\varepsilon\xi_1 M_0}\exp\left(-\frac{\varepsilon\xi_1 M_0\tau}{\nu_0}\right) \approx \left(K_0 + \frac{\nu_0^2 K_1}{\varepsilon\xi_1 M_0}\right) + \nu_0(1-K_1)\tau, \tag{2.102}$$

$$\varphi' = \nu_0 - \nu_0 K_1 \exp\left(-\frac{\varepsilon\xi_1 M_0\tau}{\nu_0}\right) \approx \nu_0(1-K_1) + K_1\varepsilon\xi_1 M_0\tau, \tag{2.103}$$

$$y = -\varepsilon\mu_1\cos\left[\left(K_0 + \frac{\nu_0^2 K_1}{\varepsilon\xi_1 M_0}\right) + \nu_0(1-K_1)\tau\right] + K_2\tau + K_3, \tag{2.104}$$

$$y' = K_2 + \varepsilon\mu_1\nu_0(1-K_1)\sin\left[\left(K_0 + \frac{\nu_0^2 K_1}{\varepsilon\xi_1 M_0}\right) + \nu_0(1-K_1)\tau\right], \tag{2.105}$$

where the constants $K_0$, $K_1$, $K_2$ and $K_3$ depend on the initial conditions (2.101). The elastic force acts when $y = (\pm\varepsilon V_k/2)$ and the motion functions are (2.97)–(2.100) with initial conditions which correspond to $\tau$ and (2.103)–(2.105) at that displacement position.

In Fig. 2.19 the $y-\tau$, $y'-\tau$, $\varphi-\tau$ and $\varphi'-\tau$ time history diagrams for $\zeta_1 = 0$, $\mu_1 = 1$, $\eta_1 = 1$, $\xi_1 = 0.5$, $\nu_0 = 1.1$, $\varepsilon = 0.1$. and initial conditions $y(0) = 0.5$,

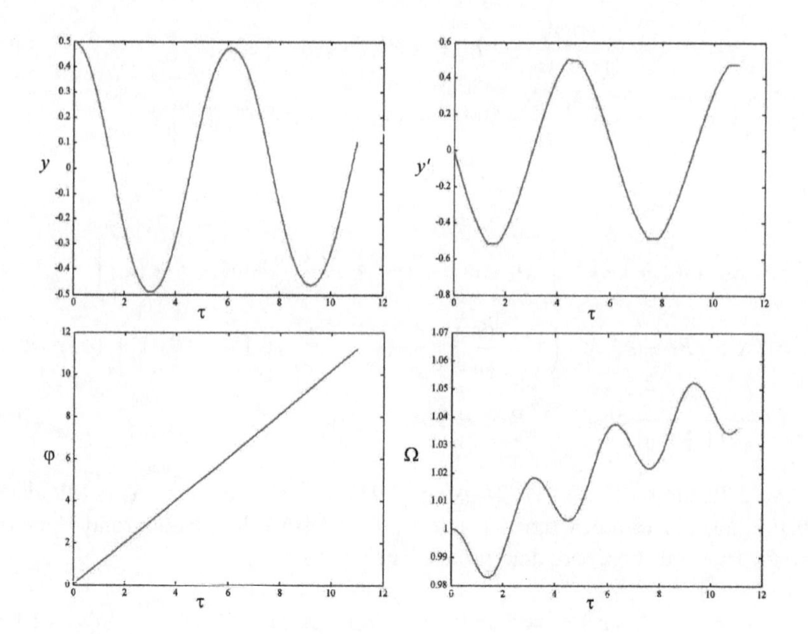


**Fig. 2.19** Time-history diagrams: **a** $y-\tau$, **b** $y'-\tau$, **c** $\varphi-\tau$, **d** $\Omega-\tau$, for $V_k = 0.1$, $\mu_1 = 1$, $\eta_1 = 1$, $\xi_1 = 0.5$, $\nu_0 = 1.1$, $\varepsilon = 0.1$ and initial conditions $y(0) = 0.5$, $y'(0) = 0$, $\varphi(0) = 0$, $\varphi'(0) = 1$

$y'(0) = 0$, $\varphi(0) = 0$, $\varphi'(0) = 1$ are plotted. Analytical solutions obtained by solving Eqs. (2.87) and (2.88) are shown. Due to small parameter values and a short time period the differences between the solutions are negligible.

### 2.3.3 *Steady-State Motion of the System*

The system of non-linear equations (2.86) is approximately solved by applying of the well known Krylov–Bogolyubov method of slow variable amplitude and phase (Bogolyubov and Mitropolskij 1974). For $\varepsilon = 0$ the solution is

$$y = a\cos(\varphi + \psi), \qquad \varphi' = \Omega = const., \tag{2.106}$$

where $a$, $\varphi$ and $\psi$ are the amplitude, frequency and phase of vibration and $\Omega$ is the constant angular velocity. Based on (2.106) the approximate solution is

$$y\,(\tau) = a(\tau)\cos\left(\varphi(\tau) + \psi(\tau)\right), \qquad \varphi' = \Omega(\tau), \tag{2.107}$$

where the amplitude $a = a\,(\tau)$, phase $\psi = \psi\,(\tau)$ and excitation frequency $\varphi'(\tau)$ are functions of slow time $\tau$. The first time derivative of (2.107) is

$$y'\,(\tau) = -a\sin\left(\varphi + \psi,\right) \tag{2.108}$$

when

$$\psi' = (1 - \Omega) + \frac{a'}{a}\frac{\cos\left(\varphi + \psi\right)}{\sin\left(\varphi + \psi\right)}. \tag{2.109}$$

The time derivative of (2.108) is

$$y'' = -a'\sin\left(\varphi + \psi\right) - a\Omega\cos\left(\varphi + \psi\right) - a\psi'\cos\left(\varphi + \psi\right). \tag{2.110}$$

Substituting (2.107)–(2.110) into (2.86) the differential equations with new variables $A$, $\psi$ and $\Omega$ are obtained

$$\begin{aligned}
a' &= -\left(\varepsilon\zeta_1 a\sin\left(\varphi + \psi\right) + \varepsilon\mu_1\Omega^2\cos\varphi\right)\sin\left(\varphi + \psi\right) + \varepsilon\,\kappa_1\,f_k\sin\left(\varphi + \psi\right) \equiv g_A,\\
\psi' &= (\Omega - 1) - \frac{\cos\left(\varphi + \psi\right)}{a}\left(\varepsilon\zeta_1 a\sin\left(\varphi + \psi\right) + \varepsilon\mu_1\Omega^2\cos\varphi\right)\\
&\quad + \frac{\cos\left(\varphi + \psi\right)}{a}\,\varepsilon\kappa_1\,f_k\\
&\equiv g_\psi,\\
\Omega' &= -\varepsilon\eta_1 a\cos\left(\varphi + \psi\right)\sin\varphi + \varepsilon\xi_1\,\mathcal{M}\,(\Omega) \equiv g_\Omega.
\end{aligned} \tag{2.111}$$

To solve the system of coupled equations (2.111) is not an easy task. Due to the fact that the functions $g_A$, $g_\psi$ and $g_\Omega$ are periodical, the averaging procedure is introduced. The averaged differential equations (2.111) are

$$\begin{aligned} a' &= \frac{1}{2\pi}\int_0^{2\pi} g_A d\varphi = -\frac{1}{2}\left(\varepsilon\zeta_1 a + \varepsilon\mu_1\Omega^2 \sin\psi\right) + G_A, \\ \psi' &= \frac{1}{2\pi}\int_0^{2\pi} g_\psi d\varphi = 1 - \Omega - \frac{\varepsilon\mu_1}{2a}\Omega^2 \cos\psi + G_\psi, \\ \Omega' &= \frac{1}{2\pi}\int_0^{2\pi} g_\Omega d\varphi = \frac{1}{2}\varepsilon\eta_1 a \sin\psi + \varepsilon\xi_1 \mathcal{M}(\Omega). \end{aligned} \tag{2.112}$$

For $a > V_k/2$

$$\begin{aligned} G_A &= \frac{1}{2\pi}\int_0^{2\pi} \left(\varepsilon\left(\sin\left(\varphi+\psi\right)\right)\left(\kappa_1 f_k\right)\right) d\varphi = 0, \\ G_\psi &= \frac{1}{2\pi}\int_0^{2\pi} \left(\varepsilon\frac{1}{a}\left(\cos\left(\varphi+\psi\right)\right)\left(\kappa_1 f_k\right)\right) d\varphi \\ &= \frac{1}{\pi}\left(-\frac{V_k}{2a}\sqrt{1-\left(\frac{V_k}{2a}\right)^2} + \arccos\frac{V_k}{2a}\right) + \left(-\frac{1}{2}\right), \end{aligned} \tag{2.113}$$

and for $a \leq V_k/2$ when the elastic force is zero

$$G_A = 0, \qquad G_\psi = -\frac{1}{2}. \tag{2.114}$$

For the steady-state motion, when $a = const.$, $\psi = const.$ and $\Omega = const.$, the differential equations (2.112) simplify to

$$\varepsilon\zeta_1 a + \varepsilon\mu_1\Omega^2 \sin\psi = 0, \tag{2.115}$$

$$\left(1 - \Omega - \frac{\varepsilon\mu_1}{2a}\Omega^2 \cos\psi\right) + G_\psi = 0, \tag{2.116}$$

$$\frac{1}{2}\varepsilon\eta_1 a \sin\psi + \varepsilon\xi_1 \mathcal{M}(\Omega) = 0. \tag{2.117}$$

From (2.115) and (2.116) the $A$ - $\Omega$ relation is obtained

$$A = \frac{\varepsilon\mu_1\Omega^2}{\sqrt{\varepsilon^2\alpha_1^2 + 4\left(1 - \Omega + G_\psi\right)^2}}. \tag{2.118}$$

For $a > V_k/2$ the amplitude-frequency function depends on the clearance, i.e.,

$$a = \frac{\varepsilon\mu_1\Omega^2}{\sqrt{\varepsilon^2\zeta_1^2 + \left(1 - 2\Omega - \frac{V_k}{a\pi}\sqrt{1 - \left(\frac{V_k}{2a}\right)^2} + \frac{2}{\pi}\arccos\frac{V_k}{2a}\right)^2}}. \tag{2.119}$$

For $a = V_k/2$, the amplitude-frequency function is

$$a = \frac{\varepsilon\mu_1\Omega^2}{\sqrt{(1 - 2\Omega)^2 + \varepsilon^2\zeta_1^2}}. \tag{2.120}$$

The relation is independent on the value of the clearance.

The amplitude has the extreme value for

$$\Omega = \frac{1 - p}{4}\left(3 \pm \sqrt{1 - \frac{8\varepsilon^2\zeta_1^2}{(1 - p)^2}}\right), \tag{2.121}$$

where

$$p = \frac{2}{\pi}\arccos\frac{V_k}{2a} - \frac{V_k}{a\pi}\sqrt{1 - \left(\frac{V_k}{2a}\right)^2}. \tag{2.122}$$

If $a = V_k/2$ and

$$\Omega = \frac{1}{4}(3 \pm \sqrt{1 - 8\varepsilon^2\zeta_1^2}), \tag{2.123}$$

the maximal amplitude is

$$a = \varepsilon\mu_1\frac{(5 - 3\sqrt{1 - 8\varepsilon^2\zeta_1^2} - 4\varepsilon^2\zeta_1^2)}{4\sqrt{2 - 2\sqrt{1 - 8\varepsilon^2\zeta_1^2} - 4\varepsilon^2\zeta_1^2}} \approx \varepsilon\mu_1\frac{1 + 4\varepsilon^2\zeta_1^2}{4\sqrt{2}\varepsilon^2\zeta_1^2}. \tag{2.124}$$

The maximal amplitude depends on the damping properties of the system and mass distribution in the system. For extremely small $\zeta_1$, when the damping is negligible, the maximal value of the amplitude $a$ tends to infinity for $\Omega = 1/2$.

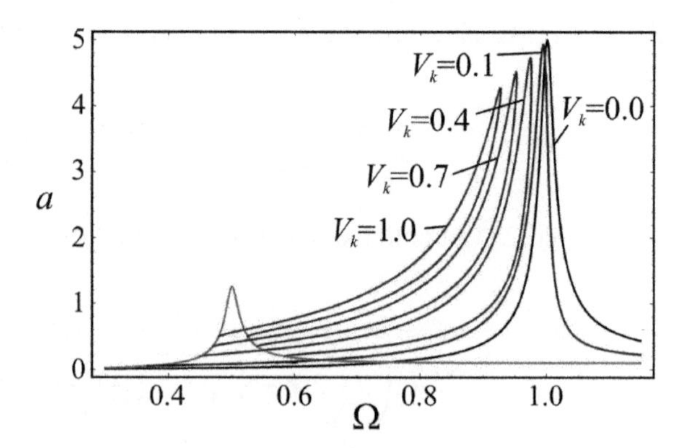


**Fig. 2.20** Amplitude-frequency curves for various values of clearance

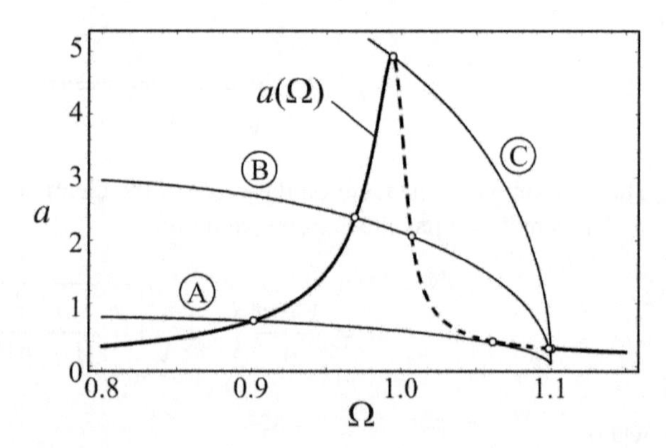


**Fig. 2.21** Frequency-response curve and characteristics of the motor: stable (——) and unstable (- - -) solutions

In Fig. 2.20 the amplitude-frequency diagrams for various values of clearance $V_k$ are plotted. The parameters of the system are $\varepsilon\zeta_1 = 0.02$, $\varepsilon\mu_1 = 0.1$, $\varepsilon\eta_1 = 0.1$, $\varepsilon\xi_1 = 0.05$, $\nu_0 = 1.1$ and $\varepsilon = 0.1$.

Using the characteristics of the motor (2.83) and the relations (2.115) and (2.117), we obtain the following relation

$$M_0\left(1 - \frac{\Omega}{\nu_0}\right)\Omega^2 = \frac{1}{2}\frac{\zeta_1\eta_1}{\mu_1\xi_1}a^2. \tag{2.125}$$

Solving the Eqs. (2.118) and (2.125), we obtain the approximate values of the steady-state amplitude $a$ and angular velocity of motor $\Omega$.

In Fig. 2.21, for parameter values $\zeta_1 = 0.2$, $\mu_1 = 1$, $\eta_1 = 1$, $\xi_1 = 0.5$, $\nu_0 = 1.1$, $\varepsilon = 0.1$ and $V_k = 0.1$, the frequency-response curve is plotted. The intersection between the motor characteristic (A, B, C) and the curve defines the number of the steady-state motions. For the two boundary curves A and C the number of steady-state solutions is two. Inside the boundary curves A and C there are three steady-state solutions (for example for the curve B). Outside the boundary curves only one steady-state solution exists: for small values of $M_0$ (below the curve A), and for very high values of $M_0$ (above curve C).

Which of steady-state motion will be realized depends on the stability and initial conditions. For stability analysis the perturbed amplitude, phase and frequency, $a = a_S + a_1$, $\psi = \psi_S + \psi_1$ and $\Omega = \Omega_S + \Omega_1$, are considered, where $a_S$, $\psi_S$ and $\Omega_s$ are the steady-state values and $a_1$, $\psi_1$ and $\Omega_1$ the perturbations. The linearized differential equations with perturbed values are

$$\begin{aligned}
a_1' &= -\frac{1}{2}\varepsilon\zeta_1 a_1 - \frac{1}{2}\varepsilon\mu_1\Omega_S^2\cos(\psi_S)\psi_1 - \varepsilon\mu_1\Omega_S\sin(\psi_S)\Omega_1,\\
\psi_1' &= \left(\frac{1}{2}\varepsilon\mu_1\Omega_S^2\frac{\cos(\psi_S)}{a_S^2} + \left(\frac{\partial G_\psi}{\partial a}\right)_S\right)a_1 + \frac{1}{2}\frac{\varepsilon}{a_S}\mu_1\Omega_S^2\sin(\psi_S)\psi_1\\
&\quad + \left(-1 - \frac{\varepsilon\mu_1}{a}\Omega_S\cos(\psi_S)\right)\Omega_1,\\
\Omega_1' &= \frac{a_1}{2}\varepsilon\eta_1\sin(\psi_S) + \frac{\psi_1}{2}\varepsilon\eta_1 a_S\cos(\psi_S) + \varepsilon\xi_1\Omega_1\left(\frac{\partial\mathcal{M}(\Omega)}{\partial\Omega}\right)_S,
\end{aligned} \tag{2.126}$$

where $\frac{\partial G_\psi}{\partial a} = \frac{1}{\pi}\frac{V_k}{a^2}\sqrt{1-\left(\frac{V_k}{2a}\right)^2}$ and $\frac{\partial\mathcal{M}(\Omega)}{\partial\Omega} = -\frac{M_0}{\nu_0}$, and $(\cdot)_S$ is the notation for steady-state condition. Using the Routh–Hurwitz procedure the stability is investigated. In Fig. 2.21, the stability regions in frequency-response diagram are plotted. The dot line is used for the unstable solutions.

To prove the correctness of the previous mentioned analytical procedure the approximate analytical solutions are compared with 'exact' numerical solutions. The system of two differential equations (2.80) is transformed into four first order differential equations

$$\begin{aligned}
y_1' &= y_2,\\
y_2' &= \frac{1}{\left(1-\mu\eta\sin^2 y_3\right)}\left(-y_1 - \zeta y_2 + \mu\left(\xi\mathcal{M}(y_4)\sin y_3 + y_4^2\cos y_3\right) - \kappa f_k\right),\\
y_3' &= y_4,\\
y_4' &= \frac{\eta\sin y_3\left(-y_1 - \zeta y_2 + \mu\left(\xi\mathcal{M}\sin y_3 + y_4^2\cos y_3\right) - \kappa f_k\right)}{\left(1-\mu\eta\sin^2 y_3\right)} + \xi\,\mathcal{M},
\end{aligned} \tag{2.127}$$

where $y_1 = y$, $y_2 = y'$, $y_3 = \varphi$, $y_4 = \varphi'$ and $\mathcal{M} = \mathcal{M}(y_4)$. The system of differential equations (2.127) is numerically solved by applying the Runge–Kutta method. The analytical and numerical solutions are plotted in Fig. 2.22.

It can be seen that the system cannot be made to respond at a frequency between $\Omega_T$ and $\Omega_H$ (grey points) and also $\Omega_R$ and $\Omega_P$ (black points) by increasing and decreasing of the control parameter $M_0$, respectively. This phenomena is called the Sommerfeld effect.

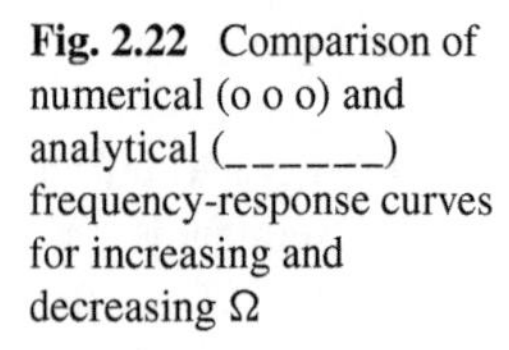

Fig. 2.22 Comparison of numerical (o o o) and analytical (______) frequency-response curves for increasing and decreasing Ω

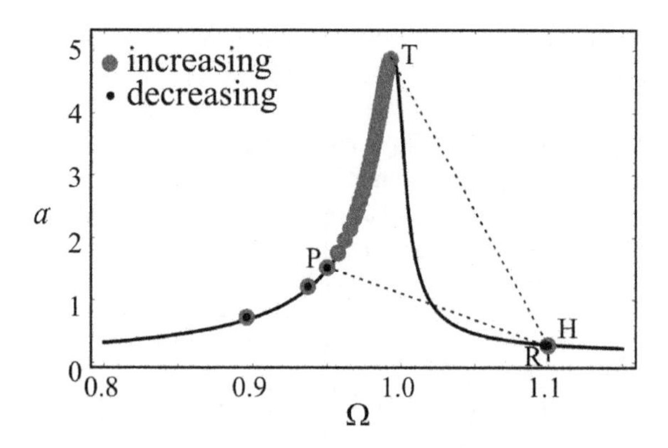


## 2.3.4 Chaotic Motion

Changing the control parameter $M_0$ the influence of motor properties on the system motion are numerically analyzed. The constant parameters of the system are: $\zeta = 0.04$, $\mu = .9375$, $\eta = 0.1$, $\xi = 0.5$, $\nu_0 = 0.2$ and $V_k = 0.01$.

For $M_0 = 0.35$ the motion of the system is periodical with period equal to excitation period (Fig. 2.23a). For control parameter $M_0 = 0.43$ the motion is periodic with period equal to double excitation period (Fig. 2.23b). Increasing the control parameter $M_0$ causes periodic motions with period doubling, as shown in bifurcation diagram (Fig. 2.24).

The multiplied bifurcations give the chaotic motion. It is concluded that for high values of control parameter ($M_0 > 75$), when the system tends to ideal one, the motion is chaotic. In Fig. 2.25 the phase plane for $M_0 = 100$ is plotted.

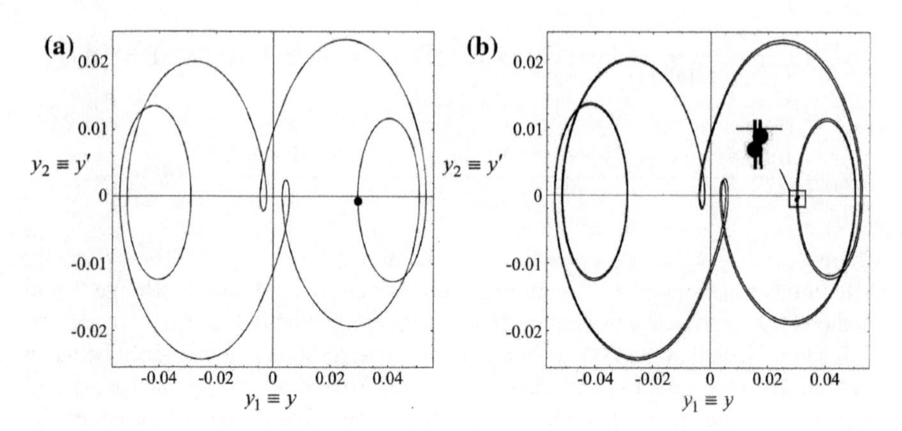


Fig. 2.23 Periodic motion: **a** period 1, **b** period 2

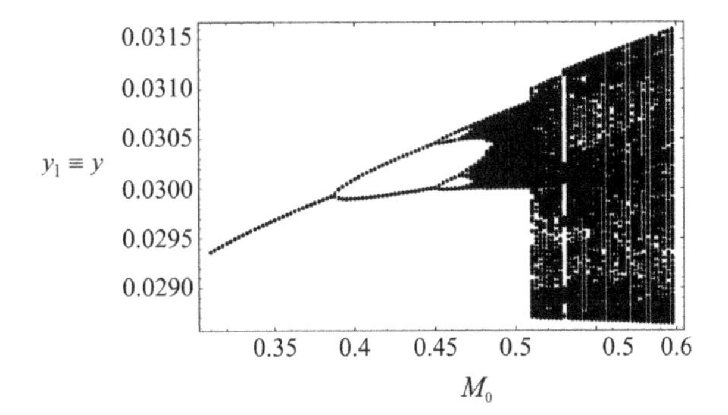


**Fig. 2.24** Bifurcation diagram for control parameter $M_0$

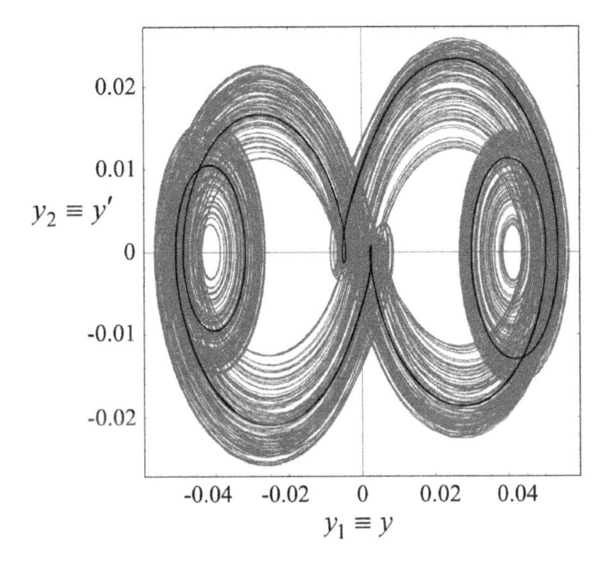


**Fig. 2.25** Phase portrait for $M_0 = 100$: before chaos control (*grey line*) and after chaos control (*black line*)

To prove the existence of the chaotic motion the Lyapunov exponent is calculated. Using the procedures suggested by Wolf (1984) and Wolf et al. (1985), and also by Sandri (1996) the maximal Lyapunov exponent for chaotic motion is $\lambda = 0.0284$ (see Fig. 2.26).

The parameter $k$ represents the number of periods of vibrations.

For the case when the control parameter is the clearance $V_k$ the bifurcation diagram is plotted (see Fig. 2.27).

The parameters of the system are: $\alpha = 0.04$, $\mu = .9375$, $\eta = 0.1$, $\xi = 0.5$, $\nu_0 = 0.2$ and $M_0 = 10$. For the small values of clearance ($V_k \leqslant 0.005$) and for high values of clearance ($V_k \geqslant 0.2$) the period one motion occurs. In some regions of the interval $0.005 < V_k < 0.2$ even chaotic motion appears.

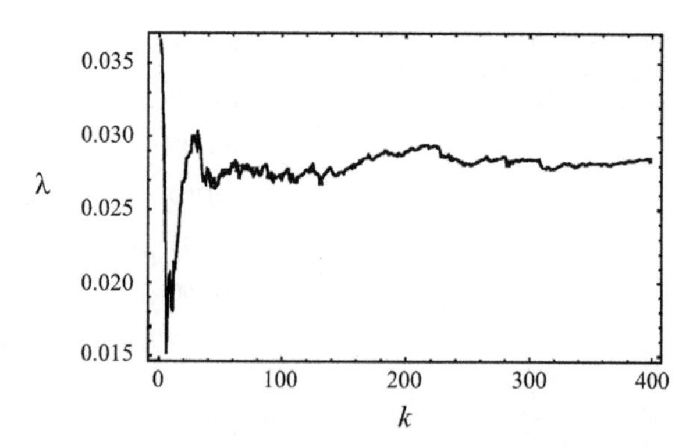


**Fig. 2.26** Distribution of Lyapunov exponent

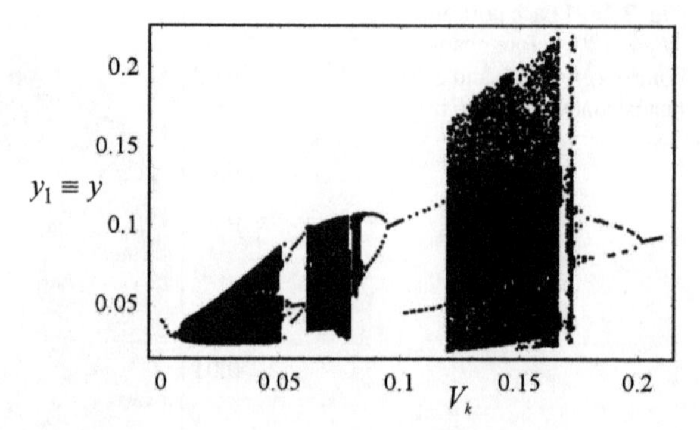


**Fig. 2.27** Bifurcation diagram for control parameter $V_k$

## 2.3.5 Chaos Control

Based on the known methods of chaos control (Ott et al. 1990; Pyragas 1992, 1996; Alvarez-Ramirez et al. 2003; Tereshko et al. 2004; Balthazar and Brasil 2004), the following control function is introduced

$$g\left(y'\right) = -h\ \tanh\left(\chi\, y'\right), \tag{2.128}$$

where $y'$ is dimensionless velocity of oscillator, $h$ is the amplitude and $\chi$ the gradient of the control function.

Solving the system of coupled differential equations of motion (2.80) with addition of the control function (2.128)

$$\begin{aligned} y'' + y &= -\alpha y' - \kappa f_k + \mu\left(\varphi'' \sin\varphi + \left(\varphi'\right)^2 \cos\varphi\right) - h\ \tanh\left(\chi\, y'\right), \\ \varphi'' &= \eta y'' \sin\varphi + \xi\, \mathcal{M}\left(\varphi'\right), \end{aligned} \tag{2.129}$$

the properties of the controlled system are obtained. The black line in Fig. 2.25 shows the motion of the system after chaos control. The chaotic attractor is transformed into periodic attractor for $\chi = 0.15$ and $h = 0.02$.

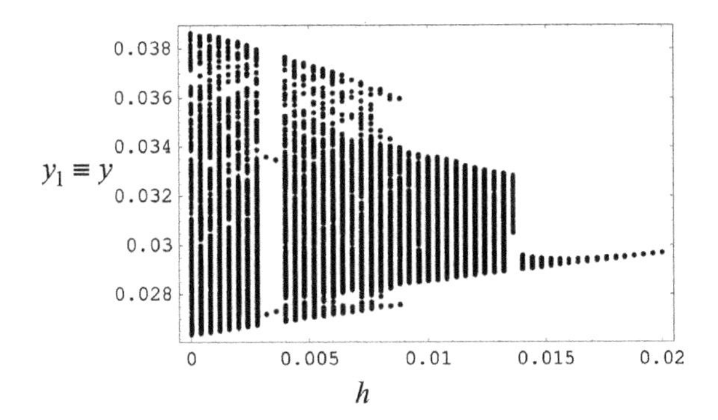


**Fig. 2.28** Bifurcation diagram after chaos control

It is worth to say that the applied control function is not an unique one, i.e., the chaos control is possible for various parameters of control function. In Fig. 2.28, the bifurcation diagram for the constant value of parameter value $\chi = 0.15$ and variable value of parameter $h$ is plotted. It is obvious that the control of chaos and its transformation to periodic solution is possible for $h > 0.014$, and to periodic motion with period 1 for $h > 0.019$.

## 2.4 Conclusion

The most important results of the chapter can be expressed as:

1. In the non-ideal mechanical system which contains a non-ideal source and a linear oscillator or an oscillator with clearance the Sommerfeld effect is evident. In amplitude-frequency diagram the jump phenomena occurs. For certain values of frequencies there are no steady-state positions.

2. In spite of the fact that the elastic force is linear, the clearance causes the bending of the amplitude-frequency curve: the higher the value of the clearance, the bending is more significant.

3. Due to clearance the motion is continual but divided into intervals with and without elastic force. It causes the disturbance of the periodic motion.

4. For certain system parameters chaotic motion occur. The type of the steady-state motion depends not only on the torque but also on the value of the clearance in the system.

5. The chaos control based on the function which depends on the velocity of oscillator vibration is very convenient for non-ideal mechanical systems with clearance. The control is directed onto the oscillator and not on the motor as it is usually done.

6. Analog model is very appropriate to give an explanation of the dynamics of the resonance capture effect and can help students and young researchers to understand this phenomenon.

## References

Alvarez-Ramirez, J., Espinosa-Paredes, G., & Puebla, H. (2003). Chaos control using small-amplitude damping signals. *Physics Letters A*, *316*, 196–205.

Balthazar, J. M., Mook, D. T., Weber, H. I., Brasil, R. M. I. R. F., Fenili, A., Beltano, D., et al. (2003). An overview on non-ideal vibrations. *Meccanica*, *38*, 613–621.

Balthazar, J. L., & Brasil, R. M. L. R. F. (2004). On saturation control of a non-ideal vibrating portal frame founded type shear - building. *Journal of Vibration and Control*, *10*, 1739–1748.

Bogolyubov, N. N., & Mitropolskij, Ju. A. (1974). *Asimptoticheskie metodi v teorii nelinejnih kolebanij*. Moscow: Nauka.

Cveticanin, L. (2010). Dynamics of the non-ideal mechanical systems: A review. *Journal of the Serbian Society for Computational Mechanics*, *4*(2), 75–86.

Cveticanin, L. (2015). *Dynamics of bodies with time-variable mass*. Berlin: Springer. ISBN 978-3-319-22055-0.

Dantas, M. J. H., & Balthazar, J. M. (2007). On the existence and stability of periodic orbits in non ideal problems: General results. *Zeitschrift fur angewandteMathematik und Physik*, *58*, 940–958.

Dimentberg, M. F., McGovern, L., Norton, R. L., Chapdelaine, J., & Harrison, R. (1997). Dynamics of an unbalanced shaft interacting with a limited power supply. *Nonlinear Dynamics*, *13*, 171–187.

Felix, J. L., Balthazar, J. M., & Brasil, R. M. L. R. F. (2009). Comments on nonlinear dynamics of a non-ideal Duffing-Rayleigh oscillator: Numerical and analytical approaches. *Journal of Sound and Vibration*, *319*, 1136–1149.

Felix, J. L. P., Balthazar, J. M., & Dantas, M. J. H. (2011). On a nonideal (MRD) damper-electro-mechanical absorber dynamics. *International Journal of Bifurcation and Chaos*, *21*(10), 2871–2882.

Goncalves, P. J. P., Silveira, M., Pontes Junior, B. R., & Balthazar, J. M. (2014). The dynamic behavior of a cantilever beam coupled to a non-ideal unbalanced motor through numerical and experimental analysis. *Journal of Sound and Vibration*, *333*, 5115–5129.

Kononenko, V. O. (1969). *Vibrating systems with a limited power supply*. London: Iliffe Books Ltd.

Kononenko, V. O. (1980). *Nelinejnie kolebanija mehanicheskih system. Izabranie trudi*. Kiev: Naukova dumka.

Kovriguine, D. A. (2012). Synchronization and sommerfeld effect as typical resonant patterns. *Archive of Applied Mechanics*, *82*(5), 591–604.

Lin, R. M., & Ewins, D. J. (1993). Chaotic vibration of mechanical systems with backlash. *Mechanical Systems and Signal Processing*, *7*, 257–272.

Nayfeh, A. H., & Mook, D. T. (1976). *Nonlinear oscillations*. New York: Wiley.

Nayfeh, A. H., & Mook, D. T. (1979). *Nonlinear oscillations*. New York: Wiley-Interscience.

Ott, E., Grebogi, C., & Yorke, Y. A. (1990). Controlling chaos. *Physical Review Letter*, *64*, 1196–1999.

Pyragas, K. (1992). Continuous control of chaos by self controlling feedback. *Physics Letters A*, *170*, 421–428.

Pyragas, K. (1996). Continuous control of chaos by self-controlling feedback. *Controlling chaos* (pp. 118–123). San Diego: Academic Press.

Samantaray, A. K., Dasgupta, S. S., & Bhattacharyya, R. (2010). Sommerfeld effect in rotationally symmetric planar dynamical systems. *International Journal of Engineering Science*, *48*, 21–36.

Sandri, M. (1996). Numerical calculation of Lyapunov exponents. *The Mathematica Journal*, *6*, 78–84.

Souza, S. L. T., Caldas, I. L., Viana, R. L., Balthazar, J. M., & Brasil, R. M. L. R. F. (2005a). Impact dampers for controlling chaos in systems with limited power supply. *Journal of Sound and Vibration*, *279*, 955–965.

Souza, S. L. T., Caldas, I. L., Viana, R. L., Balthazar, J. M., & Brasil, R. M. L. R. F. (2005b). Basins of attraction changes by amplitude constraining of oscillators with limited power supply. *Chaos, Solitons and Fractals*, *26*, 1211–1220.

Tereshko, V., Chacon, R., & Preciado, V. (2004). Controlling chaotic oscillators by altering their energy. *Physics Letters A*, *320*, 408–416.
Tusset, A. M., Balthazar, J. M., Basinello, D. G., Pntes, B. R., & Felix, J. L. P. (2012a). Statements on chaos control designs, including fractional order dynamical system, applied to a „MEMS" conb-drive actuator. *Nonlinear Dynamics*, *69*(4), 1837–1857.
Tusset, A. M., Balthazar, J. M., Chavarette, F. R. & Felix, J. L. P. (2012b). On energy transfer phenpmena, in a nonlinear ideal and nonideal essential vibrating systems, coupled to a (MR) magneto-rheological damper. *Nonlinear Dynamics*, *69*(4), 1859–1880.
Warminski, J., Balthazar, J. M., & Brasil, R. M. L. R. F. (2001). Vibrations of a non-ideal parametrically and self-excited model. *Journal of Sound and Vibration*, *245*, 363–374.
Wolf, A. (1984). In Holden, A. V. (Ed.), *Quantifying chaos with Lyapunov exponent. Nonlinear science: Theory and application*. Manchester: Manchester University Press.
Wolf, A., Swift, J., Swinney, H., & Vastano, J. (1985). Determining Lyapunov exponents from a time series. *Physica D*, *16*, 285–317.
Zukovic, M., & Cveticanin, L. (2009). Chaos in non-ideal mechanical system with clearance. *Journal of Vibration and Control*, *15*, 1229–1246.

# Chapter 3
# Nonlinear Oscillator and a Non-ideal Energy Source

In this chapter the motion of the non-ideal system which contains a nonlinear one degree-of-freedom oscillator and a non-ideal energy source is considered. In such a non-ideal oscillator-motor system there is an interaction between motions of the oscillator and of the motor as it was already explained in the previous chapter. However, due to nonlinear properties of the oscillator in the non-ideal system beside the Sommerfeld effect some additional phenomena are evident. Depending on the parameter properties of the oscillator the motion is regular or irregular. Results on motion of the non-ideal systems with nonlinear oscillators are published in Dimentberg et al. (1997), Warminski et al. (2001), Warminski and Kecik (2006) Dantas and Balthazar (2007), Felix et al. (2009a), Zukovic and Cveticanin (2007, 2009), Nbendjo et al. (2012), Cveticanin and Zukovic (2015a, b) etc.

This chapter is divided into five sections. In Sect. 3.1, a generalization of the model of the non-ideal oscillator-motor is done: a strong nonlinear oscillator is coupled with a motor with nonlinear torque property. The model of the structure-motor system is generalized by assuming that the driving torque is a nonlinear function of the angular velocity and the oscillator is with strong nonlinearity. The oscillator-motor system is assumed as a non-ideal one where not only the motor affects the motion of the oscillator but also vibrations of the oscillator have an influence on the motor motion. The model of the motor-structure system is described with two coupled strong nonlinear differential equations. An improved asymptotic analytic method based on the averaging procedure is developed for solving such a system of strong nonlinear differential equations. The steady state motion and its stability is studied. Results available the discussion of the Sommerfeld effect. A new procedure for determination of parameters of the non-ideal system for which the Sommerfeld effect does not exist is developed. For these critical values of the parameter the Sommerfeld effect is suppressed. In Sect. 3.2, the suggested theoretical consideration is applied for pure nonlinear oscillator driven by a motor with nonlinear torque characteristics. As a special case the pure nonlinear oscillator where the order of the nonlinearity is a positive rational number is investigated. The influence of the order of nonlinearity

© Springer International Publishing AG 2018

L. Cveticanin et al., *Dynamics of Mechanical Systems with Non-Ideal Excitation*,
Mathematical Engineering, DOI 10.1007/978-3-319-54169-3_3

on the dynamic properties of the system is also analyzed. The numerical simulation is done for the motor oscillator system, where the motor torque is a cubic function of the angular velocity and the oscillator is with pure nonlinearity. The obtained results are compared with those obtained analytically (Cveticanin and Zukovic 2015b). In Sect. 3.3, a pure strong nonlinear oscillator which is coupled to a non-ideal source whose torque is a linear function of the angular velocity is considered. An analytical solving procedure based on averaging is developed. The approximate solution has the form of the Ateb function. The resonant case is considered. Steady-state solution and characteristic points are determined. Special attention is given to suppression of the Sommerfeld effect. The section ends with some numerical examples (Cveticanin and Zukovic 2015a). In Sect. 3.4 the non-ideal system with the Duffing oscillator of cubic type is analyzed. The hardening Duffing oscillator with one stable fixed point which is coupled to a non-ideal energy source is mathematically modelled and analytically solved. Approximate solution of the problem is calculated. Condition for the steady-state motion are obtained. Stability of motion is investigated and phenomena of jump in the amplitude-frequency diagram is treated. Using the numerical simulation the chaotic motion is detected. A procedure for controlling chaos is introduced (Zukovic and Cveticanin 2007). Finally, in Sect. 3.5, the non-ideal system with bistable Duffing oscillator, which has three fixed points, is considered. The semi-trivial and non-trivial solutions are determined. Based on the semi-trivial solutions the conditions for quenching of the amplitude of the mechanical system are obtained. In this chapter the stability of non-trivial solutions is investigated. Based on the signs of the Lyapunov exponents (Lyapunov 1893) regions of chaos and hyperchaos are determined (Nbendjo et al. 2012).

## 3.1 Nonlinear Oscillator Coupled with a Non-ideal Motor with Nonlinear Torque

In the previous chapter the systems which have the following limitations are considered:

- the elastic force of the structure is assumed to be linear (Zukovic and Cveticanin 2009), or with small nonlinearity (Dimentberg et al. 1997; Warminski et al. 2001; Dantas and Balthazar 2006; Felix et al. 2009b),
- the torque property of the motor is assumed to be a linear function of the angular velocity (Dantas and Balthazar 2003; Tsushida et al. 2003 and 2005; Souza et al. 2005a and 2005b; Castao et al. 2010).

In this section a generalization to the model of motor-structure system is done. It is assumed that the elastic property of the system need not be linear or with a small nonlinearity but with a strong nonlinearity of any order, described with any positive rational exponent of the displacement. The motor torque is assumed to be a nonlinear function of angular velocity. No limitation to the form of the forcing torque is introduced. Such a generalization gives us an opportunity to give a more realistic view of the dynamics of the system.

In general, the model of the non-ideal system is an oscillator-motor one which has two degrees of freedom. The motion is described with a system of two coupled differential equations (see Felix et al. 2009b)

$$m\ddot{x} + \hat{f}_1(x) = \hat{f}_2(x, \dot{x}) + \hat{F}(\varphi, \dot{\varphi}, \ddot{\varphi}, \hat{q}), \qquad I\ddot{\varphi} = \mathcal{M}(\dot{\varphi}) + \hat{R}(\varphi, \dot{\varphi}, \ddot{x}, \hat{q}), \tag{3.1}$$

where $x$ and $\varphi$ are generalized coordinates of the system (displacement and angular position), $\hat{f}_1$ is the deflection function of the oscillator, $\hat{f}_2$ is the function which describes other properties of the oscillator (damping, relaxation, hysteresis...), $\hat{F}$ and $\hat{R}$ are coupling functions of the oscillator and the motor, $m$ is the mass of the oscillator, $I$ is the moment of inertia of the rotating part of the motor and $\hat{q}$ is the measure of the unbalance of the rotor. The deflection function of the oscillator $\hat{f}_1(x)$ is usually assumed to be linear or weakly nonlinear, while the function $\hat{f}_2$ is supposed to be a small one. In general, the torque of the electro-motor is

$$\mathcal{M}(\dot{\varphi}) = L(\dot{\varphi}) - H(\dot{\varphi}), \tag{3.2}$$

where $\dot{\varphi}$ is the angular velocity of the motor and $L(\dot{\varphi})$ and $H(\dot{\varphi})$ are driving and resisting torques. The torque property of the motor is usually assumed to be a linear function of the angular velocity

$$\mathcal{M}(\dot{\varphi}) = V_m^* - C_m^*\dot{\varphi}, \tag{3.3}$$

where $C_m^*$ and $V_m^*$ are characteristics of the motor (Felix et al. 2009b). Comparing the real motor torque with the (3.3) it is evident that the assumed model represents the first approximation of the real one.

The aim is to make the generalization of the problem on non-ideal systems considering the whatever any nonlinear oscillator and the improved version of the motor torque model. The function $\hat{f}_1(x)$ in (3.1) need not to be a small nonlinear function, but may be a strong nonlinear one. The elastic force in the oscillator is the function of any rational order of the displacement (integer or non-integer). Besides, the motor torque (3.2) need not to be a linear velocity function (3.3), but it may be a nonlinear one.

### 3.1.1 Nonlinear Motor Torque Property

As it is stated by Nayfeh and Mook (1979), for determination of the influence of the motion on the motor properties it is necessary to know the characteristics of the motor. Kononenko and Korablev (1959) plotted experimentally obtained torques as a function of the frequency or angular velocity of rotor for various types of DC motors, for an asynchronous and a synchronous motor. For the most of the mentioned characteristics it is common that they are nonlinear.

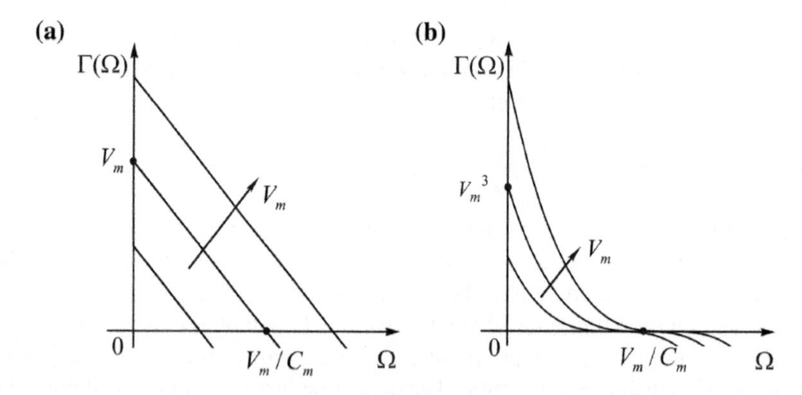


**Fig. 3.1** Torque curves for various values of control parameter $V_m$ and constant parameter $n$: **a** linear property ($n = 1$), **b** cubic property ($n = 3$)

The DC series wound motor, which is considered in this section, has a net of nonlinear characteristics for various constant control or regulator parameters (see Kononenko 1969). Mathematical model of the motor characteristics is assumed in the form

$$\mathcal{M}(\dot{\varphi}) = (V_m^* - C_m^* \dot{\varphi})^n, \tag{3.4}$$

where $V_m^*$ and $C_m^*$ are constant parameters and $n = 2, 3, 4, \ldots$ is a positive integer. It should be mentioned that the relation (3.4) includes the linear model (3.3) for $n = 1$. The DC series wound motor develops a large torque and can be operated at low speed. It is a motor that is well suited for starting heavy loads. Because of that it is often used for industrial cranes and winches, where very heavy loads must be moved slowly and lighter loads moved very rapidly. Introducing the dimensionless time parameter $\tau$, the relation (3.4) transforms into

$$\Gamma(\Omega) = (V_m - C_m \Omega)^n, \tag{3.5}$$

where $\Gamma(\Omega)$ is the dimensionless driving torque with dimensionless parameters $V_m$ and $C_m$ and angular velocity $\Omega$.

In Fig. 3.1 the torque curves for various values of control parameter $V_m$ and constant parameter $n$ are plotted: (a) linear property ($n = 1$), (b) cubic property ($n = 3$). It can be concluded that for increasing of the control parameter $V_m$ the curves move to right in the $\Gamma(\Omega) - \Omega$ plane. For the arbitrary value of the motor frequency, the higher the control parameter $V_m$, the higher the value of the torque $\Gamma(\Omega)$. In Fig. 3.2, the torque curves for constant value of parameter $V_m$ and various values of the order of nonlinearity is plotted. Increasing the order of nonlinearity $n$ the curves move to the left in the $\Gamma(\Omega) - \Omega$ plane.

Namely, for a certain constant frequency, the motor torque is higher if the order of nonlinearity $n$ is smaller. Nevertheless, the driving torque is zero and independent on the order $n$ for the angular velocity $\Omega = V_m / C_m$. The higher the control parameter

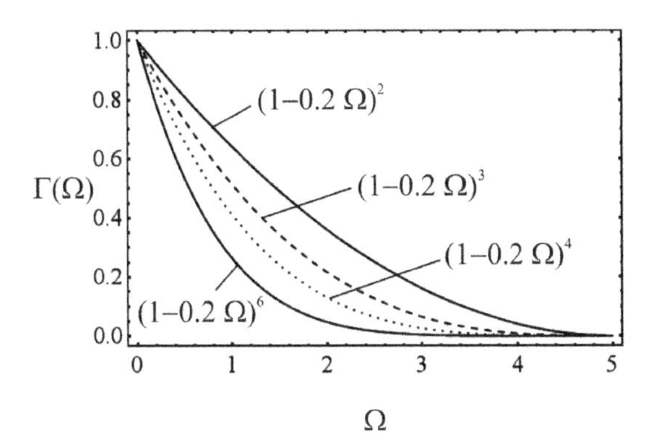


**Fig. 3.2** Torque curves for various values of the nonlinearity order $n$ and constant control parameter $V_m$

$V_m$, the higher the value of the angular velocity $\Omega$ for which the driving torque is zero.

### 3.1.2 Solution Procedure in General

Introducing the motor characteristics (3.5) into (3.1) and after some modification dimensionless differential equations follow as

$$y'' + f_1(y) = f_2(y, y') + F(\varphi, \varphi', \varphi'', q), \tag{3.6}$$

$$\varphi'' = \Gamma(\varphi') + R(\varphi, \varphi', y'', q), \tag{3.7}$$

where $y$ and $\varphi$ are the dimensionless generalized coordinates, $q$ is the dimensionless unbalance measure, $f_1$, $f_2$, $F$, $R$ and $\Gamma$ are the functions of the dimensionless coordinates and parameters and $(') = d/d\tau$, $('') = d^2/d\tau^2$ with dimensionless time $\tau$.

To classify the 'small' and 'arge' values in the system (3.6), (3.7), we introduce a small parameter $\varepsilon << 1$. Due to the physical sense of the problem the functions $f_2$, $F$, $R$ and $\Gamma$ are small and we have

$$y'' + f_1(y) = \varepsilon f_2(y, y') + \varepsilon F(\varphi, \varphi', \varphi'', q), \tag{3.8}$$

$$\varphi'' = \varepsilon^2 \Gamma(\varphi') + \varepsilon^2 R(\varphi, \varphi', y'', q). \tag{3.9}$$

The terms on the right side of Eqs. (3.8) and (3.9) are small values of the first and second order, respectively, but different from zero. It must be mentioned that, in this paper, we analyze the system with small foundation damping and in the Eq. (3.8) the damping term is of order $\varepsilon$.

In the previous investigation it was assumed that the function $f_1(y)$ is a linear one, and the Eq. (3.8) is with small perturbed terms on the right-hand side of the equation. In this paper the generalization of the problem is done, as the function $f_1(y)$ need not be linear. The suggested mathematical procedure is based on the method described for the perturbed linear differential equation.

If $\varepsilon = 0$ the Eqs. (3.8) and (3.9) simplify into two uncoupled differential equations

$$y'' + f_1(y) = 0, \qquad \varphi'' = 0. \tag{3.10}$$

For the case when $f_1(y)$ is the linear deflection function, the differential equation (3.10) is a linear one and has the exact solution in the form of the trigonometric function. Otherwise, the trigonometric function represents only the approximate solution of the nonlinear differential equation (3.10). In the papers (Cveticanin 2009; Cveticanin 2009; Cveticanin and Pogany 2012) it is already shown that the approximate solution of trigonometric type is very close to the numerical solution of (3.10) and represents a satisfactory asymptotic solution. It gives as the opportunity to assume the asymptotic solution to (3.10) in the form

$$y = a\cos(\omega(a)t + \psi), \qquad \varphi' = \Omega, \tag{3.11}$$

where $a$, $\psi$ and $\Omega$ are arbitrary constants which satisfy the initial conditions. It is worth to say that the frequency of vibration $\omega$ of the nonlinear differential equation (3.10) depends on the amplitude $a$ and has to satisfy exactly or approximately the relation

$$-a\omega^2\cos(\omega t + \psi) + f_1(a\cos(\omega t + \psi)) \approx 0. \tag{3.12}$$

It is of special interest to consider the resonant case (see Cveticanin 1995), when the difference between the frequency of vibration of the structure $\omega(a)$ and of the driving frequency $\Omega$ is small. Due to the fact that $\omega$ depends on $a$, there is a trace of frequencies which have to satisfy the relation

$$\Omega - \omega(a) = (\varepsilon\sigma)^2, \tag{3.13}$$

where $\varepsilon\sigma << 1$. The solution of (3.11) and its first time derivative for the resonant case are

$$y = a\cos(\varphi + \psi), \qquad y' = -a\Omega\sin(\varphi + \psi). \tag{3.14}$$

The method suggested in this paper requires the solution of (3.8) and (3.9) to be close to (3.14) which is the solution of (3.11). Namely, the solution of (3.8) is the perturbed version of (3.14), where the parameters are time variable. Using the procedure given by Kononenko (1969) and Cveticanin (1992) the solution of (3.8) is suggested in the form

$$y = a(\tau)\cos(\varphi(\tau) + \psi(\tau)) \equiv a\cos(\varphi + \psi), \tag{3.15}$$

and

$$y' = -a(\tau)\Omega(\tau)\sin(\varphi(\tau) + \psi(\tau)) \equiv -a\Omega\sin(\varphi + \psi). \tag{3.16}$$

The first time derivative of (3.15) is

$$y' = a' \cos(\varphi + \psi) - a(\Omega' + \psi') \sin(\varphi + \psi). \tag{3.17}$$

Comparing (3.17) with (3.16), it follows

$$a' \cos(\varphi + \psi) - a\psi' \sin(\varphi + \psi) = 0. \tag{3.18}$$

Substituting the solution (3.15) and the corresponding first (3.16) and second time derivative of (3.16) into (3.8), we obtain

$$\begin{aligned}
&-a'\Omega \sin(\varphi + \psi) - a\Omega(\Omega + \psi') \cos(\varphi + \psi) - a\Omega' \sin(\varphi + \psi) \\
&+ f_1(a \cos(\varphi + \psi)) \\
&= \varepsilon f_2(a \cos(\varphi + \psi), -a\omega \sin(\varphi + \psi)) + \varepsilon F(\psi, \Omega, \Omega', q),
\end{aligned} \tag{3.19}$$

where according to (3.11) the differential equation (3.9) transforms into two first order differential equations

$$\Omega' = \varepsilon^2 \Gamma(\Omega) + \varepsilon^2 R, \tag{3.20}$$

$$\varphi' = \Omega, \tag{3.21}$$

with

$$R = R(\varphi, \Omega, -a\Omega^2 \cos(\varphi + \psi), q).$$

Using the relation (3.12) and neglecting terms with the second order small parameter $O(\varepsilon^2)$, the relation (3.18) with (3.19) gives two first order differential equations

$$a' = -\frac{\varepsilon}{\Omega}(F + f_2) \sin(\varphi + \psi) + a(\Omega^2 - \omega^2) \sin(\varphi + \psi) \cos(\varphi + \psi) \tag{3.22}$$

$$a\psi' = -a\frac{\Omega^2 - \omega^2}{\Omega} \cos^2(\varphi + \psi) - \frac{\varepsilon(F + f_2)}{\Omega} \cos(\varphi + \psi) \tag{3.23}$$

where

$$F = F(\varphi, \Omega, \Omega', q), \qquad f_2 = f_2(a \cos(\varphi + \psi), -a\Omega \sin(\varphi + \psi)).$$

Equations (3.20)–(3.23) are four first order differential equations which correspond to two second order differential equations (3.8) and (3.9). Our task is to solve and analyze these equations.

**Averaging procedure**

Due to complexity of Eqs. (3.20)–(3.23) it is a heavy task to solve them. This is the reason that the approximate solution procedure for the system of differential equations (3.20)–(3.23) is introduced. In order to eliminate all resonances for the dynamic system described with (3.22) and (3.23), we define the resonant surface by rewriting the relation (3.13) into

$$\Omega(\tau) - \omega(a) = (\varepsilon\sigma)^2,$$

with $a = a(\tau)$. Now, we perform the averaging over the slow varying variables and apply the standard averaging procedure (see Zhuravlev and Klimov 1988; Cveticanin 1993, 2003). Averaging Eqs. (3.20)–(3.23) over the period of vibration gives

$$a' = -\frac{1}{\Omega}(\bar{F}(\psi, \Omega, q) + \bar{f}_2(a)), \tag{3.24}$$

$$a\psi' = -a\frac{\Omega^2 - \omega^2}{2\Omega} - (\bar{F}^*(\psi, \Omega, q) + \bar{f}_2^*(a)), \tag{3.25}$$

$$\Omega' = \Gamma(\Omega) + \bar{R}(\psi, \Omega, a, q), \tag{3.26}$$

where $\bar{F}$, $\bar{F}^*$, $\bar{f}_2$, $\bar{f}_2^*$ and $\bar{R}$ are averaged functions $F$, $f_2$ and $R$, respectively. Equations (3.24) and (3.25) give variations of the amplitude and initial phase of vibration of the oscillator, while (3.26) describes the variation of the averaged angular velocity of the motor. Solving these equations we obtain $a - t$, $\psi - t$ and $\Omega - t$ relations for various values of parameter. Equations describe the non-stationary motion of the system and give us very objective qualitative analysis of the problem.

It is of special interest to study the influence of the motion of the motor on the oscillator, but also of the oscillator on the motor. It requires the analysis of the coupled system of differential equations (3.24)–(3.26). Solutions $a - t$ and $\Omega - t$ have to be compared with corresponding relations for the case when there is not an interaction between the support and the motor. Then, $\bar{F} = \bar{R} = 0$ and the Eqs. (3.24)–(3.26) simplify into

$$a' = -\frac{\bar{f}_2(a)}{\omega(a)}, \qquad \xi' = \frac{\bar{f}_2^*(a)}{a}, \tag{3.27}$$

$$\Omega' = \Gamma(\Omega). \tag{3.28}$$

Separating variables in Eqs. $(3.27)_1$ and (3.28) and after some calculation, we have

$$s(a_0) - s(a) = \tau, \tag{3.29}$$

$$\Gamma_1(\Omega) - \Gamma_1(\Omega_0) = \tau, \tag{3.30}$$

where

$$s(a) = \int \frac{\bar{f}_2(a)}{\omega(a)} da, \qquad \Gamma_1(\Omega) = \int \frac{d\Omega}{\Gamma(\Omega)},$$

$a_0$ is the initial amplitude of the oscillator vibration and $\Omega_0$ is the initial angular velocity of the motor. Based on $(3.27)_2$ and (3.29) the $\psi - a$ and $\psi - \tau$ relations, for the initial phase of vibration of the oscillator, are calculated.

Comparing Eqs. (3.24)–(3.26) and (3.27), (3.28), it is evident that the most significant difference in amplitude and angular velocity is for the resonant case when the difference between the angular velocity of the motor $\Omega$ and the eigenfrequency of the oscillator $\omega$ is quite small.

### 3.1.3 Steady-State Motion and Its Stability

The steady state motion of the coupled oscillator-motor system is described with the system of three algebraic equations

$$\bar{F}(\psi, \Omega, q) + \bar{f}_2(a) = 0, \tag{3.31}$$

$$a\frac{\Omega^2 - \omega^2}{2\Omega} + \bar{F}^*(\psi, \Omega, q) + \bar{f}_2^*(a) = 0, \tag{3.32}$$

$$\Gamma(\Omega) + \bar{R}(\psi, \Omega, a, q) = 0, \tag{3.33}$$

which represent right-hand sides of Eqs. (3.24)–(3.26). Eliminating the parameter $\xi$ by combining Eqs. (3.31) and (3.32), and also (3.31) and (3.33), a system of two algebraic equations is obtained

$$Q_1(a, \Omega, q) = 0, \qquad Q_2(a, \Omega, q, V_m) = 0, \tag{3.34}$$

where equations give $a - \Omega$ relations for various values of $q$ and parameter $V_m$ of the driving torque (see Eq. (3.3)). Algebraic equations (3.34) are nonlinear and very complex. The solution of (3.34) gives the amplitude of oscillator vibration and the angular velocity of motor for the steady state motion. Very often, the solution of (3.34) is analyzed graphically by plotting of the frequency-response curves $a - \Omega$ for the oscillator (relation $(3.34)_1$) and the motor (relation $(3.34)_2$). The intersection of the curves give us the steady-state parameters of the system.

To analyze the stability of the steady-state solution the Jacobi determinant is formed

$$J = \begin{bmatrix} \frac{d\bar{f}_2(a)}{da} & \frac{\partial \bar{F}}{\partial \psi} & \frac{\partial \bar{F}}{\partial \Omega} \\ \frac{\Omega^2-\omega^2}{2\Omega} - \frac{a\omega}{\Omega}\frac{d\omega}{da} + \frac{\bar{f}_2^*(a)}{da} & \frac{\partial \bar{F}^*}{\partial \psi} & \frac{\partial \bar{F}^*}{\partial \Omega} - a \\ \frac{\partial \bar{R}}{\partial a} & \frac{\partial \bar{R}}{\partial \psi} & \frac{\partial \bar{R}}{\partial \Omega} + \frac{d\Gamma}{d\Omega} \end{bmatrix}. \tag{3.35}$$

The characteristic equation is

$$J_3\lambda^3 + J_2\lambda^2 + J_1\lambda + J_0 = 0, \tag{3.36}$$

where

$$J_3 = 1, \qquad J_2 = -\left(\frac{d\bar{f}_2(a)}{da} + \frac{\partial \bar{F}^*}{\partial \psi} + \frac{\partial \bar{R}}{\partial \Omega} + \frac{d\Gamma}{d\Omega}\right), \qquad J_0 = -\det(J),$$

$$J_1 = \left(\frac{d\bar{f}_2(a)}{da} + \frac{\partial \bar{F}^*}{\partial \psi}\right)\left(\frac{\partial \bar{R}}{\partial \Omega} + \frac{d\Gamma}{d\Omega}\right) + \frac{d\bar{f}_2(a)}{da}\frac{\partial \bar{F}^*}{\partial \psi}$$
$$+\frac{\partial \bar{F}}{\partial \psi}\left(\frac{\Omega^2 - \omega^2}{2\Omega} - \frac{a\omega}{\Omega}\frac{d\omega}{da} + \frac{\bar{f}_2^*(a)}{da}\right)$$
$$-\frac{\partial \bar{F}}{\partial \Omega}\frac{\partial \bar{R}}{\partial a} - \frac{\partial \bar{R}}{\partial \psi}\left(\frac{\partial \bar{F}^*}{\partial \Omega} - a\right). \tag{3.37}$$

Using the Routh–Hurwitz criteria it can be concluded that the stability for the steady-state solution (3.31)–(3.33) is satisfied if

$$J_0 > 0, \qquad J_1 > 0, \qquad J_2 > 0, \qquad J_1 J_2 - J_0 J_3 > 0. \tag{3.38}$$

### 3.1.4 Characteristic Points on the Steady State Curves

Let us determine the locus of characteristic points and characteristic control parameter in the frequency-response curve $(3.34)_1$ where the jump phenomena appears (see Fig. 3.3). The criteria is that at these points P and R the both curves given with (3.34) have the same direction of the joint tangent. Namely, for

$$dQ_1 = \frac{\partial Q_1}{\partial \Omega} d\Omega + \frac{\partial Q_1}{\partial a} da = 0, \qquad dQ_2 = \frac{\partial Q_2}{\partial \Omega} d\Omega + \frac{\partial Q_2}{\partial a} da = 0, \tag{3.39}$$

the equality of the direction of the tangents of the curves $(3.34)_1$ and $(3.34)_2$ in P is

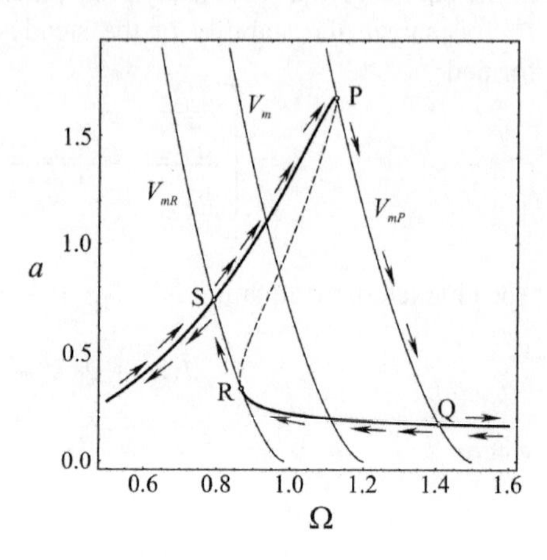


**Fig. 3.3** Characteristic points

$$\frac{da}{d\Omega} \equiv -\left(\frac{\frac{\partial Q_1}{\partial \Omega}}{\frac{\partial Q_1}{\partial a}}\right)_{\Omega_P, a_P, V_{mP}} = -\left(\frac{\frac{\partial Q_2}{\partial \Omega}}{\frac{\partial Q_2}{\partial a}}\right)_{\Omega_P, a_p, V_{mp}}. \tag{3.40}$$

Solving algebraic equations (3.40) and also (3.34), values $\Omega_P$, $a_P$, $V_{mP}$ for a peak point are obtained. The frequency $\Omega_P$ corresponds to the amplitude $a_P$ and gives the control parameter for motor torque $V_{mP}$. Using the value of the control parameter $V_{mP}$, Eq. (3.34) give us the additional pair of $(\Omega_Q, a_Q)$ values due to the fact that system of algebraic equations is nonlinear. Points P and Q correspond to the same value of control parameter $V_{mP}$.

The same procedure is applied for determining the $\Omega_R$, $a_R$, $V_{mR}$ and also $(\Omega_S, a_S)$ which corresponds to $V_{mR}$. The critical frequency $\Omega_R$ with correspondent amplitude $a_R$ gives the value of the control parameter $V_{mR}$ for which the jump phenomena to the point S appears during decreasing of the control parameter $V_m$. In the region between $V_{mP}$ and $V_{mR}$ in the amplitude-frequency diagram a gap exists.

To eliminate the Sommerfeld effect the control parameter $V_m$ has to be beyond the interval $(V_{mP}, V_{mR})$. The number of stable steady-state solutions outside this interval is only one.

### *3.1.5 Suppression of the Sommerfeld Effect*

The Sommerfeld effect does not appear if, for all of values of the driving torque, only one steady-state response of the oscillator exists. Then, the intersection between the amplitude-frequency curves of the oscillator and of the motor has only one unique solution. Using this criteria the parameters of the system have to be calculated. For technical reasons we suggest an approximate analytical method for determination of the parameters of the non-ideal system where Sommerfeld effect does not exist. The basic requirement of the method is that the bone curve $Q_3(a, \Omega)$ of the amplitude-frequency characteristic of the oscillator $(3.34)_1$ and the amplitude-frequency curve of the motor $(3.34)_2$ have the equal gradient for the extreme steady-state position $(a^*, \Omega^*)$. Namely, the following relations have to be satisfied

$$Q_1(a^*, \Omega^*, q^*) = 0, \qquad Q_2(a^*, \Omega^*, q^*, V_m^*) = 0, \qquad Q_3(a^*, \Omega^*) = 0,$$
$$\frac{da}{d\Omega} \equiv -\left(\frac{\frac{\partial Q_1}{\partial \Omega}}{\frac{\partial Q_1}{\partial a}}\right)_{\Omega^*, a^*, V_m^*, q^*} = -\left(\frac{\frac{\partial Q_3}{\partial \Omega}}{\frac{\partial Q_3}{\partial a}}\right)_{\Omega^*, a^*, V_m^*, q^*}. \tag{3.41}$$

Solving the system of algebraic equations (3.41) the parameter $q^*$ is obtained, for which only one solution for $\Omega^*$, $a^*$ and $V_m^*$ exists. Due to the fact that for either value of the control parameter $V_m$ the amplitude-frequency curve $(3.34)_2$ remains parallel to $Q_2(a, \Omega, q^*, V_m^*) = 0$ and also to the bone curve $Q_3(a, \Omega) = 0$ it can be concluded that there is only one intersection between any $Q_2(a, \Omega, q^*, V_m) = 0$ and the amplitude-frequency curve $Q_1(a, \Omega, q^*) = 0$.

*Remark 1* Using the same relations (3.41) instead of $q^*$ another critical parameter of the system can be calculated (for example $C_m^*$).

### 3.1.6 Conclusion

Analyzing the results the following is concluded:

1. The generalized non-ideal mechanical system contains the nonlinear oscillator of any order and a motor with the driving torque which need not to be the linear function of the angular velocity.
2. The approximate solution procedure of the problem suggested in the text is suitable for the near resonant case and gives the results which are possible to be used for the stability analysis and discussion of the characteristic properties of the system.
3. The approximate value of the control parameter for the non-ideal source is analytically calculated applying the method of equating the gradient of the both amplitude-frequency curves (of the oscillator and of the motor) in the intersection points. The criteria for the Sommerfeld effect is obtained.
4. It can be concluded that the method developed in the text gives the parameter values for which the Sommerfeld effect is suppressed. For these parameters there exists only one steady-state response of the oscillator for all values of the driving torque.
5. The suitable choice of non-ideal system parameters available the motion without jumps.
6. The analytically obtained results show a good agreement with numerically obtained ones. It proves the correctness of the analytic procedures.

## 3.2 Pure Nonlinear Oscillator and the Motor with Nonlinear Torque

Let us consider a motor-structure system shown in Fig. 3.4.

A motor with an unbalance is connected to a viscoelastic structure with nonlinear properties. The motion occurs in a horizontal plane and is constrained so that the motor executes a rectilinear motion along the $x$ -axis. The elastic force of the structure is assumed as a pure nonlinear displacement function

$$F_e = kx\,|x|^{\alpha-1}\,, \tag{3.42}$$

where $\alpha \geq 1$ and $\alpha \in \mathbb{R}$ is a positive rational number (integer or non-integer) which represents the order of nonlinearity and $k$ is the coefficient of rigidity. Experimental investigation on a significant number of materials, for example: aluminum, titanium

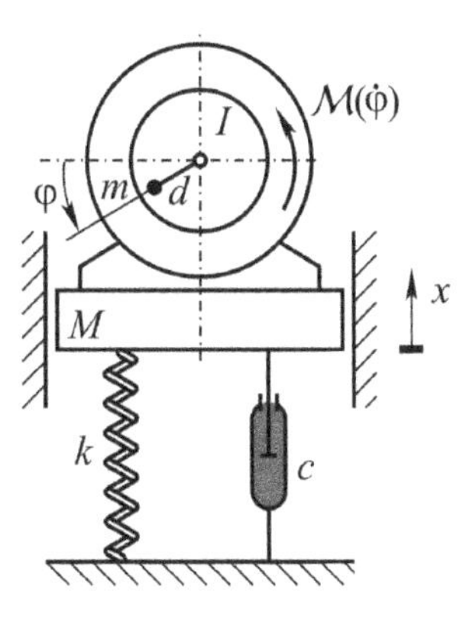


**Fig. 3.4** Model of the motor-structure non-ideal system

and other aircraft materials (Prathap and Varadan 1976), copper and copper alloys (Lo and Gupta 1978), aluminum alloys and annealed copper (Lewis and Monasa 1982), wood (Haslach 1985), ceramic materials (Colm and Clark 1988), hydrophilic polymers (Haslach 1992; Pilipchuk 2010), composites (Chen and Gibson 1998), polyurethane foam (Patten et al. 1998), felt (Russell and Rossing 1998), etc., show that the stress-strain properties of the material are nonlinear. The nonlinear dependence of the restoring force on the deflection is a polynomial whose exponent is of positive integer or non-integer order. For most of these materials the damping properties are also nonlinear. However, for the mentioned metallic materials and their alloys the order of nonlinearity in the damping force is small and the linear damping force model gives a good approximation. Thus, the damping force-velocity function is

$$F_d = c\dot{x}, \tag{3.43}$$

where $c$ is the damping constant. Mass of the system is $M$, the moment of inertia of the motor rotor is $J$, mass of the rotor unbalance is $m$ and the length of the rotor unbalance is $d$. The considered non-ideal system has two degrees-of-freedom, represented by the generalized coordinates $x$ and $\varphi$ and the motion is described with two Lagrange differential equations

$$\begin{aligned} \frac{d}{dt}\frac{\partial T}{\partial \dot{x}} - \frac{\partial T}{\partial x} + \frac{\partial U}{\partial x} + \frac{\partial \Phi}{\partial \dot{x}} &= Q_x, \\ \frac{d}{dt}\frac{\partial T}{\partial \dot{\varphi}} - \frac{\partial T}{\partial \varphi} + \frac{\partial U}{\partial \varphi} + \frac{\partial \Phi}{\partial \dot{\varphi}} &= Q_\varphi, \end{aligned} \tag{3.44}$$

where $T$ is the kinetic energy, $U$ is the potential energy, $\Phi$ is the dissipative function and $Q_x$ and $Q_\varphi$ are the generalized forces. The kinetic energy, potential energy and the dissipation function are expressed by

$$T = \frac{1}{2}M\dot{x}^2 + \frac{1}{2}m(\dot{x} - d\dot{\varphi}\cos\varphi)^2 + \frac{1}{2}m(d\dot{\varphi}\sin\varphi)^2 + \frac{1}{2}J\dot{\varphi}^2, \tag{3.45}$$

$$U = \frac{k}{\alpha + 1} x^{\alpha+1}, \qquad \Phi = \frac{1}{2} c\dot{x}^2, \tag{3.46}$$

where $\varepsilon << 1$ and a dot denotes differentiation with respect to time $t$. For the driving torque (3.2) the generalized forces are

$$Q_x = 0, \qquad Q_\varphi = \mathcal{M}(\dot{\varphi}). \tag{3.47}$$

Equations of motion (3.44) have the form

$$\begin{aligned} \ddot{x}(M+m) + kx\,|x|^{\alpha-1} + c\dot{x} - md(\ddot{\varphi}\cos\varphi - \dot{\varphi}^2 \sin\varphi) = 0, \\ (J + md^2)\ddot{\varphi} - md\ddot{x}\cos\varphi = \mathcal{M}(\dot{\varphi}), \end{aligned} \tag{3.48}$$

and the initial conditions are

$$x(0) = x_0,\ \dot{x}(0) = 0,\ \varphi(0) = 0,\ \dot{\varphi}(0) = \omega_0. \tag{3.49}$$

It is convenient to normalize the coordinates and time according to

$$x \longrightarrow y = x/l, \qquad t \longrightarrow \tau = \Omega^* t, \tag{3.50}$$

where $l$ is the initial length of the non-deformed spring and $\Omega^*$ is the synchronous angular velocity of the rotor. By introducing (3.50) the differential equations (3.48) transform into

$$\begin{aligned} y'' + \varepsilon\zeta_1 y' + p^2 y\,|y|^{\alpha-1} = \varepsilon\mu_1(\varphi''\cos\varphi - \varphi'^2 \sin\varphi), \\ \varphi'' = \varepsilon^2[\eta_2 y''\cos\varphi + \mathcal{M}(\varphi')], \end{aligned} \tag{3.51}$$

with non-dimensional initial conditions

$$y(0) = A,\ y'(0) = 0,\ \varphi(0) = 0,\ \varphi'(0) = \frac{\omega_0}{\Omega^*} = \omega, \tag{3.52}$$

where

$$\begin{aligned} p = \frac{\omega^*}{\Omega^*}, \quad \omega^{*2} = \frac{kl^{\alpha-1}}{M+m}, \quad A = \frac{x_0}{l}, \quad \varepsilon = \frac{m}{M+m}, \\ \varepsilon\mu_1 = \left(\frac{m}{M+m}\right)\left(\frac{d}{l}\right), \quad \varepsilon^2\eta_2 = \left(\frac{d}{l}\right)\frac{ml^2}{(J+md^2)}, \\ \varepsilon\zeta = \left(\frac{m}{M+m}\right)\frac{c}{\Omega^* m}, \quad \varepsilon^2\mathcal{M}(\varphi') = \frac{\mathcal{M}(\Omega^*\varphi')}{(J+md^2)\Omega^{*2}}, \end{aligned} \tag{3.53}$$

$\Omega^*$ is the synchronous angular velocity (Dimentberg et al. 1997) and prime denotes differentiation with respect to $\tau$. It is worth to say that $\varepsilon << 1$ is a small positive parameter. Using the expression (3.51) we have

$$y'' + p^2 y\,|y|^{\alpha-1} = -\varepsilon(\mu_1 \varphi'^2 \sin\varphi + \zeta_1 y'), \tag{3.54}$$

$$\varphi'' = \varepsilon^2[\mathcal{M}(\varphi') - \eta_2 p^2 y\,|y|^{\alpha-1} \cos\varphi]. \tag{3.55}$$

The Eqs. (3.54) and (3.55) represent the system of two coupled differential equations which describe the motion of the non-ideal system given in Fig. 3.4. Comparing (3.54) and (3.55) with (3.8) and (3.9) we have

$$f_1 = p^2 y\,|y|^{\alpha-1}, \quad f_2 = -\zeta_1 y', \quad F = -\mu_1 \varphi'^2 \sin\varphi, \quad R = -\eta_2 p^2 y\,|y|^{\alpha-1} \cos\varphi. \tag{3.56}$$

Now, the differential equations (3.54) and (3.55) have to be solved.

### 3.2.1 Approximate Solution Procedure

For the case when the small parameter $\varepsilon$ tends to zero, the differential equations (3.54) and (3.55) transform into

$$y'' + p^2 y\,|y|^{\alpha-1} = 0, \qquad \varphi'' = 0. \tag{3.57}$$

The two differential equations (3.57) are uncoupled and can be solved independently but according to the initial conditions (3.52). Equation $(3.57)_1$ is a second order pure nonlinear differential equation with rational order of nonlinearity. The approximate analytical solution of the $(3.57)_1$ is assumed in the form of a harmonic function $(3.11)_1$ with the frequency (see Cveticanin 2009; Cveticanin and Pogany 2012)

$$\omega = \omega_\alpha \sqrt{p^2} a^{(\alpha-1)/2}, \tag{3.58}$$

where

$$\omega_\alpha = \sqrt{\frac{\alpha+1}{2}} \frac{\sqrt{\pi}\,\Gamma(\frac{3+\alpha}{2(\alpha+1)})}{\Gamma(\frac{1}{\alpha+1})}, \tag{3.59}$$

and $\Gamma$ is the gamma function (Gradstein and Rjizhik 1971). The relation (3.58) is the exact analytically obtained frequency of vibration of the nonlinear elastic structure $(3.57)_1$. The asymptotic solution $(3.11)_1$ approximately satisfies the differential equation $(3.57)_1$, i.e.,

$$-a\omega^2 \cos(\omega t + \psi) + p^2 a^\alpha \cos^\alpha(\omega t + \psi) \approx 0. \tag{3.60}$$

Using the generalized procedure given in Sect. 3.3 and substituting the approximate solution with time variable parameters (3.15) and the corresponding derivatives into (3.54) it follows

$$-a'\Omega\sin(\varphi+\psi) - a\Omega(\Omega+\psi')\cos(\varphi+\psi) \quad (3.61)$$
$$+p^2 a^\alpha \cos^\alpha(\varphi+\psi)$$
$$= -\varepsilon\mu_1\Omega^2\sin\varphi + \varepsilon\zeta_1 a\Omega\sin(\varphi+\psi),$$

where for the resonant condition (3.13) and using the relation (3.60), the differential equation (3.61) simplifies into

$$-a'\Omega\sin(\varphi+\psi) - a(\Omega^2-\omega^2+\Omega\varphi')\cos(\varphi+\psi) = \varepsilon\mu_1\Omega^2\sin\phi + \varepsilon\zeta_1 a\Omega\sin(\varphi+\psi). \quad (3.62)$$

The relations (3.18) and (3.62) are the two first order differential equations which correspond to the second order differential equation (3.54). Solving the Eqs. (3.18) and (3.62) for $a'$ and $\psi'$ the relations (3.22) and (3.23) for (3.116) follow

$$a' = \varepsilon\mu_1\Omega\sin\varphi\sin(\varphi+\psi) - \varepsilon\zeta_1 a\sin^2(\varphi+\psi)$$
$$-a\frac{\Omega^2-\omega^2}{\Omega}\sin(\varphi+\psi)\cos(\varphi+\psi), \quad (3.63)$$
$$a\psi' = -a\frac{\Omega^2-\omega^2}{\Omega}\cos^2(\varphi+\psi) - \varepsilon\mu_1\Omega\sin\varphi\cos(\varphi+\psi)$$
$$+\varepsilon\zeta_1 a\sin(\varphi+\psi)\cos(\varphi+\psi). \quad (3.64)$$

Substituting the solution (3.15) into the differential equation (3.55) and using the relations (3.56) and (3.60) we obtain (3.20) with (3.21), i.e.,

$$\Omega' = \varepsilon^2\mathcal{M}(\Omega) - \varepsilon^2\eta_2 a\omega^2\cos(\varphi+\psi)\cos\varphi, \qquad \varphi' = \Omega. \quad (3.65)$$

Equations (3.63)–(3.65) represent the four first order differential equations which correspond to the second order differential equations (3.54) and (3.55). After averaging equations transform into

$$a' = -\frac{\zeta_1 a}{2} + \frac{\mu_1\Omega}{2}\cos\psi, \quad (3.66)$$
$$\psi' = -\frac{\Omega^2-\omega^2}{\Omega} + \frac{\mu_1\Omega}{2a}\sin\psi, \quad (3.67)$$
$$\Omega' = \mathcal{M}(\Omega) - \frac{\eta_2 a\Omega^2}{2}\cos\psi. \quad (3.68)$$

Equations (3.66)–(3.68) describe the non-stationary motion of the system.

### 3.2.2 Steady-State Motion and Its Properties

If there is no interaction between the oscillator and the motor the amplitude of vibration of the oscillator decreases exponentially from the initial amplitude $A$ (due

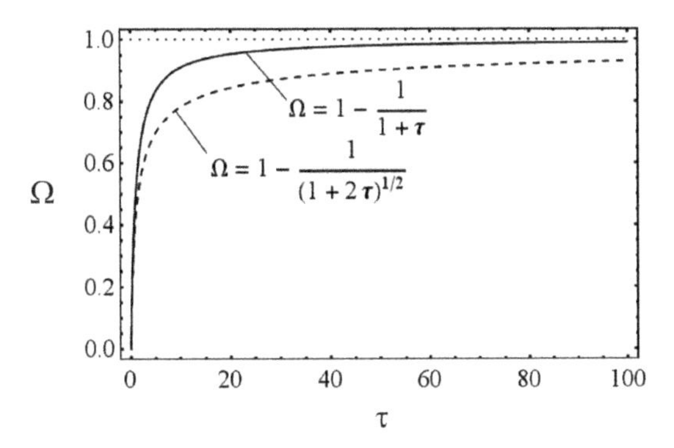


**Fig. 3.5** $\Omega - \tau$ curves for various values of the nonlinearity order $n$: $n = 2$ (*full line*), $n = 3$ (*dotted line*)

to viscous damping) as $a = A\exp(-\gamma\tau/2)$, and the variation of the angular velocity of the motor satisfies the relation

$$\Omega = \frac{V_m}{C_m} - \frac{V_m}{C_m}\frac{1}{(1 + C_m V_m^{n-1}(n-1)\tau)^{1/(n-1)}}. \tag{3.69}$$

In Fig. 3.5, $\Omega - \tau$ curves for various values of parameter $n$ are plotted.

Namely, for $V_m = 1$ and $C_m = 1$ the $\Omega - \tau$ relations according to (3.69) are: for $n = 2$

$$\Omega = 1 - \frac{1}{1 + \tau}, \tag{3.70}$$

for $n = 3$

$$\Omega = 1 - \frac{1}{(1 + 2\tau)^{1/2}}. \tag{3.71}$$

Analyzing the relations (3.69) i.e., (3.70) and (3.71), it is obvious that for any value of parameter $n$, the angular velocity is zero for $\tau = 0$, and tends to the constant steady-state value $V_m/C_m$. The smaller the value of the parameter $n$, the steady-state value is achieved in a shorter time.

For the steady-state respons, Eqs. (3.66)–(3.68) have the form

$$\frac{\zeta_1 a}{2} = \frac{\mu_1 \Omega}{2}\cos\psi, \tag{3.72}$$

$$\frac{\Omega^2 - \omega^2}{\Omega}a = \frac{\mu_1 \Omega}{2}\sin\psi, \tag{3.73}$$

$$\mathcal{M}(\Omega) = \frac{\eta_2 a \Omega^2}{2}\cos\psi. \tag{3.74}$$

Solving the algebraic equations (3.72)–(3.74), the steady-state properties of the system are determined.

Eliminating $\xi$ from (3.72) and (3.73) and also from (3.72) and (3.74) following two $a - \Omega$ relations are obtained

$$(\mu_1 \Omega)^2 = a^2 \left( \zeta_1^2 + 4 \left( \frac{\Omega^2 - \omega^2}{\Omega} \right)^2 \right), \tag{3.75}$$

$$\zeta_1 a^2 \Omega = 2 \frac{\mu_1}{\eta_2} \mathcal{M}(\Omega), \tag{3.76}$$

i.e., after substituting (3.58) and (3.4) it follows

$$(\mu_1 \Omega)^2 = a^2 \left( \zeta_1^2 + 4 \left( \frac{\Omega^2 - \omega_\alpha^2 p^2 a^{(\alpha-1)}}{\Omega} \right)^2 \right), \tag{3.77}$$

$$\zeta_1 a^2 \Omega = 2 \frac{\mu_1}{\eta_2} (V_m - C_m \Omega)^n. \tag{3.78}$$

The solution of the system of algebraic equations (3.75) and (3.76) depends on the motor torque function but also on the order of nonlinearity of the oscillator. Solving algebraic equations (3.77) and (3.78) for $a$ and $\Omega$, the steady-state phase angle $\psi$ is calculated. Namely, the relation for phase is

$$2 \frac{\Omega^2 - \omega^2}{\Omega \zeta_1} = \tan \psi, \tag{3.79}$$

and is obtained by dividing Eqs. (3.72) and (3.73).

### 3.2.3 *Characteristic Points*

According to the procedure given in Sect. 3.3, characteristic points in amplitude-frequency curves can be calculated. Solving the system of algebraic equations (3.77) and (3.78) and also the relation

$$\frac{da}{d\Omega} \equiv \frac{2\mu_1^2 \Omega^3 - \Omega a^2 \zeta_1^2 - 8a^2 \left( \Omega^2 - \omega^2 \right) \Omega}{a \zeta_1^2 \Omega^2 + 4a \left( \Omega^2 - \omega^2 \right)^2 - 4(\alpha - 1) a \left( \Omega^2 - \omega^2 \right) \omega^2}$$
$$= -\frac{\zeta_1 a^2 + 2n C_m \frac{\mu_1}{\eta_2} (V_m - C_m \Omega)^{n-1}}{2 \zeta_1 a \Omega},$$

where $\omega$ is given as (3.58), the parameters $a_P$, $\Omega_P$ and $V_{mP}$ of the characteristic point P are obtained.

Due to complexity to the suggested calculation an approximate solution procedure is recommended. Using the fact that the locus of the point P is near the position of the point P', where solutions of the Eq. (3.77) bifurcate from one to two, i.e., from three to two real solutions, it is suggested to consider the characteristics of P' instead of P. Thus, for

$$\Omega_{P'} = \omega_{p'} = \omega_\alpha a_{p'}^{(\alpha-1)/2}\sqrt{p^2}, \tag{3.80}$$

the relation (3.77) gives the peak amplitude

$$a_{P'} = \left(\frac{\zeta_1}{\omega_\alpha \mu_1 \sqrt{p^2}}\right)^{\frac{2}{\alpha-3}}. \tag{3.81}$$

Substituting (3.81) into (3.80), the locus $\Omega_{P'}$ for $a_{P'}$ is obtained

$$\Omega_{P'} = \left(\frac{\zeta_1}{\mu_1}\right)^{\frac{\alpha-1}{\alpha-3}} \left(\omega_\alpha \sqrt{p^2}\right)^{\frac{2}{3-\alpha}}. \tag{3.82}$$

Equations (3.81) and (3.82) with (3.78) give the value of a control parameter $V_{mP'}$ of the motor torque

$$V_{mP'} = C_m \left(\frac{\zeta_1}{\mu_1}\right)^{\frac{\alpha-1}{\alpha-3}} \left(\omega_\alpha \sqrt{p^2}\right)^{\frac{2}{3-\alpha}} + \left(\frac{\eta_2}{2}\right)^{\frac{1}{n}} \left(\frac{\zeta_1}{\mu_1}\right)^{\frac{2\alpha}{n(\alpha-3)}} \left(\omega_\alpha \sqrt{p^2}\right)^{\frac{6}{n(3-\alpha)}}. \tag{3.83}$$

For this approximate value of the control parameter $V_{mP'}$ the Sommerfeld effect has to appear. Analyzing the relation (3.77) it is obvious that the position of the extreme point P' is on a line

$$a = \frac{\mu_1}{\zeta_1}\Omega.$$

The gradient of the line does not depend on the order of nonlinearity $\alpha$, but only on the parameters $\mu_1$ and $\zeta_1$.

### 3.2.4 Suppression of the Sommerfeld Effect

As it is previously shown Eqs. (3.81)–(3.83) give us values $\Omega_{P'}$ and $a_{P'}$ of the point P' and also the corresponding control parameter $V_{mP'}$, which forces the amplitude-frequency torque curve through the point P'. If the amplitude-frequency torque curve and the backbone curve (3.80) have the equal gradients in the P' i.e.,

$$\frac{da}{d\Omega} \equiv \frac{2}{\omega_\alpha(\alpha-1)\sqrt{p^2}a^{\frac{\alpha-3}{2}}} = -\frac{\zeta_1 a^2 + 2nC_m \frac{\mu_1}{\eta_2}(V_m - C_m\Omega)^{n-1}}{2\zeta_1 a\Omega}, \tag{3.84}$$

the additional condition for suppression of the Sommerfeld effect in the system is obtained. Namely, solving the system of four algebraic equations (3.81)–(3.83) and (3.84) four unknown values are obtained: $a^*$, $\Omega^*$, $V_m^*$ and also $\alpha^*$, $\zeta_1^*$, $\mu_1^*$, $\eta_2^*$ or $C_m^*$. The forth mentioned parameter is the control parameter for elimination of the Sommerfeld effect.

The left side of the Eq. (3.84) is the gradient of the backbone curve in the point P'. Using (3.81) the gradient is

$$\left(\frac{da}{d\Omega}\right)_{P'} = \frac{2\mu_1}{\zeta_1(\alpha-1)}.$$

It depends on the order of nonlinearity $\alpha$: for $\alpha < 1$ it is negative, for $\alpha > 1$ it is positive, while for $\alpha = 1$ it represents an orthogonal direction. The bending is higher for $\alpha$ significantly higher or smaller than 1.

Using relations (3.81)–(3.83), Eq. (3.84) is rewritten as

$$\frac{4\zeta_1}{1-\alpha} = \gamma + C_m n \mu_1 \left(\frac{2}{\eta_2}\right)^{\frac{1}{n}} \left(\frac{\zeta_1}{\mu_1}\right)^{\frac{2}{n}\left((n-1)+\frac{n-3}{\alpha-3}\right)} \left(\omega_\alpha \sqrt{p^2}\right)^{\frac{2(n-3)}{(3-\alpha)n}}. \tag{3.85}$$

If the order of nonlinearity $\alpha$ and parameters $n$, $\gamma$, $\mu_1$ and $\eta_2$ are known, solving the relation (3.85), for example, for the parameter $C_m^*$, we have

$$C_m^* = \frac{3+\alpha}{1-\alpha}\frac{1}{n}\left(\frac{\eta_2}{2}\right)^{\frac{1}{n}} \left(\frac{\zeta_1}{\mu_1}\right)^{1-\frac{2}{n}\left((n-1)+\frac{n-3}{\alpha-3}\right)} \left(\omega_\alpha \sqrt{p^2}\right)^{\frac{2(n-3)}{(\alpha-3)n}}. \tag{3.86}$$

Substituting (3.86) into (3.87) the control parameter $V_m^*$ is

$$V_m^* = \left(\frac{3+\alpha}{1-\alpha}\frac{1}{n}+1\right)\left(\frac{\eta_2}{2}\right)^{\frac{1}{n}} \left(\frac{\zeta_1}{\mu_1}\right)^{\frac{2\alpha}{n(\alpha-3)}} \left(\omega_\alpha \sqrt{p^2}\right)^{\frac{6}{(3-\alpha)n}}. \tag{3.87}$$

It can be concluded that for parameter values (3.86) and (3.87) in the non-ideal system with known order of nonlinearity $\alpha$ the jump phenomena is excluded.

### 3.2.5 Numerical Examples

In this section, numerical examples of electro-motors connected with pure nonlinear elastic structures are considered. As is shown in the previous section, there are numerous materials whose elastic properties are strong nonlinear. We choose two of them: the copper alloy with $\alpha = 4/3$ and aluminium alloy with $\alpha = 5/3$ (Jutte 2008). The corresponding elastic forces are

$$F_{e1} = kx\,|x|^{1/3}, \qquad F_{e2} = kx\,|x|^{2/3}. \tag{3.88}$$

If the motor-structure system is driven with the motor torque of cubic type

$$\mathcal{M}(\dot\varphi) = (V_m^* - C_m^*\dot\varphi)^3. \tag{3.89}$$

differential equations of motion are:
for $\alpha = 4/3$

$$y" + p^2 y\,|y|^{1/3} = -\varepsilon(\mu_1 \varphi'^2 \sin\varphi + \zeta_1 y'), \tag{3.90}$$
$$\varphi" = \varepsilon^2[(V_m - C_m\varphi')^3 - \eta_2 p^2 y\,|y|^{1/3}\cos\varphi], \tag{3.91}$$

for $\alpha = 5/3$

$$y" + p^2 y\,|y|^{2/3} = -\varepsilon(\mu_1 \varphi'^2 \sin\varphi + \zeta_1 y'), \tag{3.92}$$
$$\varphi" = \varepsilon^2[(V_m - C_m\varphi')^3 - \eta_2 p^2 y\,|y|^{2/3}\cos\varphi]. \tag{3.93}$$

Corresponding averaged differential equations of motion obtained analytically are

$$a' = -\frac{\zeta_1 a}{2} + \frac{\mu_1 \Omega}{2}\cos\psi, \tag{3.94}$$
$$\xi' = -\frac{\Omega^2 - \omega_\alpha^2 p^2 a^{(\alpha-1)}}{\Omega} + \frac{\mu_1\Omega}{2a}\sin\psi, \tag{3.95}$$
$$\Omega' = (V_m - C_m\Omega)^3 - \frac{\eta_2 a \Omega^2}{2}\cos\psi, \tag{3.96}$$

where dependently on the order $\alpha$ the frequency constants are $\omega_{4/3} = 0.96916$ and $\omega_{5/3} = 0.940\,81$ (see Cveticanin 2009). For $\mu_1 = 0.15$, $\eta_2 = 0.05$, $p^2 = 1$, $C_m = 1$ and $\zeta_1 = 0.1$ the steady-state amplitude-frequency relations (3.77) for these systems are:
for $\alpha = 4/3$

$$0.0225\Omega^2 = a^2\left(0.01 + 4\left(\frac{\Omega^2 - 0.93927a^{1/3}}{\Omega}\right)^2\right), \tag{3.97}$$

for $\alpha = 5/3$

$$0.0225\Omega^2 = a^2\left(0.01 + 4\left(\frac{\Omega^2 - 0.885\,12a^{2/3}}{\Omega}\right)^2\right), \tag{3.98}$$

while the relation (3.78) for the motor is

$$0.1a^2\Omega = 6(V_m - \Omega)^3. \tag{3.99}$$

In Fig. 3.6, amplitude-frequency steady-state diagrams for various values of control parameter $V_m$ and order of nonlinearity $\alpha$ are plotted.

Analytically obtained (full line) curves (3.97)–(3.99) are compared with numerically (3.90)–(3.93) ones, obtained by increasing (circle) and decreasing (squares) of the control parameter $V_m$. Numerical solutions are obtained applying the Runge–Kutta

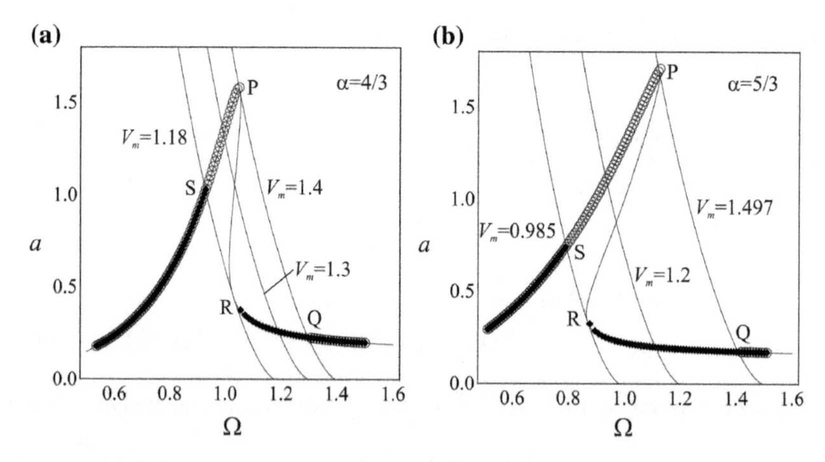


**Fig. 3.6** Frequency-response curves for various $\alpha$ obtained: **a** analytically (*full line*), **b** numerically (*circles* - for the increasing of the $V_m$, *squares* - for decreasing of the $V_m$)

**Table 3.1** Coordinates of peak in the $a_p$-$\Omega_p$ diagram for control parameter $V_{mP}$ and certain value of $\alpha$

| $\alpha$ | $a_{P'}$ | $\Omega_{P'}$ | $V_{mP'}$ | $V_{mP}$ |
|---|---|---|---|---|
| 4/3 | 1.5667 | 1.0445 | 1.3941 | 1.4000 |
| 5/3 | 1.6764 | 1.1176 | 1.4917 | 1.4970 |

procedure. We note that there is a hysteresis in diagrams in the region P-Q-R-S (see Fig. 3.6), where two stable steady-state responses exist. This phenomena, called Sommerfeld effect, depends on the orders of nonlinearity $\alpha$.

Using relations (3.81)–(3.83) coordinates of peaks ($a_{P'}$, $\Omega_{P'}$) in the diagram (3.77) and corresponding control parameters $V_{mP'}$ are calculated and shown in Table 3.1. Approximate values $V_{mP'}$ are compared with the exact numerically obtained value $V_{mP}$. It can be seen that they are in good agreement.

For the value of the control parameter $V_{mP}$ the amplitude-frequency curve (3.78) represents the boundary for which the Sommerfeld effect exists.

The question is whether the Sommerfeld effect in this non-ideal system can be suppressed. Using the procedure given in this section, the value of the parameter $\alpha$ for which the Sommerfeld effect is suppressed can be calculated. For the driving torque with cubic nonlinearity ($n = 3$) the relation (3.85) simplifies to

$$\frac{4\zeta_1}{1-\alpha} = \zeta_1 + 3C_m\mu_1 \left(\frac{2}{\eta_1}\right)^{\frac{1}{3}} \left(\frac{\zeta_1}{\mu_1}\right)^{\frac{4}{3}}. \tag{3.100}$$

Substituting the previously mentioned parameter values and $C_m = 1$ into (3.100) we obtain the critical value of the order of nonlinearity

$$\alpha^* = 0.59851. \tag{3.101}$$

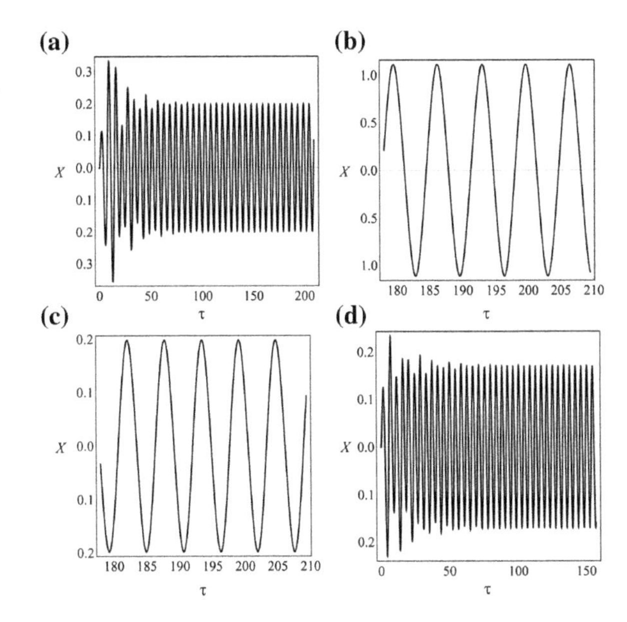


**Fig. 3.7** Time history diagrams for $\alpha = 5/3$: **a** non-stationary diagram for $V_m = 1.2$; **b, c** Steady-state diagrams for $V_m = 1.2$; **d** non-stationary motion for $V_m = 1.4$

For $\alpha = \alpha^*$ the Sommerfeld effect is suppressed. It means, that for all values of $\alpha \geq 1$ the Sommerfeld effect exists.

To prove the correctness of analytically and numerically obtained results, let us plot curves for the steady-state and of the non-stationary motion. In Fig. 3.7, amplitude-frequency diagrams of the oscillator with nonlinearity order $\alpha = 5/3$ for the control parameter $V_m = 1.2$ (non-stationary motion in Fig. 3.7a and two steady-state motions in Fig. 3.7b, c) and for the control parameter $V_m = 1.4$ (non-stationary motion in Fig. 3.7d) are shown. Comparing these diagrams with values shown in Fig. 3.7, it can be seen that they are in good agreement.

Based on the obtained results, it can be concluded that the methods developed for the general type of the non-ideal system (mentioned in the previous sections) are applicable for the pure nonlinear oscillator coupled with a non-ideal source excited with the nonlinear torque.

## 3.3 Pure Strong Nonlinear Oscillator and a Non-ideal Energy Source

Let us consider a non-ideal energy source which is settled on a foundation which is a nonlinear oscillator (Fig. 3.4) The elastic characteristic of the oscillator is pure non-linear and described with an elastic force which is proportional to the displacement $x$ with the positive rational exponent $\alpha \in \mathbb{R}$ (integer or noninteger)

$$F_e = kx\,|x|^{\alpha-1}, \tag{3.102}$$

where $k$ is the rigidity constant. The damping property of the system is supposed to be a linear velocity function and is given with the damping force

$$F_d = c\dot{x}, \tag{3.103}$$

where $c$ is the damping coefficient.

The oscillator is driven by a motor which has an unbalance $m$ which is on the distance $d$ to the rotor shaft. Position of the unbalance is varying in time and is described with the angle $\varphi$. As it is suggested in Dantas and Balthazar (2003), Tsuchida et al. (2003, 2005), Souza et al. (2005a, b), Felix et al. (2009a), Castao et al. (2010), the motor torque is a linear function of angular velocity

$$\mathcal{M}(\dot{\varphi}) = V_{\varphi m} - C_{\varphi m}\dot{\varphi}, \tag{3.104}$$

where $C_{\varphi m}$ and $V_{\varphi m}$ are constant values and $\dot{\varphi}$ is the angular velocity of the motor. The system executes a rectilinear motion and the displacement is given with the variable $x$. This motion has an effect on the rotation of the rotor of the motor.

The suggested oscillator-motor system has two-degrees-of-freedom. The two generalized coordinates are $x$ and $\varphi$. In general, the Lagrange differential equations of motion for the system are

$$\begin{aligned}
\frac{d}{dt}\frac{\partial T}{\partial \dot{x}} - \frac{\partial T}{\partial x} + \frac{\partial U}{\partial x} + \frac{\partial \Phi}{\partial \dot{x}} &= Q_x,\\
\frac{d}{dt}\frac{\partial T}{\partial \dot{\varphi}} - \frac{\partial T}{\partial \varphi} + \frac{\partial U}{\partial \varphi} + \frac{\partial \Phi}{\partial \dot{\varphi}} &= Q_\varphi,
\end{aligned} \tag{3.105}$$

where $T$ is the kinetic energy, $U$ is the potential energy, $\Phi$ is the dissipation function and $Q_x$ and $Q_\varphi$ are the corresponding generalized forces.

The kinetic energy, potential energy and the dissipation function are, respectively,

$$T = \frac{1}{2}M\dot{x}^2 + \frac{1}{2}m(\dot{x} - d\dot{\varphi}\cos\varphi)^2 + \frac{1}{2}m(d\dot{\varphi}\sin\varphi)^2 + \frac{1}{2}J\dot{\varphi}^2, \tag{3.106}$$

$$U = \frac{k}{\alpha+1}x^{\alpha+1}, \qquad \Phi = \frac{1}{2}c\dot{x}^2, \tag{3.107}$$

where a dot denotes differentiation with respect to time $t$, $M$ is mass of the oscillator, $J$ is the moment of inertia of the motor rotor, $m$ is mass of the rotor unbalance and $d$ is the length of the rotor unbalance. The virtual work of the motor torque is

$$\delta A = \mathcal{M}(\dot{\varphi})\delta\varphi, \tag{3.108}$$

and the generalized forces

$$Q_x = 0, \quad Q_\varphi = \mathcal{M}(\dot{\varphi}). \tag{3.109}$$

Substituting (3.106)–(3.109) into (3.105), the mathematical model of the system is obtained (Cveticanin 2009)

$$\ddot{x}(M+m) + kx\,|x|^{\alpha-1} + c\dot{x} - md(\ddot{\varphi}\cos\varphi - \dot{\varphi}^2\sin\varphi) = 0,$$
$$(J+md^2)\ddot{\varphi} - md\ddot{x}\cos\varphi = \mathcal{M}(\dot{\varphi}). \tag{3.110}$$

In this case the effect of gravitational force is omitted. The model (3.110) represents a system of two coupled strong nonlinear second order differential equations. To find the exact (closed form) solution of (3.110) is even impossible. An approximate solving procedure for the resonant motion of the system is suggested. It is based on the averaging procedure adopted for the system (3.110). The influence of the nonlinearity order on the Sommerfeld effect is discussed. A method for suppressing Sommerfeld effect is developed. The critical parameters for the Sommerfeld phenomena are approximately determined. As examples the steady-state resonant motions of the oscillators with non-integer order driven by a non-ideal force are considered.

### *3.3.1 Model of the System*

It is convenient to normalize the coordinates and time in (3.110) according to

$$x \longrightarrow y = x/l, \qquad t \longrightarrow \tau = \Omega^* t, \tag{3.111}$$

where $l$ is the initial length of the non-deformed spring and $\Omega^*$ is the synchronous angular velocity of the rotor (see Dimentberg et al. 1997). By introducing (3.111) and the notation $\phi \to \psi_1$, the differential equations (3.110)) transform into

$$y'' + \zeta y' + p^2 y\,|y|^{\alpha-1} = \mu(\varphi''\cos\varphi - \varphi'^2\sin\varphi),$$
$$\varphi'' = \eta y''\cos\varphi + (V_m^* - C_m^*\varphi'), \tag{3.112}$$

where

$$p^2 = \frac{kl^{\alpha-1}}{\Omega^{*2}(M+m)}, \qquad \zeta = \left(\frac{c}{M+m}\right)\frac{1}{\Omega^*}, \qquad \mu = \frac{dm}{l(M+m)},$$
$$\eta = \frac{dml}{J+md^2}, \qquad V_m^* = \frac{V_{\varphi m}}{(J+md^2)\Omega^{*2}}, \qquad C_m^* = \frac{C_{\varphi m}}{(J+md^2)\Omega^*}, \tag{3.113}$$

and $(') \equiv (d/d\tau)$, $('') \equiv (d^2/d\tau^2)$. After some modification, we have

$$y''(1-\mu\eta\cos^2\varphi)+p^2y\,|y|^{\alpha-1}+\zeta y'=\mu\left[(V_m^*-C_m^*\varphi')\cos\varphi-\varphi'^2\sin\varphi\right],$$
$$\varphi''(1-\mu\eta\cos^2\varphi)+\mu\eta\varphi'^2\cos\varphi\sin\varphi=-\eta(\zeta y'+p^2y\,|y|^{\alpha-1})\cos\varphi$$
$$+(V_m^*-C_m^*\varphi'). \quad (3.114)$$

Due to the physical properties of the system, it can be concluded that the parameters $\mu$ and $\eta$ are small in comparison to 1 and can be treated as the product of a small parameter $\varepsilon$ and constants $\mu_1$ and $\eta_1$, i.e., $\mu=\varepsilon\mu_1$ and $\eta=\varepsilon\eta_1$. The same is valid for $V_m^*$ and $C_m^*$, and also the damping parameter $\zeta$, i.e., we have $V_m^*=\varepsilon V_m$, $C_m^*=\varepsilon C_m$ and $\zeta=\varepsilon\zeta_1$. Then, the system of differential equations (3.114) is simplified into

$$y''+p^2y\,|y|^{\alpha-1}=-\varepsilon(\mu_1\varphi'^2\sin\varphi+\zeta_1y'),$$
$$\varphi''=-\varepsilon\eta_1p^2y\,|y|^{\alpha-1}\cos\varphi+\varepsilon V_m(1-K_m\varphi'), \quad (3.115)$$

where $\varepsilon<<1$ is a small positive parameter and $K_m=C_m/V_m$. In (3.115) the small terms of the second order are neglected. The Eq. (3.115) represent the system of two coupled differential equations which describe the motion of the non-ideal system given in Cveticanin and Zukovic (2015a).

### *3.3.2 Analytical Solving Procedure*

Let us rewrite the differential equations (3.115) into a system of four first order differential equations

$$y'=z,$$
$$z'=-p^2y\,|y|^{\alpha-1}+\varepsilon F_1,$$
$$\varphi'=\Omega,$$
$$\Omega'=\varepsilon F_2, \quad (3.116)$$

where

$$F_1=-(\mu_1\Omega^2\sin\varphi+\zeta z),\qquad F_2=V_m(1-K_m\Omega)-\eta_1p^2y\,|y|^{\alpha-1}\cos\varphi. \quad (3.117)$$

For $\varepsilon=0$, the Eq. (3.116) transform into

$$y'=z,\qquad z'=-p^2y\,|y|^{\alpha-1},\qquad \varphi'=\Omega,\qquad \Omega'=0. \quad (3.118)$$

The first two differential equations describe the motion of a pure integer or noninteger order nonlinear oscillator (see Cveticanin et al. 2012; Cveticanin and Pogany 2012). The exact analytical solution of (3.118) is

$$
\begin{aligned}
& y = a\mathrm{ca}(\alpha, 1, \nu t), \qquad z = -\frac{2ha^{(\alpha+1)/2}}{\alpha+1}\mathrm{sa}(1, \alpha, \nu t), \\
& \varphi = C_1 t + C_2, \qquad \Omega = C_1,
\end{aligned} \tag{3.119}
$$

where

$$
\nu = ha^{(\alpha-1)/2}, \qquad h = |p|\sqrt{\frac{\alpha+1}{2}}, \tag{3.120}
$$

$C_1$ and $C_2$ are constants of integration, $A$ is an arbitrary constant and $\mathrm{ca}(\alpha, 1, \nu t) = \mathrm{ca}$ and $\mathrm{sa}(1, \alpha, \nu t) = \mathrm{sa}$ are the cosine and sine Ateb-functions given by Droniuk et al. (1997, 2010), Droniuk and Nazarkevich (2010) (see Appendix). Namely, (3.119) is the generating solution of the generating Eq. (3.118). Based on that solution, the trial solution for (3.116) is introduced.

Let us express $y$, $z$, $\varphi$ and $\Omega$ as functions of new variables $a$, $\psi$, $\varphi$ and $\Omega$, i.e.

$$
\begin{aligned}
y &= a\mathrm{ca}(\alpha, 1, \psi) \equiv a\mathrm{ca}(\psi), \\
z &= -\frac{2h}{\alpha+1}a^{(\alpha+1)/2}\mathrm{sa}(1, \alpha, \psi) \equiv -\frac{2h}{\alpha+1}a^{(\alpha+1)/2}\mathrm{sa}(\psi),
\end{aligned} \tag{3.121}
$$

and $\psi_2 \to \Theta$. According to the expressions for the derivatives of Ateb functions (A22) and (A23) (see Appendix), the first time derivatives of (3.121) follow

$$
\begin{aligned}
y' &= a'\mathrm{ca}(\psi) - \frac{2\psi'}{\alpha+1}a\mathrm{sa}(\psi), \\
z' &= -ha^{(\alpha-1)/2}A'\mathrm{sa}(\psi) - \frac{2h\psi'}{\alpha+1}a^{(\alpha+1)/2}\mathrm{ca}^{\alpha}(\psi).
\end{aligned} \tag{3.122}
$$

Substituting (3.122) into (3.116) and using the relations (A21) (see Appendix) and (3.120), the modified equations of motion are

$$
\begin{aligned}
\psi' &= ha^{(\alpha-1)/2} - \varepsilon F_1\frac{\alpha+1}{2h}a^{-(\alpha+1)/2}\mathrm{ca}(\psi), \\
a' &= -\frac{\varepsilon F_1}{h}a^{(1-\alpha)/2}\mathrm{sa}(\psi), \\
\varphi' &= \Omega, \\
\Omega' &= \varepsilon F_2,
\end{aligned} \tag{3.123}
$$

where

$$
\begin{aligned}
F_1 &= -\mu_1\Omega^2\sin\varphi + \zeta_1\frac{2h}{\alpha+1}a^{(\alpha+1)/2}\mathrm{sa}(\psi), \\
F_2 &= V_m(1 - K_m\Omega) - \eta_1 p^2 a\mathrm{ca}(\psi)\,|a\mathrm{ca}(\psi)|^{\alpha-1}\cos\varphi.
\end{aligned}
$$

Equation (3.123) are the four first order differential equations which in the first approximation correspond to (3.166). In these equations the trigonometric and Ateb periodic functions exist. It is well known that the period of trigonometric functions $\sin\varphi$ and $\cos\varphi$ is $2\pi$, while of the Ateb functions $\mathrm{sa}(\psi)$ and $\mathrm{ca}(\psi)$ is $2\Pi_\alpha$, where the expression for $\Pi_\alpha$ is given in Appendix (see Eq. (A14))

$$\Pi_\alpha = \mathrm{B}\left(\frac{1}{\alpha+1}, \frac{1}{2}\right), \tag{3.124}$$

and B is the beta function (Abramowitz and Stegun 1964). Introducing the new variable

$$\psi = \frac{\Pi_\alpha}{2\pi}\bar{\psi}, \tag{3.125}$$

we obtain the Ateb functions $\mathrm{sa}(\frac{\Pi_\alpha}{2\pi}\bar{\psi})$ and $\mathrm{ca}(\frac{\Pi_\alpha}{2\pi}\bar{\psi})$ whose period is also $2\pi$ as is for the trigonometric functions $\sin\varphi$ and $\cos\varphi$. Substituting (3.125) into (3.123), it follows

$$\begin{aligned}
\bar{\psi}' &= \left(\frac{2\pi}{\Pi_\alpha}\right) h a^{(\alpha-1)/2} - \varepsilon F_1 \frac{\alpha+1}{2h}\left(\frac{2\pi}{\Pi_\alpha}\right) a^{-(\alpha+1)/2}\mathrm{ca}\left(\frac{\Pi_\alpha}{2\pi}\bar{\psi}\right),\\
a' &= -\frac{\varepsilon F_1}{h} a^{(1-\alpha)/2}\mathrm{sa}\left(\frac{\Pi_\alpha}{2\pi}\bar{\psi}\right),\\
\varphi' &= \Omega,\\
\Omega' &= \varepsilon F_2,
\end{aligned} \tag{3.126}$$

where

$$F_1 = -\mu_1\Omega^2 \sin\varphi + \zeta_1 \frac{2h}{\alpha+1} a^{(\alpha+1)/2}\mathrm{sa}\left(\frac{\Pi_\alpha}{2\pi}\bar{\psi}\right),$$

$$F_2 = V_m\,(1 - K_m\Omega) - \eta_1 p^2 A\mathrm{ca}\left(\frac{\Pi_\alpha}{2\pi}\bar{\psi}\right)\left|a\mathrm{ca}\left(\frac{\Pi_\alpha}{2\pi}\bar{\psi}\right)\right|^{\alpha-1} \cos\psi. \tag{3.127}$$

Differential equations (3.126) represent the mathematical model of the non-ideal system for the non-resonant case which is not of a significant interest. Much more important case is the resonant one.

### 3.3.3 *Resonant Case and the Averaging Solution Procedure*

Let us introduce the new variable $\theta$ which satisfies the relation

$$\bar{\psi} = \theta + \varphi, \tag{3.128}$$

with time derivative

$$\bar{\psi}' = \theta' + \Omega. \tag{3.129}$$

Substituting (3.129) and (3.128) into (3.126), differential equations with variables $A$, $\theta$, $\varphi$ and $\Omega$ follow as

$$\begin{aligned}
\theta' &= \left[\left(\frac{2\pi}{\Pi_\alpha}\right) h a^{(\alpha-1)/2} - \Omega\right] - \varepsilon F_1 \frac{\alpha+1}{2h}\left(\frac{2\pi}{\Pi_\alpha}\right) a^{-(\alpha+1)/2}\mathrm{ca}\left(\frac{\Pi_\alpha}{2\pi}(\theta+\varphi)\right),\\
a' &= -\frac{\varepsilon F_1}{h} a^{(1-\alpha)/2}\mathrm{sa}\left(\frac{\Pi_\alpha}{2\pi}(\theta+\varphi)\right),\\
\varphi' &= \Omega,\\
\Omega' &= \varepsilon F_2,
\end{aligned} \tag{3.130}$$

where

$$\begin{aligned}
F_1 &= -\mu_1\Omega^2 \sin\varphi + \gamma \frac{2h}{\alpha+1} a^{(\alpha+1)/2}\mathrm{sa}\left(\frac{\Pi_\alpha}{2\pi}(\theta+\varphi)\right),\\
F_2 &= V_m(1-K_m\Omega) - \eta_1 p^2 a\mathrm{ca}\left(\frac{\Pi_\alpha}{2\pi}(\theta+\psi_1)\right)\left|a\mathrm{ca}\left(\frac{\Pi_\alpha}{2\pi}(\theta+\varphi)\right)\right|^{\alpha-1}\cos\varphi.
\end{aligned} \tag{3.131}$$

For the case when

$$\left(\frac{2\pi}{\Pi_\alpha}\right) h a^{(\alpha-1)/2} - \Omega = \varepsilon\sigma, \tag{3.132}$$

the condition of nonlinear resonance is satisfied. Thus, according to $(3.130)_1$, $\theta'$ is of the order $\varepsilon$. For $\theta'$ of the order $O(\varepsilon)$, the relation (3.129) yields the difference between the frequencies $(\bar{\psi}' - \varphi') \equiv (\bar{\psi}' - \Omega)$ to be also of the $\varepsilon$ order.

To solve the system of differential equations (3.130) is not an easy task. It is the reason the averaging procedure suggested by Zhuravlev and Klimov (1988) is adopted for this special case. The averaging is done over the period $2\pi$ of the variable $\varphi$. The averaged differential equations are

$$\begin{aligned}
\theta' &= \left[\left(\frac{2\pi}{\Pi_\alpha}\right) h a^{(\alpha-1)/2} - \Omega\right] - \varepsilon \bar{F}_{1\theta} \frac{\alpha+1}{2h}\left(\frac{2\pi}{\Pi_\alpha}\right) a^{-(\alpha+1)/2},\\
a' &= -\frac{\varepsilon \bar{F}_{1A}}{h} a^{(1-\alpha)/2},\\
\varphi' &= \Omega, \qquad \Omega = \varepsilon \bar{F}_2,
\end{aligned} \tag{3.133}$$

where

$$\bar{F}_{1A} = -\mu_1 \Omega^2 \bar{f}_1 + \zeta_1 \frac{2h}{\alpha+1} a^{(\alpha+1)/2} \bar{f}_2,$$
$$\bar{F}_{1\theta} = -\mu_1 \Omega^2 \bar{f}_3 + \zeta_1 \frac{2h}{\alpha+1} a^{(\alpha+1)/2} \bar{f}_4,$$
$$\bar{F}_2 = V_m(1 - K_m \Omega) - \eta_1 p^2 A^\alpha \bar{f}_5, \tag{3.134}$$

and

$$\bar{f}_1 = \frac{1}{2\pi} \int_0^{2\pi} \sin\varphi \,\mathrm{sa}\left(\frac{\Pi_\alpha}{2\pi}(\theta+\varphi)\right) d\varphi, \quad \bar{f}_2 = \frac{1}{2\pi} \int_0^{2\pi} \mathrm{sa}^2\left(\frac{\Pi_\alpha}{2\pi}(\theta+\varphi)\right) d\varphi,$$
$$\bar{f}_3 = \frac{1}{2\pi} \int_0^{2\pi} \sin\varphi \,\mathrm{ca}\left(\frac{\Pi_\alpha}{2\pi}(\theta+\varphi)\right) d\varphi,$$
$$\bar{f}_4 = \frac{1}{2\pi} \int_0^{2\pi} \mathrm{sa}\left(\frac{\Pi_\alpha}{2\pi}(\theta+\varphi)\right) \mathrm{ca}\left(\frac{\Pi_\alpha}{2\pi}(\theta+\varphi)\right) d\varphi,$$
$$\bar{f}_5 = \frac{1}{2\pi} \int_0^{2\pi} \cos\varphi \,\mathrm{ca}\left(\frac{\Pi_\alpha}{2\pi}(\theta+\varphi)\right) \left|\mathrm{ca}^{\alpha-1}\left(\frac{\Pi_\alpha}{2\pi}(\theta+\varphi)\right)\right| d\varphi. \tag{3.135}$$

Being the sine and cosine Ateb periodic functions, they are suitable for Fourier series expansion. The finite Fourier approximation of the functions is according to Droniuk et al. (2010, 2010)

$$\mathrm{sa}(1, \alpha, \psi) = \frac{a_0}{2} + \sum_{n=1}^{\infty} a_n \sin \frac{\pi n \psi}{\Pi_\alpha},$$
$$\mathrm{ca}(\alpha, 1, \psi) = \sum_{n=1}^{\infty} b_n \cos \frac{\pi n \psi}{\Pi_\alpha}, \tag{3.136}$$

where the coefficient in the series are

$$a_0 = \frac{2}{\Pi_\alpha} \int_0^{\Pi_\alpha} sa(1, \alpha, \psi) d\psi, \qquad a_n = \frac{2}{\Pi_\alpha} \int_0^{\Pi_\alpha} sa(1, \alpha, \psi) \sin \frac{\pi n \psi}{\Pi_\alpha} d\psi,$$
$$b_n = \frac{2}{\Pi_\alpha} \int_0^{\Pi_\alpha} ca(\alpha, 1, \psi) \cos \frac{\pi n \psi}{\Pi_\alpha} d\psi \tag{3.137}$$

and $\psi \equiv \frac{\Pi_\alpha}{2\pi}(\theta + \varphi)$. In this calculation the Fourier series expansion of the function $\mathrm{ca}(\psi) \left|\mathrm{ca}^{\alpha-1}(\psi)\right|$ is also introduced as (see Mickens 2004; Cveticanin 2008)

$$ca(\psi)\left|\mathrm{ca}^{\alpha-1}(\psi)\right| = \sum_{n=1}^{\infty} c_n \cos(n\psi), \tag{3.138}$$

with

$$c_n = \frac{4}{\pi}\int_0^{\pi/2} ca(\psi)\left|\mathrm{ca}^{\alpha-1}(\psi)\right| \cos(n\psi)d\psi. \tag{3.139}$$

For practical reasons, it is suitable to determine the solution in the first approximation. Then, using the first terms of the Fourier series (3.136), the expressions (3.134) and (3.135) are transformed into

$$\bar{f}_1 = \frac{1}{2}a_1\cos\theta,\quad \bar{f}_2 = \frac{1}{2}a_1^2,\quad \bar{f}_3 = -\frac{1}{2}b_1\sin\theta,\quad \bar{f}_4 = 0,\quad \bar{f}_5 = \frac{1}{2}c_1\cos\theta, \tag{3.140}$$

and

$$\bar{F}_{1A} = -\frac{1}{2}\mu_1 a_1 \Omega^2\cos\theta + \zeta_1\frac{h}{\alpha+1}a^{(\alpha+1)/2}a_1^2,$$
$$\bar{F}_{1\theta} = \frac{1}{2}\mu_1 b_1\Omega^2\sin\theta,\quad \bar{F}_2 = V_m(1-K_m\Omega) - \frac{1}{2}\eta_1 p^2 a^\alpha c_1\cos\theta \tag{3.141}$$

where $a_1$, $b_1$ and $c_1$ are the coefficients calculated according to (3.137) and (3.139) for $n = 1$. Substituting (3.141) into (3.133) the simplified averaged differential equations are

$$\theta' = \left[\left(\frac{2\pi}{\Pi_\alpha}\right)ha^{(\alpha-1)/2} - \Omega\right] - \frac{\varepsilon\mu_1}{2}b_1\Omega^2\frac{\alpha+1}{2h}\left(\frac{2\pi}{\Pi_\alpha}\right)a^{-(\alpha+1)/2}\sin\theta,$$
$$A' = -\frac{1}{h}a^{(1-\alpha)/2}\left(-\frac{\varepsilon\mu_1}{2}a_1\Omega^2\cos\theta + \varepsilon\zeta_1 a_1^2\frac{h}{\alpha+1}a^{(\alpha+1)/2}\right),$$
$$\Omega' = \varepsilon\left(V_m(1-K_m\Omega) - \frac{1}{2}\eta_1 c_1 p^2 a^\alpha\cos\theta\right). \tag{3.142}$$

The equations describe the transient motion in the resonant case.

**Steady - state solution**

Equating the right side of the Eq. (3.142) to zero and after some modification the steady-state equations up to the first order approximation are

$$\varepsilon\mu_1\Omega^2\sin\theta = \frac{1}{b_1}\frac{4h}{\alpha+1}\left[ha^{(\alpha-1)/2} - \left(\frac{\Pi_\alpha}{2\pi}\right)\Omega\right]a^{(\alpha+1)/2}, \tag{3.143}$$

$$\varepsilon\mu_1\Omega^2\cos\theta = \frac{2\varepsilon\zeta_1 a_1 h}{\alpha+1}a^{(\alpha+1)/2}, \tag{3.144}$$

$$\eta_1 c_1 p^2 a^\alpha \cos\theta = 2V_m(1 - K_m\Omega). \tag{3.145}$$

Eliminating the variable $\theta$ in the Eqs. (3.143) and (3.144) the frequency - response relation is as follows

$$1 = \frac{a^{(\alpha+1)}}{\Omega^4 \mu_1^2}\left(\left(\frac{2\zeta_1 a_1 h}{\alpha+1}\right)^2 + \left(\frac{4h}{\varepsilon b_1}\frac{ha^{(\alpha-1)/2} - \left(\frac{\Pi_\alpha}{2\pi}\right)\Omega}{\alpha+1}\right)^2\right). \tag{3.146}$$

Dividing Eqs. (3.144) and (3.145) the $a - \theta$ expression as a function of the control parameter $V_m$ is obtained

$$V_m(1 - K_m\Omega) = \frac{\eta_1 c_1 a_1 p^2}{\mu_1(\alpha+1)}\frac{\zeta_1 h}{\Omega^2}a^{(3\alpha+1)/2}. \tag{3.147}$$

**Characteristic points**

The characteristic point P for the curves (3.72) and (3.147) corresponds to the peak amplitude and exists if the condition of the equality of the gradients $da/d\Omega$ for the both curves is satisfied. Due to complexity of the expressions, it is suggested to use the approximate procedure for obtaining of this point. Instead of the exact characteristic point P, the point P' is determined, in which the bifurcation of the solutions of the Eq. (3.72) appears. It is known that the locus of these two points (P and P') are quite close to each other.

Let us assume that in (3.66) the left hand side of the equation is zero, i.e.,

$$ha^{(\alpha-1)/2} - \left(\frac{\Pi_\alpha}{2\pi}\right)\Omega = 0. \tag{3.148}$$

Substituting (3.148) into (3.72) the peak amplitude is obtained

$$a_P = \left(\frac{1}{\mu_1}\left(\frac{\Pi_\alpha}{2\pi h}\right)^2\left(\frac{2\zeta_1 a_1 h}{\alpha+1}\right)\right)^{2/(\alpha-3)}. \tag{3.149}$$

Substituting (3.149) into (3.148), the locus $\Theta_P$ for $a_P$ is obtained

$$\Omega_P = \left(\left(\frac{\Pi_\alpha}{2\pi h}\right)^{(\alpha+1)}\left(\frac{2\zeta_1 a_1 h}{\mu_1(\alpha+1)}\right)^{(\alpha-1)}\right)^{1/(\alpha-3)}. \tag{3.150}$$

Equations (3.149) and (3.150) with (3.147) give the value of a control parameter $V_{mP'}$ of the motor torque

$$V_{mP} = \frac{q_2 c_1 a_1 p^2}{\mu_1(\alpha+1)(1 - K_m\Omega_P)}\frac{\zeta_1 h}{\Omega_P^2}a_P^{(3\alpha+1)/2}. \tag{3.151}$$

For this approximate value of the control parameter $V_{mP}$ the Sommerfeld effect has to appear.

Analyzing the relations (3.149) and (3.150) it can be seen that the relations are independent on the value of $\eta_1$.

### 3.3.4 *Suppression of the Sommerfeld Effect*

As it is previously shown the Eqs. (3.149)–(3.151) give us the values $a_{P'}$, $\Omega_{P'}$ of the point P' and also the corresponding control parameter $V_{mP}$ for which the Sommerfeld effect exists and the amplitude-frequency torque curve is forced through the point P'. Our intention is to suppress the Sommerfeld effect which is evident for P'.

Let us calculate the gradient of the backbone curve (3.148) in the bifurcation point P. Substituting $a_P$ given with the relation (3.149) into the first derivative $(da/d\Omega)$ of the relation (3.148), the required gradient is obtained

$$\left(\frac{da}{d\Omega}\right)_P = \frac{\alpha+1}{\alpha-1}\frac{\mu_1}{\zeta_1 a_1}\left(\frac{2\pi}{\Pi_\alpha}\right). \tag{3.152}$$

The gradient is the function of the order of the nonlinearity $\alpha$. For $\alpha < 1$ the direction of the tangent is such to give an obtuse angle and for $\alpha > 1$ a sharp angle. For $\alpha = 1$ the tangent is orthogonal.

The gradient of the amplitude-frequency torque curve (3.147) for the bifurcation point is also worth to be determined. The first derivative $(dA/d\Omega)$ for (3.147) is calculated and the coordinates $\Omega_P$ for $a_P$ given with relations (3.149) and (3.150) have to be substituted

$$\left(\frac{da}{d\Omega}\right)_P = V_{mP}\frac{2(2-3K_m\Omega_P)\mu_1}{\eta_1 c_1 a_1 p^2 \zeta_1 h}\frac{\alpha+1}{3\alpha+1}\frac{\Omega_P}{(a_P)^{(3\alpha-1)/2}}. \tag{3.153}$$

Equating the gradients (3.152) and (3.153) in the P' and assuming the condition (3.151), we have

$$\frac{3\alpha+1}{\alpha-1}\frac{\mu_1}{\zeta_1 a_1}\left(\frac{2\pi}{\Pi_\alpha}\right) = \frac{2(2-3K_m\Omega_P)}{(\alpha+1)(1-K_m\Omega_P)}\frac{a_P}{\Omega_P}. \tag{3.154}$$

The relation (3.154) depends on the parameter of the torque $K_m$, parameters of eccentricity $\mu_1$, and also on the order of nonlinearity $\alpha$. Due to the motor properties (3.3) and the sign of the gradient of the $V_m$ curve, it can be concluded that the relation (3.154) is valid only for $\alpha < 1$.

Let us solve the relation (3.154) for the driving parameter

$$\tilde{K}_m = \frac{2-p_1}{(3-p_1)\Omega_P}, \tag{3.155}$$

where

$$p_1 = \frac{(3\alpha+1)(\alpha+1)}{\alpha-1}\frac{\mu_1}{\zeta_1 a_1}\left(\frac{\pi}{\Pi_\alpha}\right)\frac{\Omega_P}{a_P}. \tag{3.156}$$

Then the value $\tilde{V}_m$ is the necessary one for elimination of the Sommerfeld effect. Substituting the value of $\tilde{K}_m$ (3.155) into the relation for $V_{mP}$ (3.151), the corresponding control parameter $\tilde{V}_m$ is obtained

$$\tilde{V}_m = \frac{(3-p_1)\eta_1 c_1 a_1 p^2}{\mu_1(\alpha+1)}\frac{\zeta_1 h}{\Omega_P^2} a_P^{(3\alpha+1)/2}. \tag{3.157}$$

Finally, the parameter $\tilde{C}_m$ is

$$\tilde{C}_m = \frac{\tilde{V}_m}{\tilde{K}_m} = \frac{\eta_1 c_1 a_1 p^2 (3-p_1)^2}{\mu_1(\alpha+1)(2-p_1)}\frac{\zeta_1 h}{\Omega_p} a_P^{(3\alpha+1)/2}. \tag{3.158}$$

For the parameter value (3.158) i.e., (3.155) and (3.157) the jump phenomena in the non-ideal system with known order of nonlinearity $\alpha$ is excluded. For all values of the control parameter $V_m$ there is always only one steady point in the $a-\Omega$ curve, i.e., only one intersection point between $V_m$ and $a-\Omega$ curve. Then, the Sommerfeld effect does not exist.

### 3.3.5 Numerical Examples of Non-ideal Driven Pure Nonlinear Oscillators

For physical interpretation, a motor-support system mounted in a soya extraction plant is considered: mass of the system is $(M+m)$ of 5 kg, radius of the motor rotor is 0.120 m, measure of the unbalance is $md = 0.0125$ kgm, length $l$ is 0.05 m, moment of inertia of the motor rotor $J = 0.698 \times 10^{-3}$ kgm$^2$, the synchronous speed of the rotor is $\Omega^* = 1450$ rpm and the damping coefficient is $c = 37.961$ kg/s. The three types of nonlinearity are considered: one, with the order of nonlinearity $\alpha = 2/3$ (smaller than 1) and coefficient of rigidity $k = 0.42470 \times 10^5$ N/m$^{2/3}$, second, with $\alpha = 1$ (linear case) and coefficient of rigidity $k = 1.1528 \times 10^5$ N/m and with the order of nonlinearity $\alpha = 5/3$ (higher than 1) and coefficient of rigidity $k = 8.4941 \times 10^5$ N/m$^{5/3}$. Two values of the motor parameters are considered: $C_{\varphi m} = 1.898$ Nm/s and $C_{\varphi m} = 18.98$ Nm/s. The corresponding non-dimensional parameters are according to (3.53): $\varepsilon = 0.1$, $\mu_1 = 0.5$, $\eta_1 = 0.5$, $p^2 = 1$ and $\zeta_1 = 0.5$. Two values for $C_m$ are 1 and 10. Varying the control parameter $V_m$ of the motor, the steady-state properties of the system are analyzed. For $C_m = 1$ the initial control parameter for the increasing case is $V_m = 0.3$ and for the decreasing case it is $V_m = 1.2$, while for $C_m = 10$ it is $V_m = 3$ and $V_m = 12$, respectively. The analytical results are obtained by solving the relations (3.72) and (3.147), and numerical ones by

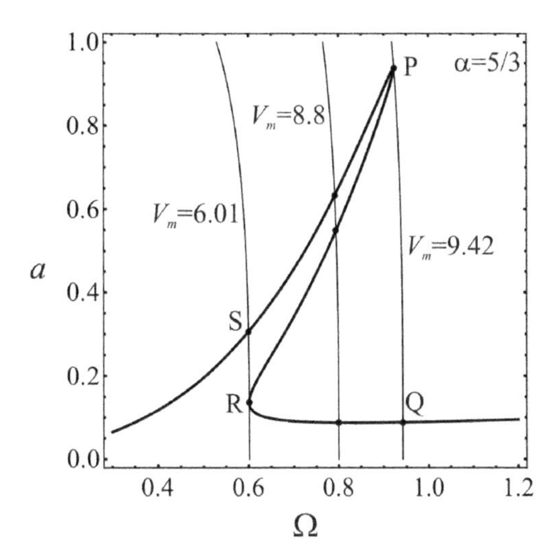


**Fig. 3.8** Frequency-response steady state curve obtained analytically for $\alpha = 5/3$ and $C_m = 10$ with characteristic points

integrating of the original differential equations of motion (3.48). For numerical calculation the Runge–Kutta procedure is applied.

The numerical procedure is as follows: for a certain value of the control parameter $V_m$ the relations (3.48) are numerically solved and the amplitude and frequency are computed. The amplitude of vibration is computed for the maximal disposition of the oscillator from the equilibrium position, while the frequency of vibration is the averaged value for $\dot{\psi}_1$ during the steady state motion. These values are the initial values for the further numerical calculation. Increasing the control parameter $V_m$ for a small value, the new steady-state parameters are computed. The numerical calculation is repeated for the new higher value of the control parameter and with initial conditions which are the previous steady-state parameters. After the significant repetition of the numerical process for increasing of the control parameter $V_m$, an amplitude-frequency curve for increasing of the control parameter is obtained. In contrary, if the same procedure is applied but the control parameter is decreased the another amplitude-frequency curve is obtained which partly differs from the first one.

If during the numerical solution of (3.48) the obtained amplitudes significantly differ for the two infinitesimal close values of the control parameter $V_m$, it is evident that the jump phenomena occurs. It gives us the chance to "recognized" the Sommerfeld effect during the numerical process.

In Figs. 3.8 and 3.9 the amplitude - frequency diagrams obtained analytically by solving (3.72) and (3.147) and numerically by solving (3.48) for $C_m = 10$ and $\alpha = 5/3$ are plotted. The control parameter $V_m$ is varied and the steady-state amplitude - frequency values are obtained. The computation time was 200 periods of vibration $T$. The steady-state numerical solution is reached after approximately $30T$. The period $T$ is computed numerical. For the certain value of control parameter $V_m$ the steady state amplitude and frequency are calculated. These values were used as the

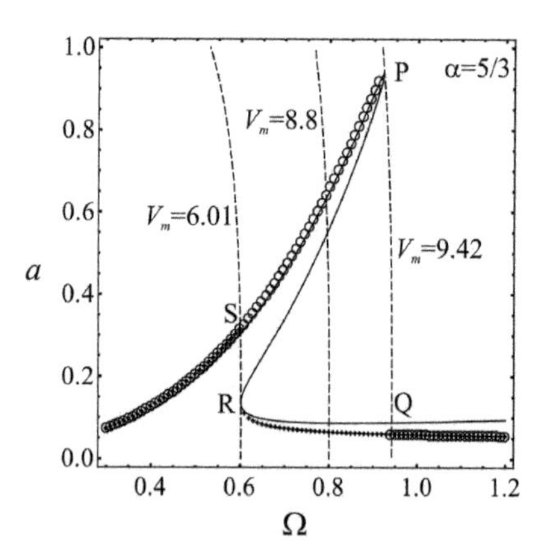


**Fig. 3.9** Frequency-response curves for $\alpha = 5/3$ and $C_m = 10$ obtained analytically (*full line*) and numerically (*circles* - for the increasing of $V_m$, *squares* - for decreasing of $V_m$)

initial conditions for computation of the steady-state values for the new value of the control parameter $V_m$.

From the both figures it is obvious that for slow increasing of $\Omega$ and variation of the motor parameter $V_m$, the amplitude of vibration increases up to the point P. Using the relations (3.149)–(3.151) the parameters of the characteristic point are obtained: $a_P = 0.9$, $\Omega_P = 0.92$ and $V_{mP} = 9.42$. Along the curve $V_{mP}$ the jump phenomena from P to Q ($a_Q = 0.098$, $\Omega_Q = 0.94$) occurs. For that position $\Omega_Q > \Omega_P$, but the amplitude $a_Q$ is significantly smaller than $a_P$. Further increase of the control parameter $V_m$ causes the further decrease of the amplitude of vibration for increase of the frequency $\Omega$. If the value of the control parameter $V_m$ is decreased, the steady-state locus in $a - \Omega$ plane moves to left along the $a - \Omega$ curve to the point R. The frequency $\Omega$ decreases and the amplitude $a$ increases slowly up to the value of $a_R = 0.1$ at the frequency $\Omega_R = 0.6$ (the values are obtained numerically). For $V_{mR} = 6.01$ a sudden change in the amplitude of vibration appears. The amplitude jumps along the curve $V_{mR}$ to the value $a_S = 0.3$ with $\Omega_S = 0.6$. The further decrease

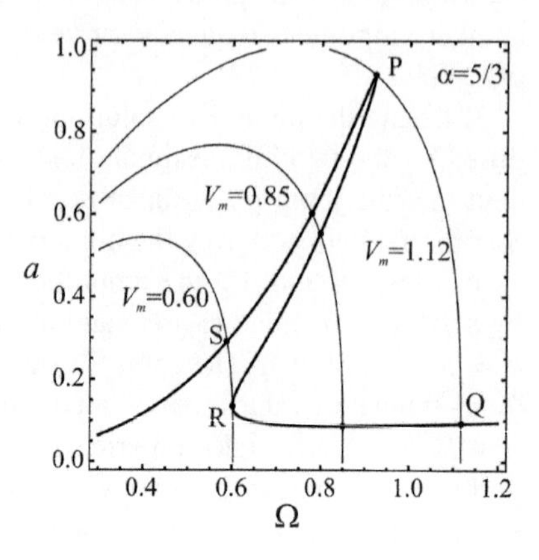


**Fig. 3.10** Frequency-response steady state curve obtained analytically for $\alpha = 5/3$ and $C_m = 1$ with characteristic points

of the control parameter $V_m$ produces amplitude and frequency decrease. The region in the $a - \Omega$ plane bounded with the part SP and RQ of the $a - \Omega$ curves and also RS and PQ parts of the $V_m - \Omega$ curves represent the hysteresis and the jump from P to Q and R to S in the system is the so called Sommerfeld effect well known in the non-ideal mechanical systems.

In Fig. 3.9 the values obtained numerically by solving of the original differential equations of motion (3.48) for increasing of the control parameter $V_m$ is plotted with circles (o) and for decreasing of the control parameter with a filled squares (■). Comparing the analytically obtained solutions (full line) with the numerically obtained values it can be seen that the results are in a good agreement.

In Figs. 3.10 and 3.11 the frequency-response curve for the oscillator with order of nonlinearity $\alpha = 2/3$ and motor parameter $C_m = 1$ is plotted. The curves in Fig. 3.10 are obtained by solving the approximate relations (3.72) and (3.147), while in Fig. 3.11 numerically solved differential equations (3.48) are plotted. Both figures show that for slow increasing of $\Omega$ and variation of the motor parameter $V_m$, the amplitude of vibration increases up to the point P whose properties are $a_P = 0.9$, $\Omega_P = 0.9$ and $V_{mP} = 1.12$.

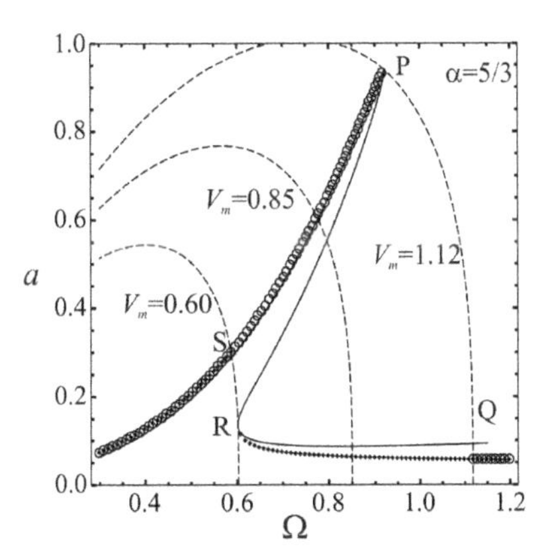


**Fig. 3.11** Frequency-response curves for $\alpha = 5/3$ and $C_m = 1$ obtained analytically (*full line*) and numerically (*circles* - for the increasing of $V_m$, *squares* - for decreasing of $V_m$)

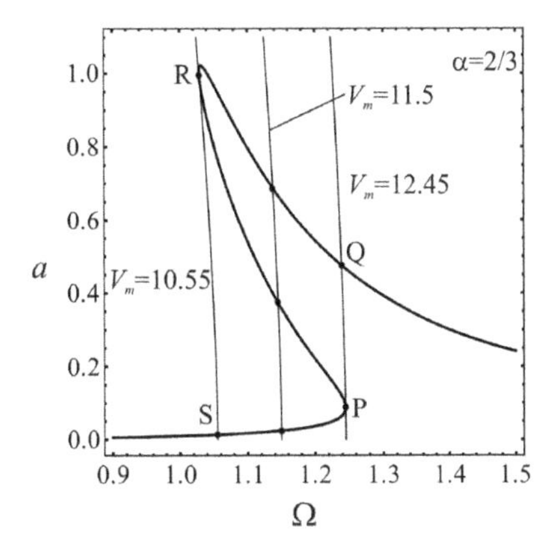


**Fig. 3.12** Frequency-response steady state curve obtained analytically for $\alpha = 2/3$ and $C_m = 10$ with characteristic points

Along the curve $V_{mP}$ the jump phenomena from P to Q ($a_Q = 0.055, \Omega_Q = 1.115$) occurs: the frequency $\Omega_Q$ is higher than $\Omega_P$, and the amplitude $a_Q$ is smaller than $a_P$. During further increase of the control parameter $V_m$, the amplitude of vibration decreases while the frequency $\Omega$ increases. If the value of the control parameter $V_m$ is decreased, the steady-state locus in $a - \Omega$ plane moves to left along the $a - \Omega$ curve to the point R. The frequency decreases to $\Omega_R = 0.6$ and the amplitude increases to $a_R = 0.125$ when $V_{mR} = 0.6$. Then a sudden change in the amplitude of vibration appears. The amplitude jumps along the curve $V_{mR}$ to the amplitude $a_S = 0.3$ and frequency $\Omega_S = 0.585$. The further decrease of the control parameter $V_m$ produces amplitude and frequency decrease. The bounded region SPQR represents the hysteresis plane. Comparing Figs. 3.8 and 3.10 it is seen that the Sommerfeld effect occurs for lower values of the motor parameter $V_m$ if $C_m = 1$ than if it is $C_m = 10$.

In Fig. 3.11 the numerically obtained $a - \Omega$ diagrams are plotted: $V_m$ is increased (o), and $V_m$ is decreased (■). These numerically obtained results by solving (3.48) are compared with analytically obtained results (full line curves). The difference between the analytical and numerical solutions is negligible.

In Figs. 3.12 and 3.13 the analytically and numerically obtained amplitude - frequency curves for the oscillator with the order of nonlinearity $\alpha = 2/3$ and motor parameter $C_m = 10$ are plotted. The analytical solutions are obtained by solving (3.72) and (3.147), and the numerical solutions by solving (3.48). The computation procedure was the same to that made in the previous calculation.

First, the control parameter $V_m$ was slowly increased. The steady-state frequency and amplitude increase, too, up to P, when the control parameter has the value $V_{mP} = 12.45$ and the amplitude and frequency are $a_P = 0.075$, $\Omega_P = 1.24$. These values are obtained analytically by solving the relations (3.149)–(3.151). For that value of control parameter ($V_{mP} = 12.45$) a sudden change in amplitude occurs for the almost

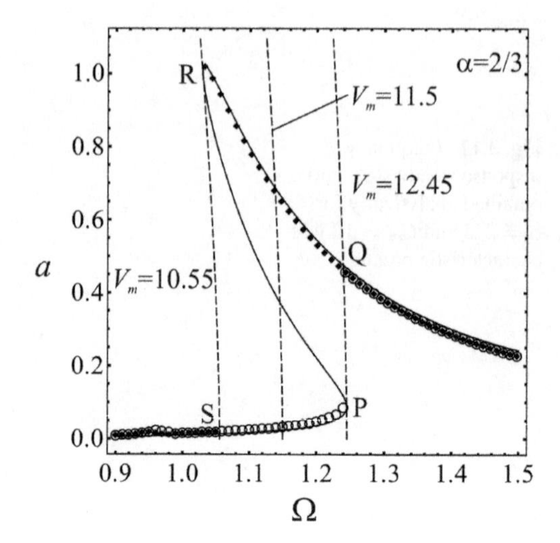


**Fig. 3.13** Frequency-response curves for $\alpha = 2/3$ and $C_m = 10$ obtained analytically (*full line*), numerically (*circles* - for the increasing of $V_m$, *squares* - for decreasing of $V_m$)

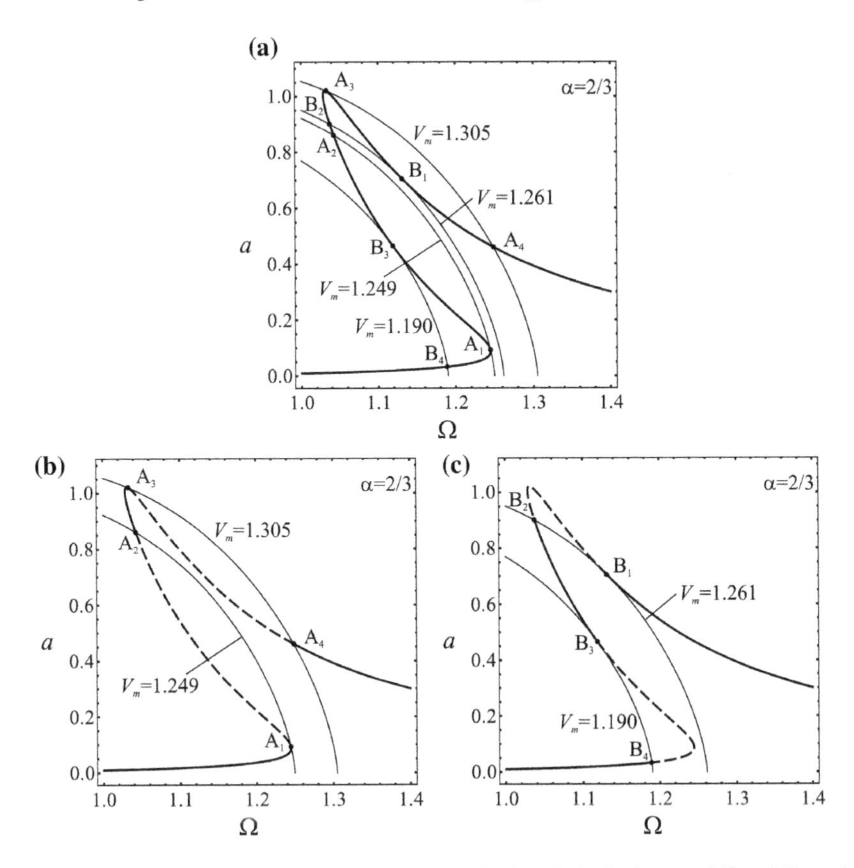


**Fig. 3.14** Frequency-response steady state curve obtained analytically for $\alpha = 2/3$ and $C_m = 1$ : **a** with characteristic points, **b** for increasing of $V_m$, **c** for decreasing of $V_m$

the same frequency. Namely, there is the jump to the point Q with the following values of the amplitude and frequency: $a_Q = 0.475$, $\Omega_Q = 1.22$. Increasing the value of $V_m$ gives decreasing of the amplitude $a$ and increasing of $\Omega$. If the control parameter is $V_m$ and the frequency $\Omega$ are decreased, the amplitude of vibration increases up to $a_R = 1.0$ for $V_{mR} = 10.55$ and $\Omega_R = 1.03$. At that value of the control parameter the jump to another steady state position occurs with amplitude $a_S = 0.02$ and frequency $\Omega_S = 1.04$. Further decrease of the control parameter gives also the decrease of the amplitude and frequency. Finally, it can be concluded that the Sommerfeld effect and the jump phenomena occur for $V_{mP} = 12.45$ and $V_{mR} = 10.55$.

Comparing the analytically obtained curve (full line) and the numerically obtained curves for the case when the control parameter increases (o) and when it decreases (■) it can be seen that the difference is negligible (see Fig. 3.13).

In Figs. 3.14 and 3.15 the frequency-response curve for the oscillator with order of nonlinearity $\alpha = 2/3$ and motor parameter $C_m = 1$ is plotted. The curves in Fig. 3.14 are obtained by solving the approximate relations (3.72) and (3.147), while

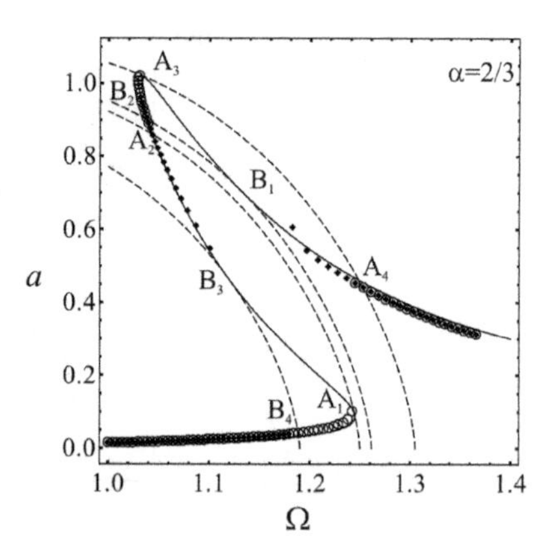


**Fig. 3.15** Frequency-response curves for $\alpha = 2/3$ and $C_m = 1$ obtained analytically (*full line*) and numerically (*circles* - for the increasing of $V_m$, *squares* - for decreasing of $V_m$)

in Fig. 3.15 numerically solved differential equations (3.48) are plotted. In Fig. 3.14b the steady-state motion for increasing value of the control parameter $V_m$ is plotted. The amplitude of vibration increases slowly with the frequency up to the point $A_1$ with coordinates $a_{A1} = 0.1, \Omega_{A1} = 1.24$ for the control parameter $V_{mA1} = 1.249$. For that value of the control parameter the other steady state motion is with the amplitude $a_{A2} = 0.875$, and frequency $\Omega_{A2} = 1.04$. For this value of the control parameter the amplitude increases significantly, and the frequency decreases. For higher values of the control parameter the tendency of increase but also decreasing of the frequency is evident. For the control parameter $V_{mA3} = 1.305$ the peak amplitude $a_{A3} = 1.025$ for $\Omega_{A3} = 1.03$ is reached. For this value of control parameter the another steady state motion is with amplitude and frequency $a_{A4} = 0.455, \Omega_{A4} = 1.25$, respectively. Further increase of the control parameter $V_m$ gives the decrease of the amplitude and increase of the frequency.

In Fig. 3.14c the procedure is repeated but in the opposite direction: the control parameter $V_m$ is decreased. Decreasing $V_m$, the amplitude is increasing up to $a_{B1} = 0.7$and the frequency is decreased to $\Omega_{B1} = 1.13$ for $V_{mB1} = 1.261$. At that value of the control parameter a jump to the amplitude $a_{B2} = 0.9$ and frequency $\Omega_{B2} = 1.04$ occurs. Further decrease of the control parameter causes decrease of the amplitude but increase of the frequency to the boundary values $a_{B3} = 0.475$, $\Omega_{B3} = 1.12$ for $V_{mB3} = 1.190$. For that value of control parameter the other steady state motion is with parameters $a_{B4} = 0.025$, $\Omega_{B4} = 1.19$. Decreasing $V_m$ the amplitude and frequency decrease. In Fig. 3.14a the four values of the control parameter $V_m$ for with the amplitude - frequency curve of the motor is the tangent of the steady state curve are shown. It means, that for $\Omega$ slowly increased two times the jump phenomena appear: from $A_1$ to $A_2$ (the amplitude jumps to a higher value) and from $A_3$ to $A_4$ (the amplitude jumps to a smaller value) as is shown in Fig. 3.14b. For slow decreasing of $\Omega$ the amplitude jumps two times in the steady-state curve, too (see Fig. 3.14c) from $B_1$ to $B_2$ (the amplitude jumps to a higher value) and from $B_3$ to $B_4$ (the amplitude jumps to a smaller value).

In Fig. 3.15 the numerically obtained $a - \Omega$ diagrams are plotted: $V_m$ is increased (o), and $V_m$ is decreased (■). These numerically obtained results by solving (3.48) are compared with analytically obtained results (full line curves). The difference between the analytical and numerical solutions is negligible.

In Fig. 3.16 the influence of the control parameter $V_m$ on amplitude-frequency curve obtained analytically by solving (3.72) and (3.147) for the linear oscillator $\alpha = 1$ and motor parameter $C_m = 1$ is analyzed. In the diagram two values of the control parameter for which the Sommerfeld effect occurs is obtained. For the case of increasing frequency, the control parameter for which two steady state motions P($a_P = 1$, $\Omega_P = 1$) and Q($a_Q = 0.14$, $\Omega_Q = 1.24$) exist is $V_{mP} = 1.251$. For the case of decreasing frequency, for the control parameter $V_{mR} = 1.101$ the jump is from R($a_R = 425$, $\Omega_R = 1.06$) to S($a_S = 0.075$, $\Omega_S = 0.7$).

In Fig. 3.17 we plot the case where the Sommerfeld effect is suppressed. The parameters of the system are $\alpha = 2/3$, $\varepsilon = 0.1$, $\mu_1 = 0.5$, $\eta_1 = 0.5$, $p^2 = 1$ and $\zeta_1 = 1.2$. Two values of $C_m$ are considered: $C_m = 0.85$ (Fig. 3.17a) and $C_m = 5$ (Fig. 3.17b) while $V_m$ is varied. The first value for $C_m$ is computed according to (3.158) and the second value is an arbitrary one. It can be seen that for $C_m = 5$ the $V_m$ intersects the $A$-$\Omega$ curve in one, two or three points giving the Sommerfeld effect.

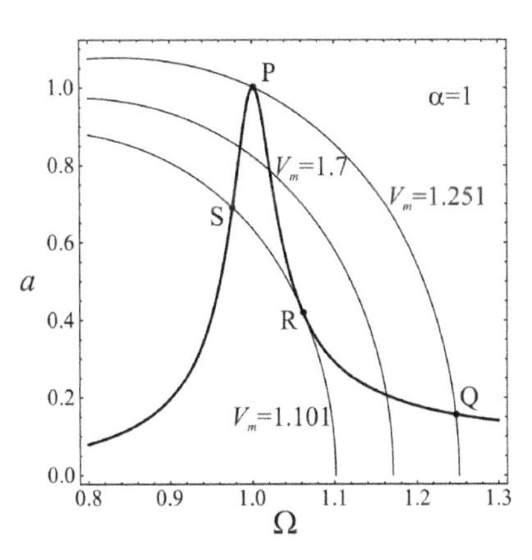


**Fig. 3.16** Frequency-response steady state curve obtained analytically for $\alpha = 1$ and $C_m = 1$ with characteristic points

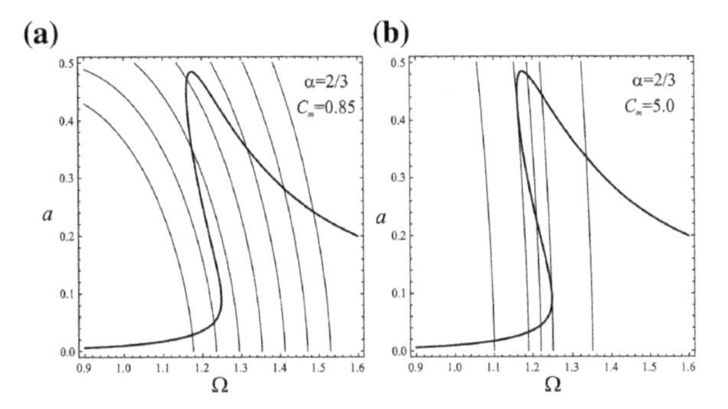


**Fig. 3.17** Frequency-response steady state curve obtained analytically for $\alpha = 2/3$ and: **a** $C_m = 0.85$, and **b** $C_m = 5$

For $C_m = 0.85$ only one intersection point of $V_m$ with the amplitude frequency exists and no amplitude jump exists. Thus, the Sommerfeld effect is eliminated.

Comparing the curves and the results plotted in Figs. 3.8, 3.9, 3.10, 3.11, 3.12, 3.13, 3.14, 3.15 and 3.16 the following is concluded:

1. The amplitude-frequency curve bends on right for $\alpha > 1$ (Figs. 3.8 and 3.9), on the left for $\alpha < 1$ (Figs. 3.12, 3.13, 3.14 and 3.15) and is straight for $\alpha = 1$ (Fig. 3.16).
2. For $C_m = 10$ the lines for $V_m$ curves in Figs. 3.8, 3.9, 3.12 and 3.13 are closed to the vertical direction and this case is close to the ideal system. For the ideal system, the oscillator has no influence on the motion of the motor and there is no coupling between these two motions. The ideal system has one-degree-of-freedom and the motion of the oscillator is forced with a time periodical excitation force. The mathematical model of the system is a second order differential equation. For $C_m = 10$ (see Figs. 3.14, 3.15 and 3.16) the $V_m$ curves are bent and the properties of non-ideal system are highly significant. For $C_m = 1$ and $\alpha = 1$ two characteristic values of the control parameter $V_m$ are evident (see Fig. 3.16), while for $C_m = 10$ and $\alpha = 2/3$ even four (Figs. 3.14 and 3.15).
3. In the system with order of nonlinearity $\alpha = 2/3$ the Sommerfeld effect can be eliminated by suitable assumption of the value of the parameter $C_m$ (see Fig. 3.17). Namely, for $C_m = 5$ the effect of jump occurs (Fig. 3.14b), while for the critical value $C_m = 0.85$ and any value of $V_m$, the Sommerfeld effect is eliminated.

### *3.3.6 Conclusion*

Analyzing the results the following is concluded:

1. In the non-ideal system which contains a pure nonlinear oscillator (order of the nonlinearity is integer or non-integer) and a motor with linear torque properties (a non-ideal source) for certain parameter values the resonant phenomena appears.
2. The approximate averaging procedure, based on the introduction of additional slow variables, as it is assumed that the angular velocity and the frequency of the structure are functions of these variables, is appropriate for analytical analysis of the problem. The suggested averaging procedure gives the equations which are suitable for analysis of the near resonant case. Results which are obtained are applicable for the analysis of the characteristic properties of the system:

   (a) The amplitude-frequency curve bends on right for the nonlinearity of the order higher than 1, i.e., the order is a positive rational number higher than 1. The amplitude-frequency curve bends on the left, if the nonlinearity is a positive rational number smaller than 1. For the linear case,the backbone curve of the amplitude-frequency steady state curve is straight.

   (b) The approximate value of the control parameter for the non-ideal source is analytically calculated applying the method of equating the gradient of the both amplitude-frequency curves (of the oscillator and of the motor) in the intersection points. The criteria for the Sommerfeld effect is analytically obtained.

(c) The Sommerfeld effect occurs not only for the linear case and when the nonlinearity is higher than 1 (as it is previously published), but also if the nonlinearity is a positive rational number smaller than 1.

(d) The parameters of the energy source effect the steady-state properties of the non-ideal mechanical system. The both parameters of the motor torque ($V_{m\phi}$ and $C_{m\phi}$) and also their rate determine the motion properties. Hence, independently on the order of nonlinearity, the higher the value of $C_{m\phi}$ the system tends to the ideal one.

(e) For the case when the nonlinearity is of the order smaller than 1, the amplitude jump effect may occur two times during increase of the motor property $V_{m\phi}$ and two times during its decrease. For certain motor property the characteristic curve may be the tangent of the amplitude-frequency curve in two points during increase of $V_{m\phi}$ and also during its decrease, and for these values the jump effect occur. In the linear non-ideal oscillator and in the oscillator with a nonlinearity higher than 1 the jump phenomena is evident once during increasing and once during decreasing of $V_{m\phi}$. Namely, the characteristic of the motor is the tangent of the amplitude-frequency curve for one $V_{m\phi}$ during increase, and for other value during decrease.

4. The method given in this text obtains motor torque parameters which can suppress the Sommerfeld effect in the non-ideal mechanical system with the positive nonlinearity of order smaller than 1. The Sommerfeld effect and the jump phenomena in the amplitude for non-ideal system may be suppressed by these critical parameter values.
5. Comparing the analytical and numerical solutions it is evident that they are in a good agreement independently on the order of nonlinearity.

## 3.4 Stable Duffing Oscillator and a Non-ideal Energy Source

Let us consider a motor operating on a structure with strong cubic nonlinearity (Fig. 3.18). Oscillator is of Duffing type and is connected with the motor with limited power supply. The motor has an eccentric mass which is the part of the non-ideal perturbation source. The driving of the system comes from the unbalanced rotor linked to the oscillator fed by an electric motor. The driven system is taken as a consequence of the dynamics of the whole system (oscillator plus rotor). Duffing oscillator, connected with non-ideal energy source, is of hardening type with strong cubic nonlinearity which was widely investigated by Krylov and Bogolubov (1943), Ueda (1985), Fang and Dowell (1987), Pezeshki and Dowell (1988), Yuste and Bejarano (1986, 1990), Cheng et al. (1991), Chen and Cheung (1996), Chen et al. (1998), Gendelman and Vakakis (2000), Mickens (2001, 2006), Andrianov (2002),

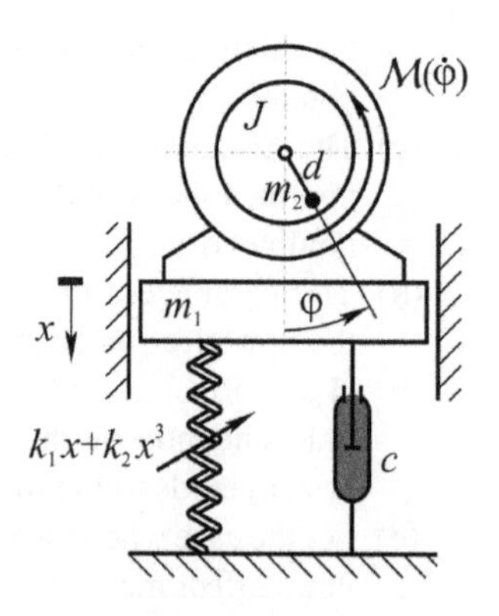


**Fig. 3.18** Model of the motor-structure non-ideal system

He (2002), Hu and Xiong (2003), Andrianov and Awrejcewicz (2003a, b), Amore and Aranda (2005), Cveticanin (2004, 2006, 2011), Ozis and Yildirm (2007), Belendez et al. (2007), Cveticanin et al. (2010), Kovacic et al. (2010) and others. The structure of mass $M$ is connected to a fixed basement by a non-linear spring and a linear viscous damper (damping coefficient $c$). The nonlinear spring stiffness is given by $k_1x + k_2x^3$, where $x$ denotes the structure displacement with respect to some equilibrium position in the absolute reference frame. The motion of the structure is due to an in-board non-ideal motor driving an unbalanced rotor. We denote by $\varphi$ the angular displacement of the rotor unbalance, and model it as a particle of mass $m$ and radial distance $d$ from the rotating axis. The moment of inertia of the rotating part is $J$. For the resonant case the structure has an influence on the motor input or output. The forcing function is dependent of the system it acts on and the source is of non-ideal type.

The non-ideal problem has two - degrees of freedom, represented by the generalized coordinates $x$ and $\varphi$. The kinetic energy $T$, potential energy $U$ and the dissipative function $\Phi$ are expressed by

$$T = \frac{1}{2}M\dot{x}^2 + \frac{1}{2}m(\dot{x}^2 + d^2\dot{\varphi}^2 - 2d\dot{x}\dot{\varphi}\sin\varphi) + \frac{1}{2}J\dot{\varphi}^2, \tag{3.159}$$

$$U = \frac{1}{2}k_1x^2 + \frac{1}{4}k_2x^4 - (M+m)gx - mgd\cos\varphi, \quad \Phi = \frac{1}{2}c\dot{x}^2. \tag{3.160}$$

A dot denotes differentiation with respect to time $t$. Lagrange's equations of motion for the system are in general

$$\frac{d}{dt}\left(\frac{\partial T}{\partial \dot{x}}\right) - \frac{\partial T}{\partial x} + \frac{\partial U}{\partial x} + \frac{\partial \Phi}{\partial \dot{x}} = Q_x,$$
$$\frac{d}{dt}\left(\frac{\partial T}{\partial \dot{\varphi}}\right) - \frac{\partial T}{\partial \varphi} + \frac{\partial U}{\partial \varphi} + \frac{\partial \Phi}{\partial \dot{\varphi}} = Q_{\dot{\varphi}},$$

where $Q_x$ and $Q_{\dot{\varphi}}$ are generalized forces.

The differential equations of motion have the form

$$\ddot{x}(M+m) + c\dot{x} - md(\ddot{\varphi}\sin\varphi + \dot{\varphi}^2\cos\varphi) + k_1 x + k_2 x^3 = (M+m)g,$$
$$(J+md^2)\ddot{\varphi} - md\ddot{x}\sin\varphi + mgd\sin\varphi = \mathcal{M}(\dot{\varphi}), \quad (3.161)$$

where $\mathcal{M}(\dot{\varphi})$ is the motor torque. The most often used model of the torque is the linear moment-speed relation (see Dimentberg et al. 1997)

$$\mathcal{M}(\dot{\varphi}) = M_0\left(1 - \frac{\dot{\varphi}}{\Omega}\right), \quad (3.162)$$

where $M_0$ and $\Omega$ are constant values.

It is convenient to normalize the coordinates and time according to

$$x \longrightarrow y = \frac{\Omega^2}{g}x, \qquad t \longrightarrow \tau = \Omega t, \quad (3.163)$$

where $g$ is the gravity constant. By introducing (3.163) the differential equations (3.161) transform into

$$y'' + \zeta y' + p^2 y + \gamma y^3 = 1 + \mu(\varphi''\sin\varphi + \varphi'^2\cos\varphi),$$
$$\varphi'' = \eta y''\sin\varphi - \eta\sin\varphi + F(1-\varphi'), \quad (3.164)$$

where

$$p = \frac{\omega^*}{\Omega}, \qquad \omega^{*2} = \frac{k_1}{M+m}, \qquad \gamma = \frac{k_3 g^2}{(M+m)\Omega^6}, \qquad \zeta = \frac{c}{\Omega(M+m)},$$
$$\mu = \frac{m}{M+m}\frac{d\Omega^2}{g}, \qquad \eta = \frac{gmd}{\Omega^2(J+md^2)}, \qquad F = \frac{M_0}{\Omega^2(J+md^2)}, \quad (3.165)$$

and prime denotes differentiation with respect to $\tau$. The differential equations (3.164) are non-linear and coupled.

### 3.4.1 Asymptotic Solving Method

In the regime near resonant the difference between the excitation frequency is close to the natural frequency. For the near resonant case one can write

$$\varphi' - p = \varepsilon\sigma, \quad (3.166)$$

where $\varepsilon\sigma$ is the detuning parameter with the small parameter $\varepsilon << 1$. Expressing parameters of equations (3.165) by

$$\zeta = \varepsilon\zeta_1, \qquad \gamma = \varepsilon\gamma_1, \qquad \mu = \varepsilon\mu_1, \qquad \eta = \varepsilon\eta_1, \qquad F = \varepsilon F_1, \quad (3.167)$$

the differential equations of motions have the form

$$z'' + p^2 z = \varepsilon\mu_1\left(\varphi''\sin\varphi + \varphi'^2\cos\varphi\right) - \varepsilon\zeta_1 z' - \varepsilon\gamma_1\left(z + \frac{1}{p^2}\right)^3\Bigg],$$

$$\varphi'' = \varepsilon\eta_1 z''\sin\varphi - \varepsilon\eta_1\sin\varphi + \varepsilon F_1(1-\varphi'), \tag{3.168}$$

where the new variable is

$$z = y - \frac{1}{p^2}. \tag{3.169}$$

Expressing $z$" and $\varphi$" from the Eq. (3.168) and assuming only the terms to $O(\varepsilon^2)$ we obtain

$$z" + p^2 z = \varepsilon\left[\mu_1\varphi'^2\cos\varphi - \zeta_1 z' - \gamma_1\left(z + \frac{1}{p}\right)^3\right] + \varepsilon^2\ldots,$$

$$\varphi" = \varepsilon\left[F_1(1-\varphi') - \eta_1(1+p^2 z)\sin\varphi\right] + \varepsilon^2\ldots. \tag{3.170}$$

Following the reference (Warminski et al. 2001)

$$z = a\cos(\varphi + \psi), \tag{3.171}$$

where $a$ and $\psi$ are the new coordinates. The first derivative is

$$z' = -ap\sin(\varphi + \psi), \tag{3.172}$$

when

$$a'\cos(\varphi+\psi) - a(\omega + \psi' - p)\sin(\varphi+\psi) = 0. \tag{3.173}$$

Determining the second derivative of $z$ and substituting (3.171) and (3.172) into (3.170) we obtain

$$-a'p\sin(\varphi+\psi) - ap(\omega+\psi')\cos(\varphi+\psi) + p^2 a\cos(\varphi+\psi)$$

$$= \varepsilon\mu_1\omega^2\cos\varphi + ap\varepsilon\zeta_1\sin(\varphi+\psi) - \varepsilon\gamma_1\left[a\cos(\varphi+\psi) + \frac{1}{p}\right]^3,$$

$$\omega' = \varepsilon F_1(1-\omega) - \varepsilon\eta_1[1 + ap^2\cos(\varphi+\psi)]\sin\varphi. \tag{3.174}$$

Equations (3.173) and (3.174) lead to the derivatives $a'$, $\psi'$, $\varphi'$ and $\omega'$

$$a' = -\sin(\varphi+\psi)\left\{\frac{\varepsilon\mu_1}{p}\omega^2\cos\varphi + a\varepsilon\zeta_1\sin(\varphi+\psi)\right.$$

$$\left. -\frac{1}{p}\varepsilon\gamma_1\left[a\cos(\varphi+\psi) + \frac{1}{p}\right]^3\right\},$$

$$a\psi' = \varepsilon\sigma a - \cos(\varphi+\psi)\left\{\frac{\varepsilon\mu_1}{p}\omega^2\cos\varphi + a\varepsilon\zeta_1\sin(\varphi+\psi)\right.$$
$$\left.-\frac{1}{p}\varepsilon\gamma_1\left[a\cos(\varphi+\psi)+\frac{1}{p}\right]^3\right\},$$
$$\omega' = \varepsilon F_1(1-\omega) - \varepsilon\eta_1[1+ap^2\cos(\varphi+\psi)]\sin\varphi, \qquad \varphi' = \omega. \quad (3.175)$$

Unfortunately, the closed form analytical solution for (3.175) is complicate to be obtained and the numerical methods are convenient.

### 3.4.2 Stability of the Steady State Solution and Sommerfeld Effect

Introducing the averaging procedure (see Bogolyubov and Mitropolskij 1974)

$$a' = \frac{\varepsilon}{2\pi}\int_0^{2\pi} f_a d\varphi, \qquad a\psi' = \frac{\varepsilon}{2\pi}\int_0^{2\pi} f_\psi d\varphi, \qquad \omega' = \int_0^{2\pi} f_\omega d\varphi, \quad (3.176)$$

where

$$fa = -\frac{1}{p}\sin(\varphi+\psi)\left\{\mu_1\omega^2\cos\varphi + ap\zeta_1\sin(\varphi+\psi) - \gamma_1\left[a\cos(\varphi+\psi)+\frac{1}{p}\right]^3\right\},$$
$$f_\psi = a\sigma - \frac{1}{p}\cos(\varphi+\psi)\{\mu_1\omega^2\cos\varphi + ap\zeta_1\sin(\varphi+\psi)$$
$$-\gamma_1\left[a\cos(\varphi+\psi)+\frac{1}{p}\right]^3\right\},$$
$$f_\omega = F_1(1-\omega) - \eta_1[1+ap^2\cos(\varphi+\psi)]\sin\varphi, \quad (3.177)$$

the differential equations (3.175) are transformed into

$$\omega' = \varepsilon\left[F_1(1-\omega) + \frac{1}{2}\eta_1 ap^2\sin\psi\right],$$
$$a' = -\frac{\varepsilon}{2}\left(a\zeta_1 + \frac{\mu_1\omega^2}{p}\sin\psi\right),$$
$$\psi' = \varepsilon\sigma - \frac{\varepsilon}{2p}\left(\frac{\mu_1\omega^2}{a}\cos\psi - \frac{3\gamma_1 a^2}{4} - \frac{3\gamma_1}{p^2}\right). \quad (3.178)$$

For the steady-state response, Eq. (3.178) have the form

$$F_1(1-\omega_S)+\frac{1}{2}\eta_1 a_S p^2 \sin\psi_S = 0,$$
$$a_S + \frac{\mu_1\omega_S^2}{p}\sin\psi_S = 0,$$
$$\sigma - \frac{1}{2p}\left(\frac{\mu_1\omega_S^2}{a_S}\cos\psi_S - \frac{3\gamma_1 a_S^2}{4} - \frac{3\gamma_1}{p^2}\right) = 0, \tag{3.179}$$

where $S$ denotes the steady state values. Combining the second and the third Eq. (3.179) we obtain

$$a_S^2\left\{\frac{\zeta_1^2}{4}+\left[\sigma+\frac{3\gamma_1}{2p}\left(\frac{a_S^2}{4}+\frac{1}{p^2}\right)\right]^2\right\} = \left(\frac{\mu_1\omega_S^2}{2p}\right)^2, \tag{3.180}$$

while combining the first and the second Eq. (3.179) yields

$$F_1\mu_1\omega_S^2(1-\omega_S) = \frac{1}{2}\zeta_1\eta_1 a_S^2 p^3. \tag{3.181}$$

Expressing $a_S$ in (3.181) and substituting into (3.180) the $\omega_S(F_1)$ function is obtained

$$8F_1(1-\omega_S)\left\{\frac{\zeta_1^2}{4}+\left[\omega_S - p + \frac{3\gamma_1}{2p^5}\left(\frac{\mu_1\omega_S^2 F_1(1-\omega_S)}{2\zeta_1\eta_1 p}+1\right)\right]^2\right\} - \mu_1\zeta_1\eta_1 p\omega_S^2 = 0. \tag{3.182}$$

The number of real solution is one, two or three and it depends on the control parameter $F_1$. To determine which of the solutions actually correspond to a realizable motion, we need to consider the stability of the solutions. We determine the stability by determining the nature of the singular points of (3.179). To accomplish this, we let

$$a = a_S + a_1, \qquad \psi = \psi_S + \psi_1, \qquad \omega = \omega_S + \omega_1. \tag{3.183}$$

Substituting (3.183) into (3.179) and neglecting all but the linear terms in $a_1, \psi_1, \omega_1$ we obtain

$$\omega_1' = \varepsilon\left[-F_1\omega_1 + \frac{1}{2}\eta_1 a_1 p^2 \sin\psi_S + \frac{\psi_1}{2}\eta_1 a_S p^2\cos\psi_S\right],$$
$$a_1' = -\frac{\varepsilon}{2}\left(a_1\zeta_1 + \psi_1\frac{\mu_1\omega_S^2}{p}\cos\psi_S + 2\frac{\mu_1\omega_S\omega_1}{p}\sin\psi_S\right),$$
$$\psi_1' = -\omega_1 - \frac{\varepsilon}{2p}\left(2\frac{\mu_1\omega_S\omega_1}{a_S}\cos\psi_S - a_1\frac{\mu_1\omega_S^2}{a_S^2}\cos\psi_S\right. \tag{3.184}$$
$$\left.-\psi_1\frac{\mu_1\omega_S^2}{a_S}\sin\psi_S - \frac{3\gamma_1 a_S a_1}{2}\right).$$

Equations (3.184) are linear and have the solution in the form

$$(a_1, \psi_1, \omega_1) = (a_{10}, \psi_{10}, \omega_{10}) \exp(\lambda \tau),$$

where $\lambda$ is an eigenvalue of the coefficient of matrix. The solutions are stable and hence corresponding motions reliable, if the real part of each eigenvalue is negative or zero.

For the parameter values

$$\varepsilon = 0.1, \quad \zeta_1 = 0.2, \quad \gamma_1 = 0.1, \quad \eta_1 = 1.0, \quad \mu_1 = 0.5, \tag{3.185}$$

using the relations (3.180) and (3.181), i.e., (3.182) the response-control parameter $F_1$ and the frequency of vibration-control parameter $F_1$ diagrams are plotted (Fig. 3.19).

In Fig. 3.20 the frequency-response curves obtained numerically and analytically by solving the Eqs. (3.180) and (3.181) are plotted. Comparing the solutions it can be concluded that the difference between "exact" numerical solution and approximate analytical solution is negligible. Analyzing the obtained curves and the relation (3.181) it is evident that the curve depends on the control parameter $F_1$. For $\omega_S = 2/3$ the maximal response

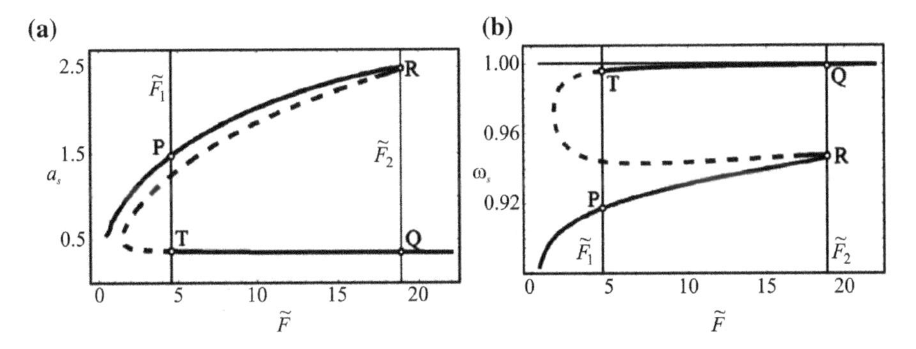


**Fig. 3.19** Jump effect for: **a** amplitude-control parameter curve, **b** frequency-control parameter curve

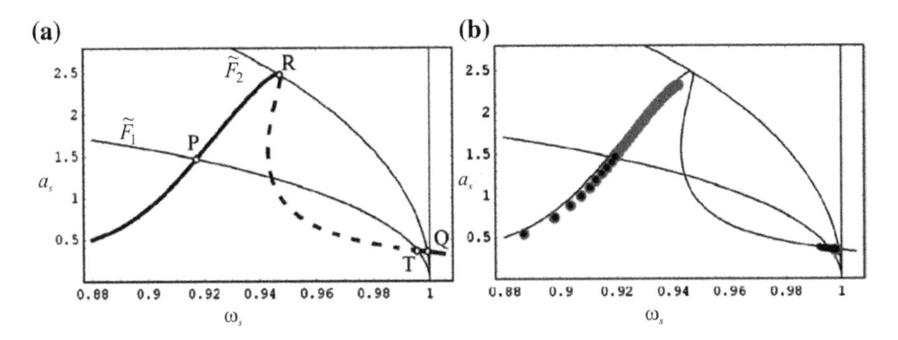


**Fig. 3.20** Frequency-response curve obtained: **a** analytically, **b** numerically

$$a_{S\max} = \frac{2}{3p}\sqrt{\frac{2\mu_1}{3\zeta_1\eta_1 p}F_1},$$

is a function of $F_1$, but is not dependent on the parameter of non-linearity $\gamma_1$.

Figures 3.19 and 3.20 show the characteristic points in the diagrams: point R and point T. Between these points the solution is unstable while all those outside this region are stable. Point R is the peak in $a_S - \omega_S$, $a_S - F_1$ and $F_1 - \omega_S$ diagrams. Analyzing the relation (3.182) and equating the first derivative to zero $(dF_1/d\omega_S = 0)$

$$0 = -8F_1\left\{\frac{\zeta_1^2}{4} + \left[\omega_S - p + \frac{3\gamma_1}{2p^5}\left(\frac{\mu_1\omega_S^2 F_1(1-\omega_S)}{2\zeta_1\eta_1 p} + 1\right)\right]^2\right\}$$
$$+2F_1(1-\omega_S)\left[\omega_S - p + \frac{3\gamma_1}{2p^5}\left(\frac{\mu_1\omega_S^2 F_1(1-\omega_S)}{2\zeta_1\eta_1 p} + 1\right)\right]$$
$$\left[1 + \frac{3\gamma_1}{2p^5}\frac{\mu_1 F_1(2\omega_S - 3\omega_S^2)}{2\zeta_1\eta_1 p}\right] - 2\mu_1\zeta_1\eta_1 p\omega_S.$$

the peak is obtained. The locus of the peak (point R) is according to (3.180)

$$\omega_{SR} = p - \frac{3\varepsilon\gamma_1}{2p}\left(\frac{a_{SR}^2}{4} + \frac{1}{p^2}\right), \tag{3.186}$$

and the amplitude is $a_{SR} = \mu_1\omega_{SR}^2/\tilde{\alpha}p$. Substituting (3.186) into the relation (3.181) we obtain the value of the control parameter for the peak amplitude

$$F_{1R} = \frac{\eta_1 p\mu_1\omega_{SR}^2}{2\zeta_1(1-\omega_{SR})}. \tag{3.187}$$

Substituting the parameter values (3.185) into (3.187) the numeric value of the control parameter is $F_{1R} = 18.246$. For that calculated value of the control parameter and the known value of the parameter of nonlinearity ($\gamma_1 = 0.1$) the amplitude and frequency of Q are obtained by solving the relations (3.180) and (3.181). For the control parameter $F_{1T} = 4.57$ two stable solutions are obtained: T with $a_{ST}$ and $\omega_{ST}$ and P with $a_{SP}$ and $\omega_{SP}$. We note that there are gaps in the diagrams (see Figs. 3.19 and 3.20) where no steady state response exists. The gaps are not the same for increasing and decreasing the control parameter $F_1$. Increasing the control parameter causes the increase of the amplitude and frequency of vibrations to R. Then the effect of jump to smaller amplitude and higher frequency in Q appears. Decreasing the control parameter $\xi_1$ decreases the frequency and increases the amplitude to T and then jump into P with higher amplitude and smaller frequency occurs. The same hysteresis is seen in Fig. 3.20. This phenomena of jump is called the Sommerfeld effect in non-ideal systems.

To eliminate the Sommerfeld effect for the certain value of the control parameter $F_{1R}$ the parameter of nonlinearity has to be calculated. The Eq. (3.180) indicates the peak amplitude

$$a_{SR} = \frac{\mu_1}{\zeta_1 p}\omega_{SR}^2. \tag{3.188}$$

The function $a_{SR}(\omega_{SR})$ is independent on the parameter of non-linearity $\gamma_1$. The relation (3.181) gives the dependence of amplitude $a_S$ on $\omega_S$ for various values of control parameter $\xi_1$. Using (3.188) and (3.181) we obtain

$$F_{1R}(1-\omega_{SR}) - \frac{1}{2}\eta_1\frac{\mu_1}{\zeta_1}\omega_{SR}^2 p = 0. \tag{3.189a}$$

$\omega_{SR}$ is the solution of (3.189a) for the control parameter $F_{1R}$. Substituting $\omega_{SR}$ into (3.188) the amplitude $a_{SR}$ is calculated. Introducing the so obtained values $\omega_{SR}$ and $a_{SR}$ into (3.180) the value of nonlinear parameter $\gamma_{1R}$ is determined

$$\gamma_{1R} = \frac{8}{3}\frac{p^3}{a_{SR}^2 p^2 + 4}(p-\omega_{SR}). \tag{3.190}$$

For $\gamma_1 < \gamma_{1R}$ only one stable solution exists for the control parameter $F_{1R}$ and the motion is without jump.

In Fig. 3.21 the $a_S - F_1$ curve for two values of parameter of nonlinearity $\gamma_1$ is shown. It is seen that for $\gamma_1 = 0.1$ the bending of the curve is smaller than for $\gamma_1 = 0.15$. For $F_{1R} = 42.1$ and $\gamma_{1R} = 0.15$ two real values of amplitude exist and for $\gamma_1 = 0.1$ only one. It can be concluded that for the real system with parameters (3.185) which works with control parameter $F_{1R} = 42.1$ the parameter of non-linearity of the structure has to be $\gamma_1 < 0.15$. Then, the Sommerfeld effect is eliminated.

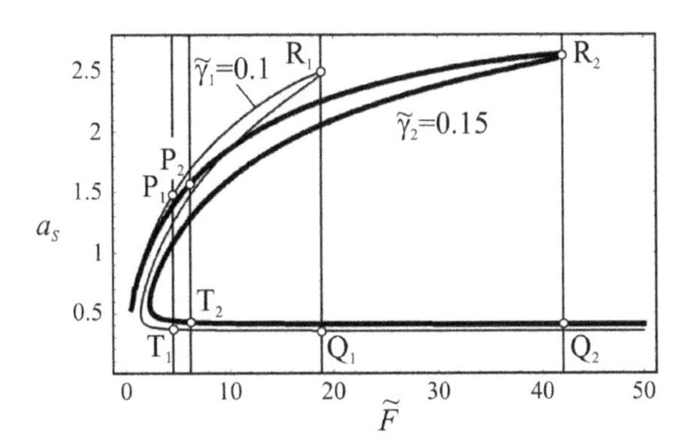


**Fig. 3.21** Frequency-control parameter curve for various values of parameter of non-linearity

### 3.4.3 Numerical Simulation and Chaotic Behavior

To prove the analytically obtained results the numerical experiment is done. The system (3.51) is rewritten in the form

$$
\begin{aligned}
y_1' &= y_2,\\
y_2' &= \frac{1}{1-\mu\eta\sin^2 y_3}(-\zeta y_2 - p y_1 - \gamma y_1^3 + 1\\
&\quad +\mu\sin y_3(\mu y_4^2\cos y_3 - \eta\sin y_3 + F(1-y_4))),\\
y_3' &= y_4,\\
y_4' &= \frac{\eta\sin y_3}{1-\mu\eta\sin^2 y_3}(-\zeta y_2 - p y_1 - \gamma y_1^3 + 1 + \mu y_4^2\cos y_3\\
&\quad +\mu\sin y_3\,(-\eta\sin y_3 + F(1-y_4)))\\
&\quad -\eta\sin y_3 + F(1-y_4).
\end{aligned}
\tag{3.191}
$$

A number of numerical simulations are done for various control parameter $F$. Applying the Runge–Kutta numerical procedure with the fixed step length the system of four first order differential equations (3.191) is solved. The results are plotted in Fig. 3.22. The phase space of the system has out of four dimensions, but we were chiefly interested in position of the oscillator itself. We also plotted the Poincaré map which represents the surface of section $(y_1(\tau_n), y_2(\tau_n))$. The points $(y_1(\tau_n), y_2(\tau_n))$

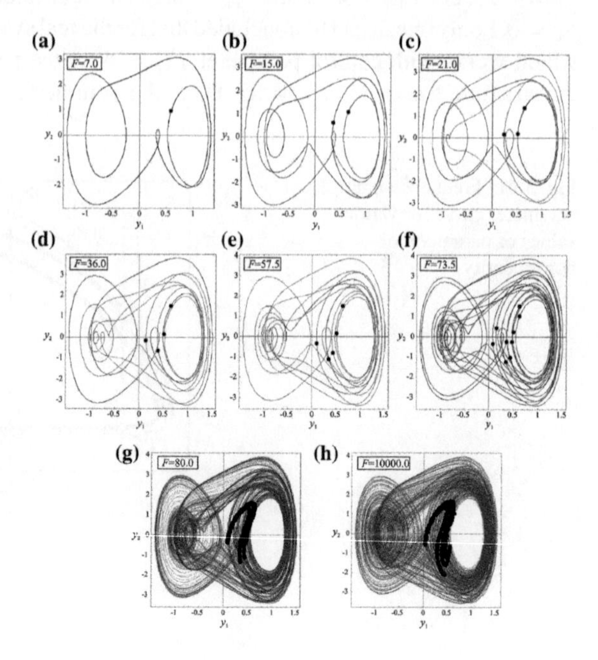


**Fig. 3.22** Trajectories in the phase space for various values of the control parameter: **a** periodic solution with period 1, **b** periodic solution with period 2, **c** periodic solution with period 3, **d** periodic solution with period 4, **e** periodic solution with period 5, **f** periodic solution with period 9, **g**, **h** chaotic solution

are captured for $\tau_n = nT$, where $n = 1, 2, \ldots$, with period $T = 2\pi/\bar{\Omega}$. The angular velocity $\overline{\Omega}$ is obtained numerically

$$\overline{\Omega} = \frac{\varphi(\tau_1) - \varphi(0)}{\tau_1} = \frac{y_3(\tau_1) - y_3(0)}{\tau_1}, \tag{3.192}$$

where $\tau_1$ is a long time period for numerical calculation.

For the parameter values of the system, mentioned in the previous section, and the control parameter $F = 7.0$ the trajectory in the phase space and the Poincaré map are plotted (Fig. 3.22a). The motion is periodic with period 1 and frequency $\overline{\Omega} = 0.999$. For $F = 15$ the periodic motion is with period 2 (Fig. 3.22b) and for $F = 21$ with period 3 (Fig. 3.22c). Increasing the value of the control parameter the motion is periodic but the period number increases, too (see Fig. 3.22d–f). For high values of control parameter ($F = 80$ and $F = 10000$) the strange attractor is obtained (Fig. 3.22g, h). The existence of strange attractor signifies chaos which is evident only if the certain criteria for the maximal (local) Lyapunov exponent are satisfied. For computational reasons the vector notation for the system of Eq. (3.191) is introduced

$$Y' = G(Y, P), \tag{3.193}$$

where $Y = [y_1, y_2, y_3, y_4]^T$ is the state space vector, $G = [g_1, g_2, g_3, g_4]$, $P(p, \alpha, \eta, \mu, F)$ is set of parameters and $[...]^T$ is denoting transpose. The equations for small deviations $\delta Y$ from the trajectory $Y(t)$ are

$$\delta Y' = L_{ij}(Y(t))\delta Y, \qquad i, j = 1, 2, \ldots 4, \tag{3.194}$$

where $L_{ij} = \partial g_i/\partial y_j$ is the Jacobian matrix of derivatives

$$\begin{aligned}
L_{11} &= L_{13} = L_{14} = 0, \qquad L_{12} = 1,\\
L_{21} &= -\frac{p + 3\alpha y_1^2}{1 - \mu\eta \sin^2 y_3}, \qquad L_{22} = -\frac{\zeta}{1 - \mu\eta \sin^2 y_3},\\
L_{23} &= \frac{-\mu y_4^2 \sin y_3 + \mu\eta(1 - y_4)\cos y_3}{1 - \mu\eta \sin^2 y_3} - \frac{Q_1}{(1 - \mu\eta \sin^2 y_3)^2},\\
L_{24} &= \frac{2\mu y_4 \cos x_3 - \mu\eta \sin y_3}{1 - \mu\eta \sin^2 y_3}, \quad L_{44} = \frac{\mu\eta y_4 \sin(2y_3) - F}{1 - \mu\eta \sin^2 y_3},\\
L_{31} &= L_{32} = L_{33} = 0, \qquad L_{34} = 1,\\
L_{41} &= -\frac{\eta p \sin y_3 + 3\eta\gamma y_1^2 \sin y_3}{1 - \mu\eta \sin^2 y_3}, \qquad L_{42} = -\frac{\zeta\eta \sin y_3}{1 - \mu\eta \sin^2 y_3},\\
L_{43} &= \frac{\mu\eta y_4^2 \cos(2y_3) - \eta \cos x_3(\zeta y_2 + p y_1 + \gamma y_1^3)}{1 - \mu\eta \sin^2 y_3}\\
&\quad + \frac{Q_2}{(1 - \mu\eta \sin^2 y_3)^2},
\end{aligned} \tag{3.195}$$

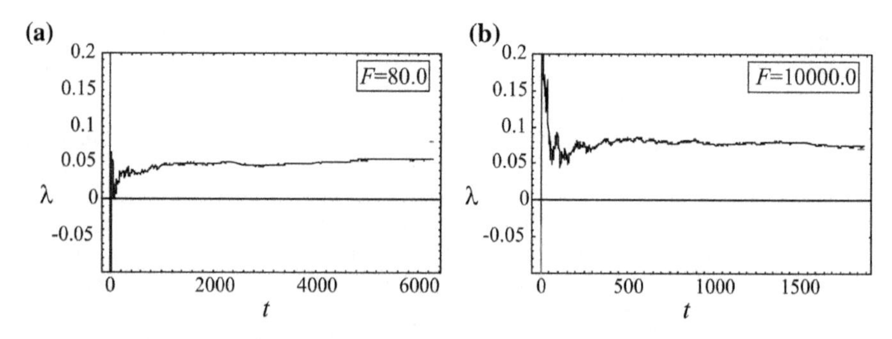


**Fig. 3.23** Positive Lyapunov exponent for the control parameters: **a** $F = 80$ and **b** $F = 10000$

where

$$Q_1 = -\mu\eta \sin(2y_3)[\zeta y_2 + py_1 + \gamma y_1^3 - \mu y_4^2 \cos y_3 - \mu F(1 - y_4) \sin y_3],$$
$$Q_2 = \mu\eta \sin(2y_3)[F(1 - y_4) + 0.4\mu\eta y_4^2 \sin(2y_3) - \eta \sin y_3(\zeta y_2 + py_1 + \gamma y_1^3)]. \tag{3.196}$$

The maximal Lyapunov exponent of the system is then as defined by Wolf et al. (1985)

$$\lambda = \lim \frac{1}{t} \log \frac{\|\delta Y(t)\|}{\|\delta Y(0)\|}. \tag{3.197}$$

In Fig. 3.23 we present the local Lyapunov exponent for the control parameters $F = 80$ and $F = 10000$. For the both parameter values the positive Lyapunov exponent is calculated. As it is discussed by Wolf et al. (1985), if the system contains at least one positive Lyapunov exponent the motion is chaotic. The two initially nearby orbits (or trajectories) diverge from each other and the separation of two nearby trajectories increases exponentially with time due to sensitive dependence on initial conditions.

The transition from periodic motion to chaos is by periodic doubling. In Fig. 3.24 the $y_1 - F$ bifurcation diagram is plotted. The bifurcation diagram is obtained for the long time integration of the differential equations (3.191) of the motion. After decay of transient motion a steady state motion is established. The fixed values of the parameters are $\alpha = 0.1, \gamma = 9, p = 1, \mu = 8.373, \eta = 0.05$ and the control parameter we choose to work with is $F$. By increasing parameter $F$ we found sequences of period doubling bifurcations (Fig. 3.24). Diagram starts from $F = 6$ where periodic motion with period $n = 1$ exists. This periodic solution bifurcates on $F \approx 8$ onto period $2T$. Further period doubling bifurcation (period 4, 8, 16,...) leads to chaotic motion for $F \approx 20$. After this interval of chaos the periodic motion with period 3 exists. Further increase of parameter $F$ leads to the new period doubling bifurcation with periods $5, 9, \ldots$ and finally to chaos for $F \in [28, 33]$. The next regions of chaotic motion are for $F \in [40, 56]$ and $F > 62$.

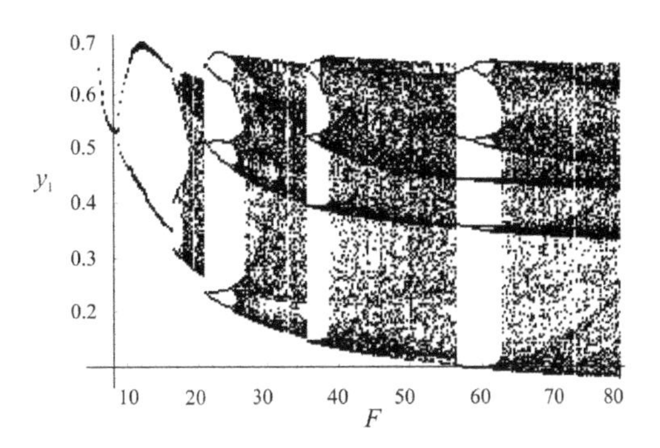


**Fig. 3.24** Period doubling bifurcation diagram

Comparing the bifurcation parameter values in Fig. 3.24 with the results of numerical experiment (see Fig. 3.22) it can be concluded that for the control parameter $F = 80$ chaotic motion exists.

### *3.4.4 Chaos Control*

It is of interest to control the chaos in the motion of the system (3.191) and specially of oscillator. There are many methods for chaos control. Dantas and Balthazar (2003) show that if we use an appropriate damping coefficient the chaotic behavior is avoided. The method of Pyragas (1992, 1995) is based on the addition of a special kind of time-continuous perturbation (external force control), which does not change the form of the desired unstable periodic solution, but under certain conditions can stabilize it. The method of Pyragas was applied in the paper. For the external force $\Phi(t)$ the model (3.191) becomes

$$
\begin{aligned}
y_1' &= y_2 + \Phi(t),\\
y_2' &= \frac{1}{1-\mu\eta\sin^2 y_3}\{-\zeta y_2 - p y_1 - \gamma y_1^3 + 1 + \mu\sin y_3[\mu y_4^2\cos y_3\\
&\quad -\eta\sin y_3 + F(1-y_4)\},\\
y_3' &= y_4,\\
y_4' &= \frac{\eta\sin y_3}{1-\mu\eta\sin^2 y_3}(-\zeta y_2 - p y_1 - \gamma y_1^3 + 1 + \mu y_4^2\cos y_3\\
&\quad +\mu\sin y_3\,(-\eta\sin y_3 + F(1-y_4))) - \eta\sin y_3 + F(1-y_4).
\end{aligned}
\tag{3.198}
$$

The external force $\Phi(t)$ is defined as

$$\Phi(t) = K[y_{up}(t) - y_1(t)], \tag{3.199}$$

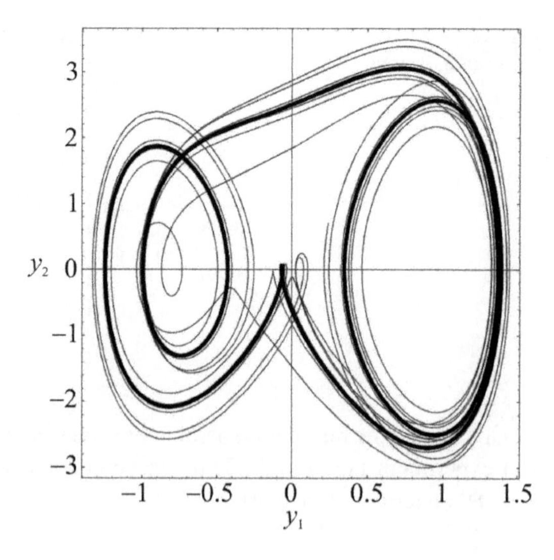


**Fig. 3.25** The two-periodic solution: before stabilization (–) and after stabilization (–)

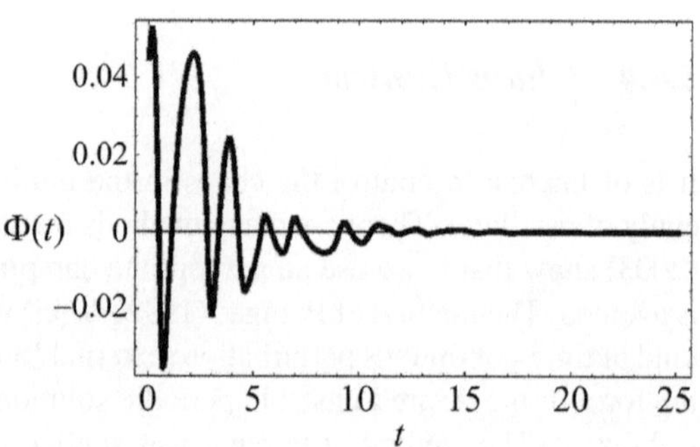


**Fig. 3.26** External force-time history diagram

where $K$ is an adjustable weight of the perturbation $\Phi(t)$ and $y_{up}(t)$ is a component of the unstable periodic solution of (3.191) which we wish to stabilize. The function $y_{up}(t)$ is time periodical with period $T$. For $\Phi(t)$ zero the system has a strange attractor.

However, by selecting the constant $K$ one can achieve the desired stabilization. Using the shooting method suggested by Van Dooren and Janssen (1996) the unstable two periodic unstable solution is detected (Fig. 3.25). Varying the value of the constant $K$ in the interval $[0.1, 2]$ it is concluded that for $K \in [0.3, 2]$ the stabilization is achieved (Fig. 3.25).

For $K = 2$ the function $\Phi(t)$ tends to a very small value (Fig. 3.26) and the component $y_1$, which is the solution after control, is very close to $y_{up}(t)$.

### 3.4.5 Conclusion

During passage through resonance of the motor-structure system which is modeled as a stable Duffing oscillator with non-ideal excitation severe vibrations appear. The energy of the system is not used for increasing of the rotation velocity, but is spent for vibrations which are harmful. Very often in the motion of the system near resonance the jump phenomena occurs: at the same value of the control parameter of the motor the amplitude of vibration skips to a higher value with lower frequency or to smaller amplitude with higher frequency. The manifestation depends on the direction of variation of the control setting. The jump phenomena and the increase in power required by a source operating near resonance are manifestations of a non-ideal energy source and are referred to as Sommerfeld effect. The Sommerfeld effect contributes to transform a regular vibration to an irregular chaotic one. It is concluded that in spite of the fact that the structure is modeled as the stable Duffing oscillator chaos appears. In the system the chaos is achieved by period doubling bifurcation. From engineering point of view it is necessary to eliminate the jump effect and the chaotic motion. The elimination of the jump phenomena for the certain control parameter is achieved by using the structure with coefficient of nonlinearity which is smaller than the critical value (3.190). Chaos is controlled using the external force control procedure where the added force does not change the form of the desired unstable periodic solution, but under certain conditions can stabilize it.

Comparing the results obtained applying the approximate analytic methods with those obtained numerically it is concluded that the difference is negligible. It proves the correctness of the used analytic procedure.

## 3.5 Bistable Duffing Oscillator Coupled with a Non-ideal Source

The model, shown in Fig. 3.27, represents an one degree-of-freedom cart connected to a fixed frame by a nonlinear spring and a dashpot (Warminski et al. 2001). Motion of the cart is due to a non-ideal motor with unbalanced rotor. In the absolute reference frame $x$ denotes the cart displacement and $\varphi$ denotes the angular displacement of the rotor. Elastic force of the spring is a cubic function of cart position $x$.

Motion of the system is described with the following equations (Kononenko 1976)

$$\ddot{x}(M+m)+c\dot{x}-md(\ddot{\varphi}\sin\varphi+\dot{\varphi}^2\cos\varphi)-k_1x+k_2x^3=0,$$
$$(J+md^2)\ddot{\varphi}-md\ddot{x}\sin\varphi=\mathcal{M}(\dot{\varphi}), \quad (3.200)$$

where $\mathcal{M}(\dot{\varphi})$ is the motor torque. For further consideration, let us introduce the dimensionless length and dimensionless time

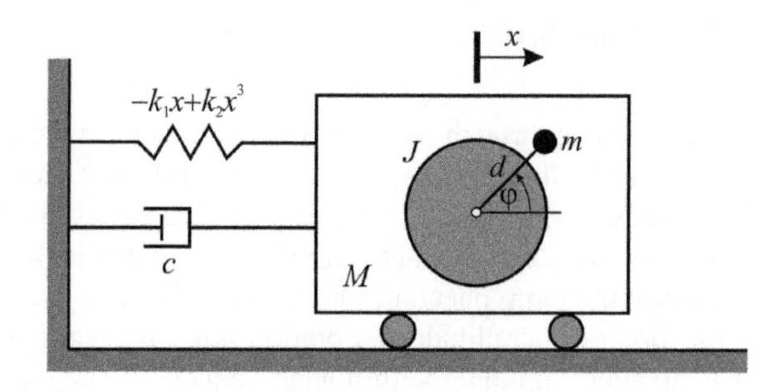


**Fig. 3.27** Model of the system

$$y = \frac{x}{d}, \quad \tau = \omega t, \tag{3.201}$$

where $d$ is the distance of the unbalanced mass to the shaft center and

$$\omega = \sqrt{\frac{k_1}{m+M}}. \tag{3.202}$$

Using (3.201) and (3.202), the dimensionless equations of motion are

$$\begin{aligned} y'' + \zeta y' - y + \gamma y^3 &= \frac{m}{M+m}(\varphi'' \sin\varphi + \varphi'^2 \cos\varphi), \\ \varphi'' &= \eta y'' \sin\varphi + \mathcal{M}(\varphi'), \end{aligned} \tag{3.203}$$

where

$$\begin{aligned} &\zeta = \frac{c}{\omega(M+m)}, \quad \gamma = \frac{d^2 k_2}{\omega^2(M+m)}, \quad \mu = \frac{m}{M+m}, \\ &\eta = \frac{md^2}{(J+md^2)}, \quad \mathcal{M}(\varphi') = \frac{\mathcal{M}(\dot{\varphi})}{\omega^2 (J+m_2 d^2)}, \end{aligned} \tag{3.204}$$

and $(') = d/d\tau$, $('') = d^2/d\tau^2$, $\zeta$ is the damping coefficient, $\gamma$ is the nonlinear parameter of the potential, while $\mu$ and $\eta$ are physical characteristics of the system. The torque is assumed to be linear, i.e.,

$$\mathcal{M}(\varphi') = E_1 - E_2\varphi', \tag{3.205}$$

with voltage or the strength of the motor $E_1$ and with characteristic parameter of the motor $E_2$. For (3.205) the equations of motion are

$$\begin{aligned} y'' + \zeta y' - y + \gamma y^3 &= \mu(\varphi'' \sin\varphi + \varphi'^2 \cos\varphi), \\ \varphi'' &= \eta y'' \sin\varphi + E_1 - E_2\varphi'. \end{aligned} \tag{3.206}$$

If this system has a principal parametric resonance 2:1, we assume that the oscillator vibrates with frequency $\Omega$, but the frequency of non-ideal force is equal to $\Omega/2$. As the voltage source is alternated, $E_1$ is periodic and has the form

$$E_1 = u_0 \cos\left(\frac{\upsilon_0}{2}\tau\right), \tag{3.207}$$

where $u_0$ is the amplitude of the voltage source and $\upsilon_0 = \Omega/\omega$. Substituting (3.207) into (3.206) we obtain

$$\begin{aligned} y'' + \zeta y' - y + \gamma y^3 &= \mu(\varphi'' \sin\varphi + \varphi'^2 \cos\varphi), \\ \varphi'' &= \eta y'' \sin\varphi + u_0 \cos\left(\frac{\upsilon_0}{2}\tau\right) - E_2\varphi'. \end{aligned} \tag{3.208}$$

Due to the strong nonlinearity there is no exact analytical solution for (3.208). Let us assume the approximate solution in the form

$$\begin{aligned} y &= A_0 + A_1\cos(\upsilon_0\tau) + A_2\sin(\upsilon_0\tau), \\ \varphi &= B_0 + B_1\cos\frac{(\upsilon_0\tau)}{2} + B_2\sin\frac{(\upsilon_0\tau)}{2}, \end{aligned} \tag{3.209}$$

where $A_0$ is the amplitude of the structure, $B_0$ is the amplitude of the rotor at rest, $A = \sqrt{A_1^2 + A_2^2}$ is the amplitude of oscillator and $B = \sqrt{B_1^2 + B_2^2}$ is the rotor amplitude. To obtain approximate solutions we expand the nonlinear function sin$\varphi$ and cos$\varphi$ in the Taylor series until third order around the lower steady state for $\varphi = 0$. Substituting (3.209) into (3.208), balancing the harmonics and neglecting the derivatives of the second order and terms having derivatives in a power higher than one, the set of first order approximate differential equations follows as

$$\begin{aligned} 0 = {} & \zeta A_1' + 2\omega A_2' - \upsilon_0\mu(B_2B_1' + B_1B_2')\left(1 - \frac{B_1^2}{4} - \frac{B_2^2}{12} + \frac{5}{4}B_0^2\right) + \zeta\upsilon_0 A_2 \\ & + \left(-1 - \upsilon_0^2 + \frac{3}{4}\gamma(A_1^2 + A_2^2) + 3\gamma A_0^2\right)A_1 + \frac{\upsilon_0^2}{4}\mu\left(1 - \frac{B_0^2}{2}\right)(B_1^2 - B_2^2), \end{aligned}$$

$$\begin{aligned} 0 = {} & \gamma A_2' + 2\omega A_1' + \upsilon_0\mu(B_1B_1' - B_2B_2')\left(1 - \frac{B_1^2}{6} - \frac{B_0^2}{2}\right) - \zeta\upsilon_0 A_1 \\ & + \left(-1 - \upsilon_0^2 + \frac{3}{4}\gamma(A_1^2 + A_2^2) + 3\gamma A_0^2\right)A_2 + \frac{\upsilon_0^2}{2}\mu B_1B_2\left(1 - \frac{B_0^2}{2}\right), \end{aligned}$$

$$0 = \omega\varepsilon_2 B_2(A_1' + \frac{\upsilon_0}{2}A_2)\left(1 - \frac{B_1^2}{4} - \frac{B_2^2}{12} - \frac{B_0^2}{2}\right) + E_2 B_1'$$
$$+\upsilon_0 B_2' + \frac{\upsilon_0}{2}E_2 B_1 - \frac{\upsilon_0^2}{4}B_1 - u_0$$
$$+\upsilon_0\eta(B_1 A_2' + \frac{\upsilon_0}{2}A_1 B_1)\left(1 - \frac{B_1^2}{6} - \frac{B_0^2}{2}\right),$$

$$0 = \upsilon_0\eta B_1(A_1' + \frac{\upsilon_0}{2}A_2)\left(1 - \frac{B_2^2}{4} - \frac{B_1^2}{12} - \frac{B_0^2}{2}\right) + E_2 B_2'$$
$$-\upsilon_0 B_1' - \frac{\upsilon_0}{2}E_2 B_1 - \frac{\upsilon_0^2}{4}B_2$$
$$-\upsilon_0\eta(B_2 A_2' + \frac{\upsilon_0}{2}A_1 B_2)\left(1 + \frac{B_2^2}{6} - \frac{B_0^2}{2}\right),$$

$$0 = A_0\left(\frac{3}{2}\gamma(A_1^2 + A_2^2) + \gamma A_0^2 - 1\right),$$
$$0 = \frac{\upsilon_0^2}{8}\eta B_0\left(A_1(B_2^2 - B_1^2) - 2A_2 B_1 B_2\right). \tag{3.210}$$

Relations $(3.210)_5$ and $(3.210)_6$ are satisfied for

$$\gamma A_0^2 = 1 - \frac{3}{2}\gamma A, \quad B_0 = 0. \tag{3.211}$$

Besides, let us assume for small oscillations that $B^2 = 0$.Using this assumption and (3.211) the Eqs. $(3.210)_1$–$(3.210)_4$ give the steady states when

$$A_1' = 0, \quad A_2' = 0, \quad B_1' = 0, \quad B_2' = 0,$$

as

$$0 = \zeta\upsilon_0 A_2 + \left(-1 - \upsilon_0^2 + \frac{3}{4}\gamma(A_1^2 + A_2^2) + 3\gamma A_0^2\right)A_1 + \frac{\upsilon_0^2}{4}\mu(B_1^2 - B_2^2),$$
$$0 = -\zeta\upsilon_0 A_1 + \left(-1 - \upsilon_0^2 + \frac{3}{4}\gamma(A_1^2 + A_2^2) + 3\gamma A_0^2\right)A_2 + \frac{\upsilon_0^2}{2}\mu B_1 B_2,$$
$$u_0 = \eta\frac{\upsilon_0^2}{2}(A_1 B_1 + A_2 B_2) + \frac{\upsilon_0}{2}E_2 B_2 - \frac{\upsilon_0^2}{4}B_1 - \eta\frac{\upsilon_0^2}{8}A_2 B_2 B_1^2,$$
$$0 = -\eta B_1\frac{\upsilon_0^2}{8}A_2 B_2^2 - \frac{\upsilon_0}{2}E_2 B_1 - \frac{\upsilon_0^2}{4}B_2 - \eta\frac{\upsilon_0^2}{2}(A_1 B_2 - A_2 B_1). \tag{3.212}$$

Equation (3.212) represent the set of four coupled equations. According to some specific considerations the system of Eq. (3.212) have semi-trivial and non-trivial solutions.

### 3.5.1 Semi-trivial Solutions and Quenching of Amplitude

The semi-trivial solution means physically that one part of the system is oscillating while the other one is at rest. This gives the condition for quenching of amplitude of oscillation of one part of the system: of the mechanical part and of the non-ideal force.

Quenching of amplitude in the mechanical part satisfies the following requirements

$$A_1 = A_2 = 0, \quad B_1 \neq 0, \quad B_2 \neq 0. \tag{3.213}$$

For these values the mechanical part of the system does not vibrate. This case can be used as a technique of control whose objective is to cancel vibration of the mechanical system. According to (3.210) the condition of quenching phenomenon in the space of the parameters of the system is derived and given as

$$B = \frac{2u_0}{\upsilon_0\sqrt{E_2^2 + (\frac{\upsilon_0}{2})^2}}. \tag{3.214}$$

In Fig. 3.28, the amplitude of non-ideal force $B$ as function of the frequency $\upsilon_0$, given with (3.211), is plotted. For numerical purposes, the following set of parameters (Souza et al. 2005a, b) is considered

$$\zeta = 0.02, \quad \gamma = 0.1, \quad \mu = 0.1, \quad \eta = 0.25, \quad E_2 = 1.5.$$

Analytical solution is compared with numerical solution of (3.210) using the fourth-order Runge Kutta algorithm (see Fig. 3.28).

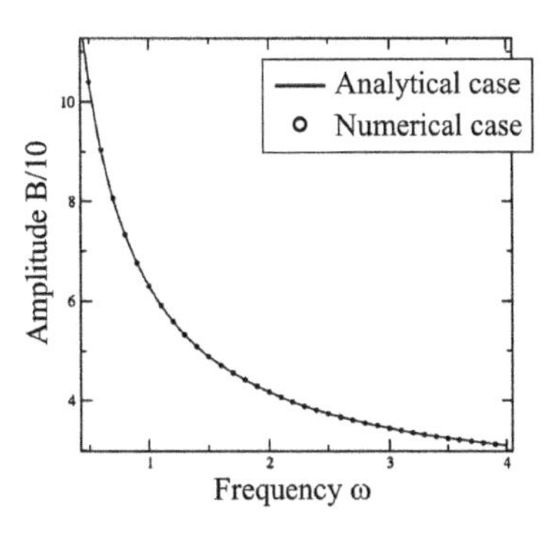


**Fig. 3.28** Amplitude-frequency curve of non-ideal system for semi-trivial case. (Nbendjo et al. 2012)

It appears that the amplitude of the non-ideal force decreases as the frequency increases. To analyze the stability of this semi-trivial solution, the amplitude modulation equations are given by

$$\begin{aligned}\dot{A}_1 &= f_1(A_1, A_2, B_1, B_2),\\ \dot{A}_2 &= f_2(A_1, A_2, B_1, B_2),\\ \dot{B}_1 &= f_3(A_1, A_2, B_1, B_2),\\ \dot{B}_2 &= f_4(A_1, A_2, B_1, B_2).\end{aligned} \tag{3.215}$$

Perturbing the solution $A_1, A_2, B_1$ and $B_2$ with $\delta A_1, \delta A_2, \delta B_1$ and $\delta B_2$ and substituting into (3.215), after linearization a set of differential equations is obtained. Stability conditions are based on the eigenvalues of the Jacobian (Warminski and Kecik 2006). If the solution is complex, and the real part of the eigenvalue is always negative, the system is stable.

Quenching of amplitude of the non-ideal force occurs for

$$B_1 = 0, \quad B_2 = 0, \quad A_1 \neq 0, \quad A_2 \neq 0. \tag{3.216}$$

This situation represents the case where the non-ideal forces does not swing and the structure vibrates. Using the set of differential equations for steady-state motion (3.212) and assuming the stationarity of solutions leads after some calculations to the following non-linear equation of the amplitude

$$A = \sqrt{\frac{(\frac{15}{4}\gamma)^2 + \zeta^2 \upsilon_0^2 + (2 - \upsilon_0^2)^2}{\frac{15}{2}\gamma(2 - \upsilon_0^2)}}. \tag{3.217}$$

Analyzing of this equation shows the evidence of mechanical part which is at equilibrium and thus the system remains stable.

### 3.5.2 Non-trivial Solutions and Their Stability

Non-trivial solutions represent the case where both systems vibrate and

$$A_1 \neq 0, \quad A_2 \neq 0, \quad B_1 \neq 0, \quad B_2 \neq 0. \tag{3.218}$$

Analytically, we were supposed to use the set of Eq. (3.210) and to derive the amplitude of both systems. Unfortunately, the system is strongly nonlinear and it is quite impossible to obtain an analytical expression of amplitude.

Moreover, to deal with such a question, we solved directly the base equations (3.208) numerically using the fourth-order Runge–Kutta algorithm and discuss the amplitude resonance curves. Afterwards, we explore the stability of the system using

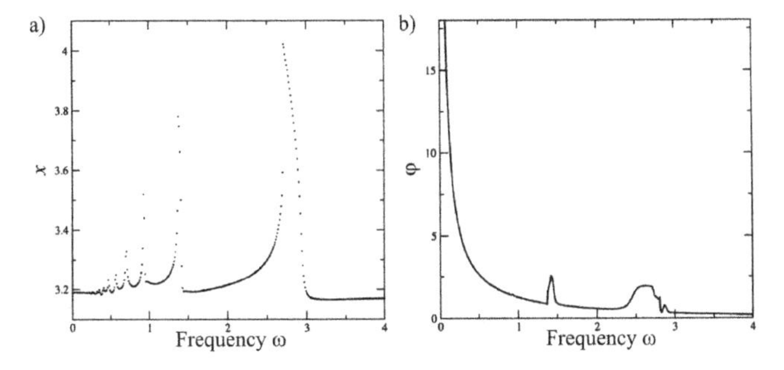


**Fig. 3.29** **a** Amplitude of the mechanical part as function of the frequency; **b** Amplitude of the non-ideal source as function of the frequency (Nbendjo et al. 2012)

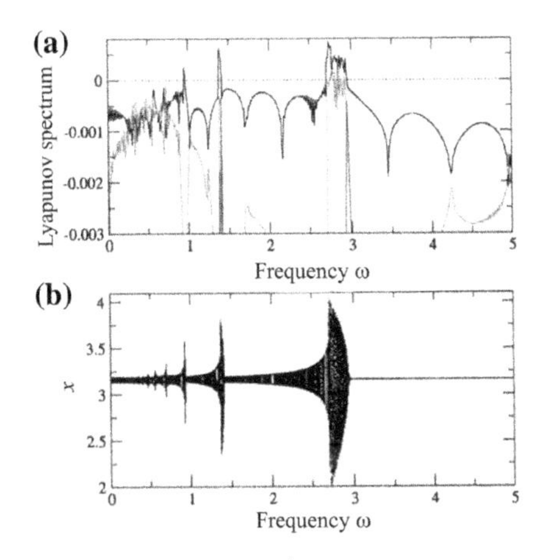


**Fig. 3.30** **a** Lyapunov spectrum and **b** bifurcation diagram as function of $v_0$ for $u_0 = 1$ (Nbendjo et al. 2012)

Lyapunov spectrum and bifurcation sequences. The frequency response curve is thus obtained from (3.208) and presented in Fig. 3.29a for the evaluation $x$ and in Fig. 3.29b for evaluation of $\varphi$ for $u_0 = 1$.

Figure 3.29a reveals a set of subharmonic resonances instead of the internal resonance 2:1 as expected. Concerning the amplitude of non-ideal forces, it decreases as frequency increases and the effect of internal resonance is visible by their appearance.

Focusing on the stability of the system allows one to display the Lyapunov spectrum as function of frequency for the specify value of $u_0 = 1$.

It appears, for example in Fig. 3.30a, that the Lyapunov spectrum is positive for certain values of frequency which is necessary for the presence of chaos in the system. On the other hand, for another set of frequencies, Fig. 3.30a shows more than one positive Lyapunov exponent, indicating hyperchaos on the system. These effects are confirmed via the corresponding bifurcation diagram (see Fig. 3.30b). Setting $v_0 = 1.5$ allows one to observe the stability of the system as $u_0$ increases.

It appears in Fig. 3.31a that the system shows a transient from periodicity to quasiperiodicity and later to chaos. To view how these transient arise, the corresponding

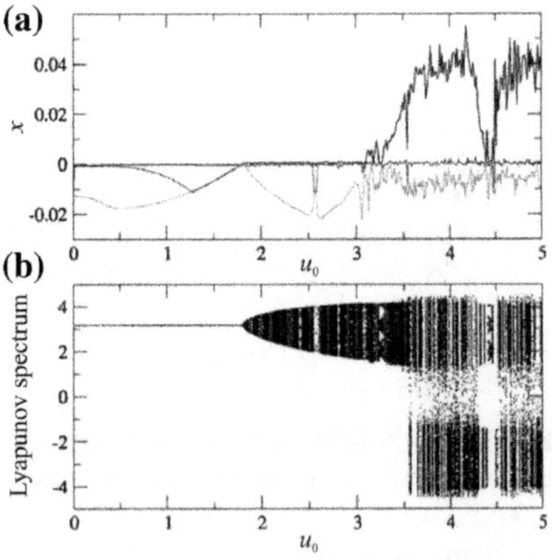

**Fig. 3.31** Lyapunov spectrum and corresponding bifurcation diagram as function of $u_0$ for $v_0 = 1.5$ (Nbendjo et al. 2012)

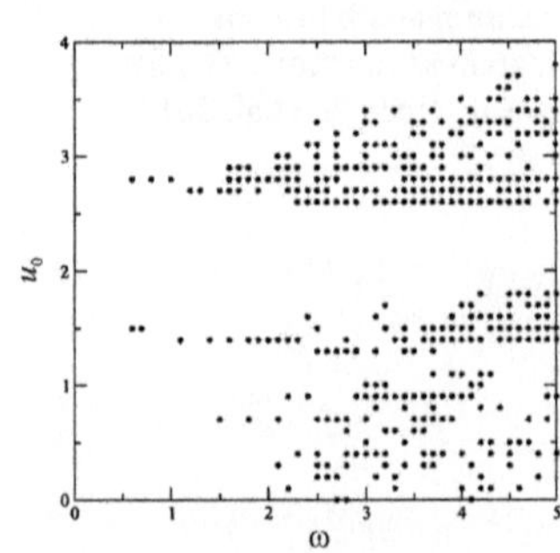


**Fig. 3.32** Region in parameter space of $(u_0, v_0)$ where hyperchaos is detected (Nbendjo et al. 2012)

bifurcation diagram is presented in Fig. 3.31b, which confirms the results obtained from the Lyapunov spectrum. It shows that for $u_0 = 0.36$ there is a crisis in which sudden change in chaotic attractors occurs. Summarizing, it is concluded that there are regions in which all Lyapunov exponents are less or equal to zero for periodic or quasiperiodic orbits. If one of the Lyapunov exponents is positive, chaos is evident. If more than one positive Lyapunov spectrum exists, the presence of the hyperchaos for a specific set of parameters is indicated (Fig. 3.32).

Based on these consideration it can be concluded that the stability of the systems deeply depends on the voltage of the energy source.

### 3.5.3 Conclusion

Alternating strength of the voltage of the source may be the reason for the limited power supply for the bistable Duffing oscillator. Thus, in the oscillator - non-ideal source system for certain parameters quenching phenomena occurs: quenching of amplitude in the mechanical part or quenching of amplitude of the non-ideal force. Besides, in the system an explosion of resonances 2:1 to a set of subharmonic resonances is revealed. Based on stability analysis of the non-trivial solutions we concluded that the periodicity of the voltage source perturbs the limited power supply and increases the possibility of the appearance of chaos and hyperchaos of the systems. Consequently, when the voltage source is alternated, if the choice of the characteristic is bad, the system can become unstable.

## Appendix: Ateb Functions

### *Sine Ateb Function*

The incomplete Beta function is defined as (see Abramowitz and Stegun 1964, 1979)

$$B_x(p,q) = \int_0^{0\le x\le 1} t^{p-1}(1-t)^{q-1}dt, \tag{A1}$$

while for the case $x = 1$, the complete Beta function is obtained

$$B(p,q) = \int_0^1 t^{p-1}(1-t)^{q-1}dt, \tag{A2}$$

where $p$ and $q$ are real numbers. For the positive integrands in (A1) and (A2), the interval for the incomplete beta function is

$$0 \le B_x(p,q) \le B(p,q). \tag{A3}$$

For the functions $B(p,q)$ and $B_x(p,q)$ the following identities are evident

$$B(p,q) = B(q,p), \quad B_x(p,q) = B(p,q) - B_y(p,q), \tag{A4}$$

where $x + y = 1$.

Let us determine the inverse of the half of the incomplete beta function (A1)

$$x \to \frac{1}{2}B_x(p,q) = \frac{1}{2}\int_0^{0\le x\le 1} t^{p-1}(1-t)^{q-1}dt. \tag{A5}$$

Introducing the notation (Cveticanin 2014)

$$\frac{1}{2}B_x(p,q) = w, \tag{A6}$$

the half of the incomplete beta function is

$$\frac{1}{2}\int_{0}^{0\leq x\leq 1} t^{p-1}(1-t)^{q-1}dt = w. \tag{A7}$$

For the the new variables

$$t = \bar{v}^{1/p}, \quad x = v^{1/p}, \tag{A8}$$

the boundary of integration $0 \leq x \leq 1$ transforms into $0 \leq |v| \leq 1$ and the integral (A7) has the forms

$$\frac{1}{2p}\int_{0}^{0\leq v\leq 1} (1-\bar{v}^{1/p})^{q-1}d\bar{v} = w, \quad \frac{1}{2p}\int_{0}^{-1\leq v\leq 0} (1-\bar{v}^{1/p})^{q-1}d\bar{v} = w. \tag{A9}$$

Let us consider the first integral in (A9). For $v = 0$ the integral is zero, and for $v = 1$ it is according to (A5)

$$\frac{1}{2p}\int_{0}^{1}(1-\bar{v}^{1/p})^{q-1}d\bar{v} = \frac{1}{2}B(p,q). \tag{A10}$$

Thus, the value of the function $w$ (A9) is bounded

$$0 \leq w \leq \frac{1}{2}B(p,q), \tag{A11}$$

as $B(p,q)$ is finite. Now, the inverse for the integral (A9) is constructed. This inverse depends on the three parameters $p, q$ and $w$. With notation given by Rosenberg (1963, 1966) we have

$$v = sa\left(\frac{1-p}{p}, \frac{1-q}{q}, w\right), \tag{A12}$$

The second integral in (A9) gives

$$-v = sa\left(\frac{1-p}{p}, \frac{1-q}{q}, -w\right). \tag{A13}$$

It follows that sa$(\frac{1-p}{p}, \frac{1-q}{q}, w)$ is the inverse of (A9) on the interval $-\frac{1}{2}B(p,q) \leq w \leq \frac{1}{2}B(p,q)$. Formula (A9) defines $w$ uniquely as a function of $v$ in this interval. Using the odd property (A13) of inverse of (A9), Rosenberg (1963) named the function 'sine Ateb function' and noted as sa.

The period of the function is

$$2\Pi(p,q) = 2B(p,q). \tag{A14}$$

## *Cosine Ateb Function*

The change of variables

$$t = 1 - \bar{u}^{1/q}, \quad x = 1 - u^{1/q}, \tag{A15}$$

transforms the interval $0 \leq x \leq 1$ into $0 \leq |u| \leq 1$. Substituting (A15) into (A7), for $1 \geq u \geq 0$, it yields

$$-\frac{1}{2q}\int_{1}^{0\leq u\leq 1}(1-\bar{u}^{1/q})^{p-1}d\bar{u} = w. \tag{A16}$$

Due to the property of the beta function (A4), we have

$$-\frac{1}{2q}\int_{0}^{0\leq u\leq 1}(1-\bar{u}^{1/q})^{p-1}d\bar{u} + \frac{1}{2q}\int_{0}^{1}(1-\bar{v}^{1/q})^{p-1}d\bar{u} = w, \tag{A17}$$

i.e.,

$$-\frac{1}{2q}\int_{0}^{0\leq u\leq 1}(1-\bar{u}^{1/q})^{p-1}d\bar{u} = -\frac{1}{2}B(p,q) + w. \tag{A18}$$

Using the notation of the inverse function (A12) i.e. (A13) and the period $\Pi(p,q) = B(p,q)$ we obtain

$$u = \pm sa\left(\frac{1-q}{q}, \frac{1-p}{p}, \frac{1}{2}\Pi(p,q) \mp w\right). \tag{A19}$$

The inverse function $u$ Rosenberg (1963) called 'cosine Ateb function' and noted it with ca:

$$ca\left(\frac{1-q}{q}, \frac{1-p}{p}, w\right) = sa\left(\frac{1-p}{p}, \frac{1-q}{q}, \frac{1}{2}\Pi(p,q) \pm w\right). \tag{A20}$$

The ca function is an even function (see Senik (1969a, b), Drogomirecka (1997)) with period $2\Pi(p,q)$ given with (A14).

### *Properties of Ateb Functions*

The relation which satisfy the sa and the ca Ateb functions is (Gottlieb 2003, Gricik and Nazarkevich 2007)

$$\mathrm{sa}^2\left(\frac{1-p}{p},\frac{1-q}{q},w\right)+\mathrm{ca}^{\alpha+1}\left(\frac{1-q}{q},\frac{1-p}{p},w\right)=1. \tag{A21}$$

The first derivatives of sa and ca functions are

$$\frac{d}{dw}\mathrm{sa}\left(\frac{1-p}{p},\frac{1-q}{q},w\right)=\mathrm{ca}^{\alpha}\left(\frac{1-q}{q},\frac{1-p}{p},w\right), \tag{A22}$$

$$\frac{d}{dw}\mathrm{ca}\left(\frac{1-q}{q},\frac{1-p}{p},w\right)=-\frac{2}{\alpha+1}\mathrm{sa}\left(\frac{1-p}{p},\frac{1-q}{q},w\right). \tag{A23}$$

The Fourier series expansion of the ca Ateb function (Gricik et al. 2009) is as follows

$$\mathrm{ca}\left(\frac{1-q}{q},\frac{1-p}{p},w\right)=\sum_{n=1}^{\infty}a_n\cos\frac{\pi n w}{\Pi(p,q)}, \tag{A24}$$

where $a_0 = 0$ and

$$a_n=\frac{2}{\Pi(p,q)}\int_0^{\Pi(p,q)}\mathrm{ca}\left(\frac{1-q}{q},\frac{1-p}{p},w\right)\cos\frac{\pi n w}{\Pi(p,q)}dw. \tag{A25}$$

Namely, the coefficient is enough to be calculated for $w \in [0, \Pi(p,q)/2]$.

## References

Abramowitz, M., & Stegun, I. A. (1964). *Handbook of mathematical functions*. National Bureu of Standards: Applied Mathematics Series, vol. 55. National Bureau of Standards, Washington.

Abramowitz, M., & Stegun, I. A. (1979). *Handbook of mathematical functions with formulas, graphs and mathematical tables*. Moscow: Nauka. (in Russian).

Amore, P., & Aranda, A. (2005). Improved Lindstedt-Poincaré method for the solution of nonlinear problems. *Journal of Sound and Vibration*, *283*, 1115–1136.

Andrianov, I. V. (2002). Asymptotics of nonlinear dynamical systems with high degree of nonlinearity. *Doklady RAN*, *386*, 165–168.

Andrianov, I. V., & Awrejcewicz, J. (2003a). Asymptotic approaches to strongly nonlinear dynamical systems. *Journal on Systems Analysis Modelling Simulation*, *43*, 255–268.

Andrianov, I. V., & Awrejcewicz, J. (2003b). Asymptotical behaviour of a system with damping and high power-form nonl-linearity. *Journal of Sound and Vibration*, *267*, 1169–1174.

Belendez, A., Pascual, C., Gallego, S., Ortuño, M., & Neipp, V. (2007). Application of a modified He's homotopy perturbation method to obtain higher-order approximations of an $x^{1/3}$ force nonlinear oscillator. *Physics Letters A*, *371*, 421–426.

Bogolyubov, N. N., & Mitropolskij, Ju A. (1974). *Asimptoticheskie metodi v teorii nelinejnih kolebanij*. Moscow: Nauka.

Castao, K. A. L., Goes, C. S., & Balthazar, J. M. (2010). A note on the attenuation of the sommerfeld effect of a non-ideal system taking into account a MR damper and the complete model of a DC motor. *Journal of Vibration and Control*, *17*(7), 1112–1118.

Chen, S. H., & Cheung, Y. K. (1996). An elliptic perturbation method for certain strongly non-linear oscillators. *Journal of Sound and Vibration*, *192*, 453–464.

Chen, W. H., & Gibson, R. F. (1998). Property distribution determination for nonuniform composite beams from vibration response measurements and Galerkin's method. *Journal of Applied Mechanics, ASME*, *65*, 127–133.

Chen, S. H., Yang, X. M., & Cheung, Y. K. (1998). Periodic solutions of strongly quadratic nonlinear oscillators by the elliptic perturbation method. *Journal of Sound and Vibration*, *212*, 771–780.

Cheng, Y. K., Chen, S. H., & Lau, S. L. (1991). A modified Lindstedt-Poincaré method for certain strongly non-linear oscillators. *International Journal of Nonlinear Mechanics*, *26*, 367–378.

Colm, I. J., & Clark, N. J. (1988). *Forming, shaping and working of high-performace ceramics*. New York: Blackie.

Cveticanin, L. (1992). An approximate solution for a system of two coupled differential equations. *Journal of Sound and Vibration*, *153*, 375–380.

Cveticanin, L. (1993). An asymptotic solution to weak nonlinear vibrations of the rotor. *Mechanism and Machine Theory*, *28*, 495–506.

Cveticanin, L. (1995). Resonant vibrations of nonlinear rotors. *Mechanism and Machine Theory*, *30*, 581–588.

Cveticanin, L. (2003). Vibrations of the system with quadratic non-linearity and a constant excitation force. *Journal of Sound and Vibration*, *261*(1), 169–176.

Cveticanin, L. (2004). Vibrations of the nonlinear oscillator with quadratic nonlinearity. *Physica A*, *341*, 123–135.

Cveticanin, L. (2006). Homotopy-perturbation method for pure non-linear differential equation. *Chaos, Solitons and Fractals*, *30*, 1221–1230.

Cveticanin, L. (2008). Analysis of oscillators with non-polynomial damping terms. *Journal of Sound and Vibration*, *317*, 866–882.

Cveticanin, L. (2009). Oscillator with fraction order restoring force. *Journal of Sound and Vibration*, *320*, 1064–1077.

Cveticanin, L. (2011). Pure odd-order oscillators with constant excitation. *Journal of Sound and Vibration*, *330*, 976–986.

Cveticanin, L. (2014). On the truly nonlinear oscillator with positive and negative damping. *Applied Mathematics and Computation*, *243*, 433–445.

Cveticanin, L., & Pogany, T. (2012). Oscillator with a sum of non-integer order non-linearities. *Journal of Applied Mathematics*, Article ID 649050, 20 p. doi:10.1155/2012/649050.

Cveticanin, L., & Zukovic, M. (2015a). Non-ideal mechanical system with an oscillator with rational nonlinearity. *Journal of Vibration and Control*, *21*(11), 2149–2164.

Cveticanin, L., & Zukovic, M. (2015b). Motion of a motor-structure non-ideal system. *European Journal of Mechanics A/Solids*, *53*, 229–240 ($2015_2$).

Cveticanin, L., Kalami-Yazdi, M., Saadatnia, Z., & Askari, H. (2010). Application of hamiltonian approach to the generalized nonlinear oscillator with fractional power. *International Journal of Nonlinear Sciences and Numerical Simulation*, *11*, 997–1002. ($(2010)_1$).

Cveticanin, L., Kovacic, I., & Rakaric, Z. (2010). Asymptotic methods for vibrations of the pure-non-integer order oscillator. *Computers and Mathematics with Applications*, *60*, 2616–2628.

Cveticanin, L., Kalami-Yazdi, M., & Askari, H. (2012). Analytical approximations to the solutions for a generalized oscillator with strong nonlinear terms. *Journal of Engineering Mathematics*, *77*(1), 211–223.

Dantas, M. H., & Balthazar, J. M. (2003). On the appearance of a Hopf bifurcation in a non-ideal mechanical system. *Mechanics Research Communications*, *30*, 493–503.

Dantas, M. H., & Balthazar, J. M. (2006). A comment on a non-ideal centrifugal vibrator machine behavior with soft and hard springs. *International Journal of Bifurcation and Chaos*, *16*, 1083–1088.

Dantas, M. J. H., & Balthazar, J. M. (2007). On the existence and stability of periodic orbits in non ideal problems: General results. *Zeitschrift fur angewandte Mathematik und Physik*, *58*, 940–958.

Dimentberg, M. F., McGovern, L., Norton, R. L., Chapdelaine, J., & Harrison, R. (1997). Dynamics of an unbalanced shaft interacting with a limited power supply. *Nonlinear Dynamics*, *13*, 171–187.

Drogomirecka, H. T. (1997). Integrating a special Ateb-function. *Visnik Lvivskogo Universitetu. Serija mehaniko-matematichna*, *46*, 108–110. (in Ukrainian).

Droniuk, I. M., Nazarkevich, M. A., & Thir, V. (2010). Evaluation of results of modelling Ateb-functions for information protection. *Visnik nacionaljnogo universitetu Lvivska politehnika*, *663*, 112–126. (in Ukrainian).

Droniuk, I., & Nazarkevich, M. (2010). Modeling nonlinear oscillatory system under disturbance by means of Ateb-functions for the Internet. *Proceedings of the 6th International working Conference on Performance Modeling and Evaluation of Heterogeneous Networks (HET-NETs'10), Zakopane, Poland* (pp. 325–334).

Droniuk, I. M., Nazarkevich, M. A., & Thir, V. (1997). Evaluation of results of modelling Ateb-functions for information protection. Visnik Nacionaljnogo Universitatu Lvivskogo Universitetu. *Serija Mehanika-matematichna*, *46*, 108–110.

Fang, T., & Dowell, E. H. (1987). Numerical simulations of periodic and chaotic responses in a stable Duffing system. *International Journal of Non-Linear Mechanics*, *22*, 401–425.

Felix, J. L. P., Balthazar, J. M., & Dantas, M. J. H. (2009a). On energy pumping, synchronization and beat phenomenon in a nonideal structure coupled to an essentially nonlinear oscillator. *Nonlineaer Dynamics*, *56*, 1–11.

Felix, J. L., Balthazar, J. M., & Brasil, R. M. L. R. F. (2009b). Comments on nonlinear dynamics of a non-ideal Duffing-Rayleigh oscillator: Numerical and analytical approaches. *Journal of Sound and Vibration*, *319*, 1136–1149.

Gendelman, O., & Vakakis, A. F. (2000). Transitions from localization to nonlocalization in strongly nonlinear damped oscillators. *Chaos, Solitons and Fractals*, *11*, 1535–1542.

Gottlieb, H. P. W. (2003). Frequencies of oscillators with fractional-power non-linearities. *Journal of Sound and Vibration*, *261*, 557–566.

Gradstein, I. S., & Rjizhik, I. M. (1971). *Tablicji integralov, summ, rjadov i proizvedenij*. Moscow: Nauka.

Gricik, V. V., & Nazarkevich, M. A. (2007). Mathematical models algorythms and computation of Ateb-functions. *Dopovidi NAN Ukraini Seriji A*, *12*, 37–43. (in Ukrainian).

Gricik, V. V., Dronyuk, I. M., & Nazarkevich, M. A. (2009). Document protection information technologies by means of Ateb-functions I. Ateb-function base consistency for document protection. *Problemy upravleniya i avtomatiki*, *2*, 139–152. (in Ukrainian).

Haslach, H. W. (1985). Post-buckling behavior of columns with non-linear constitutive equations. *International Journal of Non-Linear Mechanics*, *20*, 53–67.

Haslach, H. W. (1992). Influence of adsorbed moisture on the elastic post-buckling behavior of columns made of non-linear hydrophilic polymers. *International Journal of Non-Linear Mechanics*, *27*, 527–546.

He, J.-H. (2002). Modified Lindstedt-Poincaré methods for some strongly non-linear oscillations Part I: Expansion of a constant. *International Journal of Non-Linear Mechancis*, *37*, 309–314.

http://functions.wolfram.com/GammaBetaErf/Beta3/26/01/02/0001 (2002a).

http://functions.wolfram.com/HypergeometricFunctions/Hypergeometric2F1/03/09/19/02/0017 (2002b).

http://functions.wolfram.com/EllipticIntegrals/EllipticF/16/01/02/0001 (2002c).

http://functions.wolfram.com/EllipticFunctions/Hypergeometric2F1/03/07/17/01/0012 (2002d).

http://functions.wolfram.com/EllipticFunctions/JacobiAmplitude/16/01/01/0001.

Hu, H., & Xiong, Z. G. (2003). Oscillations in an $x^{(2m+2)/(2n+1)}$ potential. *Journal of Sound and Vibration, 259*, 977–980.

Jutte, C. V. (2008). Generalized synthesis methodology of nonlinear springs for prescribed load-displacement functions. Ph.D. Dissertation, Mechanical Engineering, The University of Michigan.

Kononenko, V. O. (1969). *Vibrating system with a limited power supply*. London: Illife.

Kononenko, V. O., & Korablev, S. S. (1959). An experimental investigation of the resonance phenomena with a centrifugal excited alternating force. *Trudji Moskovskog Teknichkog Instutita, 14*, 224–232.

Kovacic, I., Rakaric, Z., & Cveticanin, L. (2010). A non-simultaneous variational approach for the oscillators with fractional-order power nonlinearity. *Applied Mathematics and Computation, 217*, 3944–3954.

Krylov, N., & Bogolubov, N. (1943). *Introduction to nonlinear mechanics*. New Jersey: Princeton University Press.

Lewis, G., & Monasa, F. (1982). Large deflections of a cantilever beams of nonlinear materials of the Ludwick type subjected to an end moment. *International Journal of Non-Linear Mechancis, 17*, 1–6.

Lo, C. C., & Gupta, S. D. (1978). Bending of a nonlinear rectangular beam in large deflection. *Journal of Applied Mechanics, ASME, 45*, 213–215.

Lyapunov, A. M. (1893). An investigation of one of the singular cases of the theory of the stability of motion. *II Mathematicheski Sbornik, 17*, 253–333. (in Russian).

Mickens, R. E. (2001). Oscillations in an $x^{4/3}$ potential. *Journal of Sound and Vibration, 246*, 375–378.

Mickens, R. E. (2004). *Mathematical methods for the natural and engineering sciences*. New Jersey: World Scientific.

Mickens, R. E. (2006). Iteration method solutions for conservative and limit-cycle $x^{1/3}$ force oscillators. *Journal of Sound and Vibration, 292*, 964–968.

Nayfeh, A. H., & Mook, D. T. (1979). *Nonlinear oscillations*. New York: Wiley.

Nbendjo, B. R. N., Caldas, I. L., & Viana, R. L. (2012). Dynamical changes from harmonic vibrations of a limited power supply driving a Duffing oscillator. *Nonlinear Dynamics, 70*(1), 401–407.

Ozis, T., & Yildirm, T. A. (2007). Determination of periodic solution for a $u^{1/3}$ force by He's modified Lindstedt-Poincaré method. *Journal of Sound and Vibration, 301*, 415–419.

Patten, W. N., Sha, S., & Mo, C. (1998). A vibration model of open celled polyurethane foam automative seat cuchions. *Journal of Sound and Vibration, 217*, 145–161.

Pezeshki, C., & Dowell, E. H. (1988). On chaos and fractal behaviour in a general Duffing's system. *Physica D, 32*, 194–209.

Pilipchuk, V. N. (2010). *Nonlinear dynamics: Between linear and impact limits*. New York: Springer.

Prathap, G., & Varadan, T. K. (1976). The inelastic large deformation of beams. *ASME Journal of Applied Mechanics, 43*, 689–690.

Pyragas, K. (1992). Continuous control of chaos by self controlling feedback. *Physics Letters A, 170*, 421–428.

Pyragas, K. (1995). Control of chaos via extended delay feedback. *Physics Letters A, 206*, 323–330.

Rosenberg, R. M. (1963). The ateb(h)-functions and their properties. *Quarterly of Applied Mathematics, 21*, 37–47.

Rosenberg, R. M. (1966). On nonlinear vibrating systems with many degrees of freedom. *Advances in Applied Mechanics, 32*, 155–242.

Russell, D., & Rossing, T. (1998). Testing the nonlinearity of piano hammers using residual shock spectra. *Acta Acustica, 84*, 967–975.

Senik, P. M. (1969a). Inversion of the incomplete beta function. *Ukrainski Matematicheskii Zhurnal, 21*, 271–278.

Senik, P. M. (1969b). Inversion of the incomplete Beta-function. *Ukrainski matematicheski zhurnal 21, 325-333 and Ukrainian Mathematical Journal, 21*, 271–278.

Souza, S. L. T., Caldas, I. L., Viana, R. L., Balthazar, J. M., & Brasil, R. M. L. R. F. (2005a). Impact dampers for controlling chaos in systems with limited power supply. *Journal of Sound and Vibration, 279*, 955–965.

Souza, S. L. T., Caldas, I. L., Viana, R. L., Balthazar, J. M., & Brasil, R. M. L. R. F. (2005b). Basins of attraction changes by amplitude constraining of oscillators with limited power supply. *Chaos, Solitons and Fractals, 26*, 1211–1220.

Tsuchida, M., Guilherme, K. L., Balthazar, J. M., & Silva, G. N., (2003). On regular and irregular vibrations of a non-ideal system with two degrees of freedom. 1:1 resonance. *Journal of Sound and Vibration, 260*, 949–960.

Tsuchida, M., Guilherme, K. L., & Balthazar, J. M. (2005). On chaotic vibrations of a non-ideal system with two degree of freedom. 1:2 resonance and Sommerfeld effect. *Journal of Sound and Vibration, 282*, 1201–1207.

Ueda, Y. (1985). Random phenomena resulting from non-linearity in the system described by Duffing's equation. *International Journal of Non-Linear Mechanics, 20*, 481–491.

Van Dooren, R., & Janssen, H. (1996). A continuation algorithm for discovering new chaotic motions in forced Duffing systems. *Journal of Computational and Applied Mathematics, 66*, 527–541.

Warminski, J., & Kecik, K. (2006). Autoparametric vibrations of a nonlinear system with pendulum. Mathematical Problems in Engineering (No 80705).

Warminski, J., Balthazar, J. M., & Brasil, R. M. L. R. F. (2001). Vibrations of a non-ideal parametrically and self-excited model. *Journal of Sound and Vibration, 245*, 363–374.

Wolf, A., Swift, J., Swinney, H., & Vastano, J. (1985). Determining Lyapunov exponents from a time series. *Physica D, 16*, 285–317.

Yuste, S. B., & Bejarano, J. D. (1986). Construction of approximate analytical solution to a new class of a non-linear oscillator equations. *Journal of Sound and Vibration, 110*, 347–350.

Yuste, S. B., & Bejarano, J. D. (1990). Improvement of a Krylov-Bogolubov method that uses Jacobi elliptic functions. *Journal of Sound and Vibration, 139*, 151–163.

Zhuravlev, V. F., & Klimov, D. M. (1988). *Applied methods in oscillation theory*. Moscow: Nauka.

Zukovic, M., & Cveticanin, L. (2007). Chaotic responses in a stable Duffing system on non-ideal type. *Journal of Vibration and Control, 13*, 751–767.

Zukovic, M., & Cveticanin, L. (2009). Chaos in non-ideal mechanical system with clearance. *Journal of Vibration and Control, 15*, 1229–1246.

# Chapter 4
# Two Degree-of-Freedom Oscillator Coupled to a Non-ideal Source

In this chapter the two degree-of-freedom structure excited with a non-ideal source is considered. The model corresponds to real energy harvester system (Felix et al. 2009), centrifugal vibration machine (Dantas and Balthazar 2006), tuned liquid column damper mounted on a structural frame (Felix et al. 2005a), portal frame (Felix et al. 2013) and portal frame foundation type shear building (Felix et al. 2005b), rotor-structure system which moves in-plane (Quinn 1997), etc. These systems are usually modelled as two mass systems with visco-elastic connection and excited with non-ideal motor. The main attention is given to resonance capture (Balthazar et al. 2001) in the presence of a 1:1 (Zniber and Quinn 2006) and 1:2 (Tsuchida et al. 2005) frequency ratio. However, we suggest to model the aforementioned real systems as an one-mass system with two degrees-of-freedom in two orthogonal directions as the mass moves in-plane. Such model is given in the paper of Goncalves et al. (2016) and treated numerically and experimentally. The model consists of a concentrated mass which is supported by a set of linear springs and dampers positioned in two orthogonal direction, such that the mass can move horizontally and vertically in a plane. A non-ideal motor is attached to the mass such that the phenomena of resonance capture can occur. In the paper of Goncalves et al. (2016) it is concluded that the resonance can occur in both directions, in only one direction or can not occur. Limits between these cases are determined in Cveticanin et al. (2017). The system is described with the set of three coupled second order differential equations: two of them describing the vibrations of the structure in two directions and one, which gives the motion of the motor. An analytical procedure for solving the equations is developed and the constraints for resonance are given.

The chapter is divided into six sections. In Sect. 4.2, the motion of the system which contains a mass supported in two directions and excited with non-ideal energy source is modeled. Mathematical model of the system is solved analytically in Sect. 4.3. The steady-state solution and the stability conditions for solutions are determined. In Sect. 4.4, two special cases are considered: first, the case when the frequencies of the system in x and y directions are equal and then, the case when the frequency

© Springer International Publishing AG 2018

L. Cveticanin et al., *Dynamics of Mechanical Systems with Non-Ideal Excitation*, Mathematical Engineering, DOI 10.1007/978-3-319-54169-3_4

in y direction is two times higher than that in x direction. The resonant motions are investigated. The influence of vertical and horizontal stiffness on the regions of double resonance motion are considered. The obtained results are compared with numerical results in Sect. 4.5. The chapter ends with conclusions.

## 4.1 Model of the System

The system considered in this paper consists of a mass $M$ supported by springs and viscous dampers in two orthogonal directions ($x$ and $y$). The spring constants are defined with $k_x$ and $k_y$, while the damping coefficients are $c_x$ and $c_y$. The subscripts $x$ and $y$ indicate the displacement directions. To the mass $M$ a motor is attached with unbalanced mass $m$ at the distance $d$ from the center of the motor shaft (Fig. 4.1).

The motor shaft has moment of inertia defined by $J$. The motor - structure system shown in Fig. 4.1 has three degrees-of-freedom defined with three generalized coordinates $x$, $y$ and $\varphi$. where the first two define the motion in two orthogonal directions and the third is the angle position of the unbalance. Equations of motion of such system are in general

$$\frac{d}{dt}\frac{\partial T}{\partial \dot{x}} - \frac{\partial T}{\partial x} + \frac{\partial U}{\partial x} + \frac{\partial \Phi}{\partial \dot{x}} = Q_x,$$
$$\frac{d}{dt}\frac{\partial T}{\partial \dot{y}} - \frac{\partial T}{\partial y} + \frac{\partial U}{\partial y} + \frac{\partial \Phi}{\partial \dot{y}} = Q_y,$$

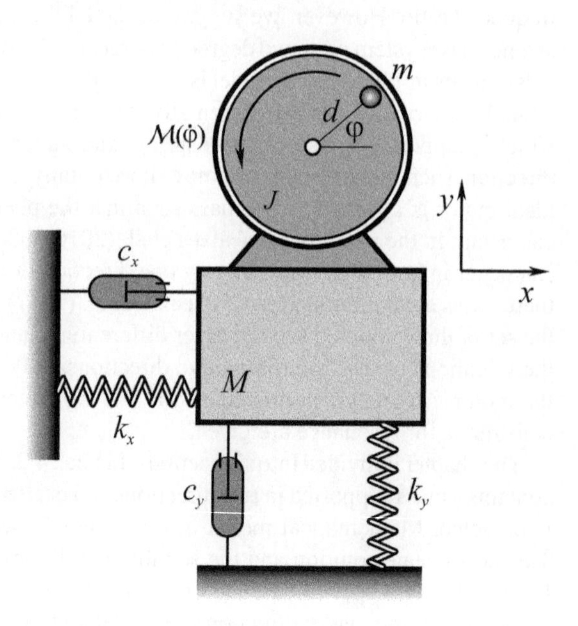


**Fig. 4.1** Model of a two-degree-of-freedom oscillator coupled with a non-ideal unbalanced motor

$$\frac{d}{dt}\frac{\partial T}{\partial \dot{\varphi}} - \frac{\partial T}{\partial \varphi} + \frac{\partial U}{\partial \varphi} + \frac{\partial \Phi}{\partial \dot{\varphi}} = Q_{\varphi}, \tag{4.1}$$

where $T$ and $U$ are the kinetic and potential energy of the system, $\Phi$ is the dissipative function and $Q_x$, $Q_y$ and $Q_{\varphi}$ are generalized forces.

Motion of the system is excited with the motor torque $\mathcal{M}$ which depends on the angular velocity of the rotor $\dot{\varphi}$. For the DC motor the motor torque model is assumed to be a linear function of $\dot{\varphi}$, i.e.,

$$\mathcal{M}(\dot{\varphi}) = \mathcal{M}_0 \left(1 - \frac{\dot{\varphi}}{\Omega_0}\right), \tag{4.2}$$

where $\mathcal{M}_0$ and $\Omega_0$ are constant values. Then the generalized force due to motor torque is

$$Q_{\varphi} = \mathcal{M}(\dot{\varphi}). \tag{4.3}$$

For the model given in Fig. 4.1 the kinetic energy is

$$T = \frac{1}{2}M(\dot{x}^2 + \dot{y}^2) + \frac{1}{2}J\dot{\varphi}^2 + \frac{1}{2}m(\dot{x}_2^2 + \dot{y}_2^2), \tag{4.4}$$

where the position of unbalance mass $m$ is

$$x_m = x + d\cos\varphi, \quad y_m = y + d\sin\varphi. \tag{4.5}$$

Substituting the time derivative of (4.5) into (4.4), we have

$$T = \frac{1}{2}(M+m)(\dot{x}^2 + \dot{y}^2) + \frac{1}{2}(J + md^2)\dot{\varphi}^2 + md\dot{\varphi}(\dot{y}\cos\varphi - \dot{x}\sin\varphi). \tag{4.6}$$

If the gravity potential energy is neglected, the potential energy of the system is

$$U = \frac{1}{2}k_x x^2 + \frac{1}{2}k_y y^2. \tag{4.7}$$

The dissipative function of the system is

$$\Phi = \frac{1}{2}c_x \dot{x}^2 + \frac{1}{2}c_y \dot{y}^2. \tag{4.8}$$

Substituting (4.3)–(4.6) into (4.1), we obtain the cart's equations in $x$ and $y$ direction

$$(M+m)\ddot{x} + k_x x + c_x \dot{x} = md(\dot{\varphi}^2\cos\varphi + \ddot{\varphi}\sin\varphi), \tag{4.9}$$

$$(M+m)\ddot{y} + k_y y + c_y \dot{y} = md(\dot{\varphi}^2\sin\varphi - \ddot{\varphi}\cos\varphi), \tag{4.10}$$

and the equation of motion of the unbalanced mass

$$(J+md^2)\ddot{\varphi}=md(\ddot{x}\sin\varphi-\ddot{y}\cos\varphi)+\mathcal{M}(\dot{\varphi}). \tag{4.11}$$

Using the parameter values

$$\omega_x=\sqrt{\frac{k_x}{M+m}},\quad \omega_y=\sqrt{\frac{k_y}{M+m}},\quad \eta=\frac{md}{J+md^2},$$
$$\zeta_x=\frac{c_x}{M+m},\quad \zeta_y=\frac{c_y}{M+m},\quad \mu=\frac{md}{M+m},\quad \varepsilon=\frac{1}{J+md^2}, \tag{4.12}$$

Equations (4.9)–(4.11) are rewritten as

$$\begin{aligned}\ddot{x}+\omega_x^2x+\zeta_x\dot{x}&=\mu(\dot{\varphi}^2\cos\varphi+\ddot{\varphi}\sin\varphi),\\ \ddot{y}+\omega_y^2y+\zeta_y\dot{y}&=\mu(\dot{\varphi}^2\sin\varphi-\ddot{\varphi}\cos\varphi),\\ \ddot{\varphi}&=\eta(\ddot{x}\sin\varphi-\ddot{y}\cos\varphi)+\varepsilon\mathcal{M}(\dot{\varphi}).\end{aligned} \tag{4.13}$$

It has to be mention that the parameter $\varepsilon$ is a small one, i.e., $\varepsilon << 1$ and the input motor torque is small. Our aim is to solve the Eq. (4.13).

## 4.2 Analytical Solution

Let us give the equations of motion (4.13) in terms of uncoupled accelerations

$$\begin{aligned}\ddot{x}=&-\varepsilon\frac{\mu\mathcal{M}(\dot{\varphi})\sin\varphi}{\mu\eta-1}+\frac{\mu\eta\sin(2\varphi)(\mu\dot{\varphi}^2\sin\varphi-F_y)}{2(\mu\eta-1)}\\ &+\frac{(\mu\eta\cos^2\varphi-1)(\mu\dot{\varphi}^2\cos\varphi-F_x)}{\mu\eta-1},\end{aligned} \tag{4.14}$$

$$\begin{aligned}\ddot{y}=&\,\varepsilon\frac{\mu\mathcal{M}(\dot{\varphi})\cos\varphi}{\mu\eta-1}+\frac{\mu\eta\sin(2\varphi)(\mu\dot{\varphi}^2\cos\varphi-F_x)}{2(\mu\eta-1)}\\ &+\frac{(\mu\eta\sin^2\varphi-1)(\mu\dot{\varphi}^2\sin\varphi-F_y)}{\mu\eta-1},\end{aligned} \tag{4.15}$$

$$\begin{aligned}\ddot{\varphi}=&-\frac{\varepsilon\mathcal{M}(\dot{\varphi})}{\mu\eta-1}+\frac{\eta\cos\varphi(\mu\dot{\varphi}^2\sin\varphi-F_y)}{\mu\eta-1}\\ &-\frac{\eta\sin\varphi(\mu\dot{\varphi}^2\cos\varphi-F_x)}{\mu\eta-1},\end{aligned} \tag{4.16}$$

where

$$F_x = x\omega_x^2 + \zeta_x \dot{x}, \quad F_y = y\omega_y^2 + \zeta_y \dot{y}. \tag{4.17}$$

For the case when the parameters $\zeta_x$, $\zeta_y$, $\mu$ and $\eta$ are small, i.e.,

$$\begin{aligned} &\zeta_x = \varepsilon\zeta_{x1}, \quad \zeta_y = \varepsilon\zeta_{y1}, \quad \mu = \varepsilon\mu_1, \\ &\eta = \varepsilon\eta_1, \end{aligned} \tag{4.18}$$

where $\varepsilon << 1$ is a small parameter, Eq. (4.13) transform into the form

$$\begin{aligned} (\varepsilon^2\mu_1\eta_1 - 1)\ddot{x} = &-\varepsilon^2\mu_1\mathcal{M}(\dot{\varphi})\sin\varphi \\ &+ \frac{\varepsilon^2\mu_1\eta_1\sin(2\varphi)(\varepsilon\mu_1\dot{\varphi}^2\sin\varphi - y\omega_y^2 - \varepsilon\zeta_{y1}\dot{y})}{2} \\ &+ (\varepsilon^2\mu_1\eta_1\cos^2\varphi - 1)(\varepsilon\mu_1\dot{\varphi}^2\cos\varphi - x\omega_x^2 - \varepsilon\zeta_{x1}\dot{x}), \end{aligned} \tag{4.19}$$

$$\begin{aligned} (\varepsilon^2\mu_1\eta_1 - 1)\ddot{y} = \varepsilon^2\mu_1\mathcal{M}(\dot{\varphi})\cos\varphi + \frac{\varepsilon^2\mu_1\eta_1\sin(2\varphi)(\mu\dot{\varphi}^2\cos\varphi - x\omega_x^2 - \varepsilon\zeta_{x1}\dot{x})}{2} \\ + (\varepsilon^2\mu_1\eta_1\sin^2\varphi - 1)(\varepsilon\mu_1\dot{\varphi}^2\sin\varphi - y\omega_y^2 - \varepsilon\zeta_{y1}\dot{y}), \end{aligned} \tag{4.20}$$

$$\begin{aligned} (\varepsilon^2\mu_1\eta_1 - 1)\ddot{\varphi} = &-\varepsilon\mathcal{M}(\dot{\varphi}) + \varepsilon\eta_1\cos\varphi(\varepsilon\mu_1\dot{\varphi}^2\sin\varphi - y\omega_y^2 - \varepsilon\zeta_{y1}\dot{y}) \\ &- \varepsilon\eta_1\sin\varphi(\varepsilon\mu_1\dot{\varphi}^2\cos\varphi - x\omega_x^2 - \varepsilon\zeta_{x1}\dot{x}). \end{aligned} \tag{4.21}$$

Eliminating the terms with the small parameter $\varepsilon$ of higher order than one, i.e., $O(\varepsilon^2)$, we have the following simplified equations

$$\ddot{x} + \omega_x^2 x = \varepsilon\mu_1\dot{\varphi}^2\cos\varphi - \varepsilon\zeta_{x1}\dot{x}, \tag{4.22}$$

$$\ddot{y} + \omega_y^2 y = \varepsilon\mu_1\dot{\varphi}^2\sin\varphi - \varepsilon\zeta_{y1}\dot{y}, \tag{4.23}$$

$$\ddot{\varphi} = \varepsilon\mathcal{M}(\dot{\varphi}) + \varepsilon\eta_1(y\omega_y^2\cos\varphi - x\omega_x^2\sin\varphi). \tag{4.24}$$

Using the notation

$$x_1 = x, \quad x_2 = \dot{x}, \quad y_1 = y, \quad y_2 = \dot{y}, \quad \Omega = \dot{\varphi}, \tag{4.25}$$

the Eqs. (4.22)–(4.24) are rewritten in the following system of first order differential equations

$$\begin{aligned} \dot{x}_1 &= x_2, \\ \dot{x}_2 &= -\omega_x^2 x_1 + \varepsilon\mu_1\Omega^2\cos\varphi - \varepsilon\zeta_{x1}x_2, \\ \dot{y}_1 &= y_2, \\ \dot{y}_2 &= -\omega_y^2 y_1 + \varepsilon\mu_1\Omega^2\sin\varphi - \varepsilon\zeta_{y1}y_2, \end{aligned}$$

$$\dot{\varphi} = \Omega,$$
$$\dot{\Omega} = \varepsilon\mathcal{M}(\Omega) + \varepsilon\eta_1(y_1\omega_y^2\cos\varphi - x_1\omega_x^2\sin\varphi), \tag{4.26}$$

where according to (4.2) the motor torque is

$$\varepsilon\mathcal{M}(\Omega) = M_0\left(1 - \frac{\Omega}{\Omega_0}\right), \tag{4.27}$$

where $M_0$ and $\Omega_0$ motor constants. Equation (4.26) represent a system of coupled differential equations whose solution is not easy to be obtained. For simplification, let us introduce the new variables

$$\begin{aligned} x_1 &= a_1\cos(\varphi + \psi_1), \\ x_2 &= -a_1\Omega\sin(\varphi + \psi_1), \\ y_1 &= a_2\cos(\varphi + \psi_2), \\ y_2 &= -a_2\Omega\sin(\varphi + \psi_2), \end{aligned} \tag{4.28}$$

where $\Omega$, $a_1$, $a_2$, $\psi_1$ and $\psi_2$ are time dependent functions. Substituting (4.28) into (4.26) and using the relation $\dot{\varphi} = \Omega$ we obtain

$$\begin{aligned} 0 &= \dot{a}_1\cos(\varphi + \psi_1) - a_1\dot{\psi}_1\sin(\varphi + \psi_1), \\ &\quad - \dot{a}_1\Omega\sin(\varphi + \psi_1) - a_1\Omega\dot{\psi}_1\cos(\varphi + \psi_1) \\ - a_1\dot{\Omega}\sin(\varphi + \psi_1) &= -\omega_x^2 a_1\cos(\varphi + \psi_1) + a_1\Omega^2\cos(\varphi + \psi_1) \\ &\quad + \varepsilon\mu_1\Omega^2\cos\varphi + \varepsilon\zeta_{x1}a_1\Omega\sin(\varphi + \psi_1), \\ 0 &= \dot{a}_2\cos(\varphi + \psi_2) - a_2\dot{\psi}_2\sin(\varphi + \psi_2), \\ &\quad -\dot{a}_2\Omega\sin(\varphi + \psi_2) - a_2\dot{\Omega}\sin(\varphi + \psi_2) \\ -a_2\Omega\dot{\psi}_2\cos(\varphi + \psi_2) &= -\omega_y^2 a_2\cos(\varphi + \psi_2) + a_2\Omega^2\cos(\varphi + \psi_2) \\ &\quad + \varepsilon\mu_1\Omega^2\sin\varphi + \varepsilon\zeta_{y1}a_2\Omega\sin(\varphi + \psi_2), \\ \dot{\Omega} &= \varepsilon\eta_1[a_2\cos(\varphi + \psi_2)\omega_y^2\cos\varphi - a_1\cos(\varphi + \psi_1)\omega_x^2\sin\varphi] \\ &\quad + \varepsilon\mathcal{M}(\Omega), \end{aligned} \tag{4.29}$$

and after some modification

$$\begin{aligned} \dot{a}_1 + a_1\frac{\dot{\Omega}}{\Omega}\sin^2(\varphi + \psi_1) &= \frac{\omega_x^2 - \Omega^2}{2\Omega}a_1\sin 2(\varphi + \psi_1) \\ &\quad - \varepsilon\mu_1\Omega\cos\varphi\sin(\varphi + \psi_1) \\ &\quad -\varepsilon\zeta_{x1}a_1\sin^2(\varphi + \psi_1), \\ a_1\left(\dot{\psi}_1 + \frac{1}{2}\frac{\dot{\Omega}}{\Omega}\sin 2(\varphi + \psi_1)\right) &= \frac{\omega_x^2 - \Omega^2}{\Omega}a_1\cos^2(\varphi + \psi_1) \\ &\quad - \varepsilon\mu_1\Omega\cos\varphi\cos(\varphi + \psi_1) \end{aligned}$$

$$
\begin{aligned}
& -\frac{1}{2}\varepsilon\zeta_{x1}a_1\sin 2(\varphi+\psi_1),\\
\dot{a}_2 + a_2\frac{\dot{\Omega}}{\Omega}\sin^2(\varphi+\psi_2) &= \frac{\omega_y^2-\Omega^2}{2\Omega}a_2\sin 2(\varphi+\psi_2)\\
&\quad -\varepsilon\mu_1\Omega\sin\varphi\sin(\varphi+\psi_2)\\
&\quad -\varepsilon\zeta_{y1}a_2\sin^2(\varphi+\psi_2),\\
a_2\left(\dot{\psi}_2 - \frac{\dot{\Omega}}{2\Omega}\sin 2(\varphi+\psi_2)\right) &= \frac{\omega_y^2-\Omega^2}{\Omega}a_2\cos^2(\varphi+\psi_2)\\
&\quad -\varepsilon\mu_1\Omega\sin\varphi\cos(\varphi+\psi_2)\\
&\quad -\frac{1}{2}\varepsilon\zeta_{y1}a_2\sin 2(\varphi+\psi_2),\\
\dot{\Omega} &= \varepsilon\eta_1[a_2\cos(\varphi+\psi_2)\omega_y^2\cos\varphi\\
&\quad -a_1\cos(\varphi+\psi_1)\omega_x^2\sin\varphi]\\
&\quad +\varepsilon\mathcal{M}(\Omega).
\end{aligned}
\tag{4.30}
$$

The Eq. (4.30) are the first order differential equations which correspond to second order equations (4.22)–(4.24). The system of equations (4.30) has to be solved for $\Omega$, $a_1$, $a_2$, $\psi_1$ and $\psi_2$. As the Eq. (4.30) are coupled, to find the solution is not an easy task. It is at this moment where the simplification is done. Averaging the equations over the period $2\pi$ of the function $\varphi$, we obtain the averaged equations

$$
\begin{aligned}
\dot{a}_1 + a_1\frac{\dot{\Omega}}{2\Omega} &= -\frac{1}{2}\varepsilon\mu_1\Omega\sin\psi_1 - \frac{1}{2}\varepsilon\zeta_{x1}a_1,\\
a_1\dot{\psi}_1 &= \frac{\omega_x^2-\Omega^2}{2\Omega}a_1 - \frac{1}{2}\varepsilon\mu_1\Omega\cos\psi_1,\\
\dot{a}_2 + a_2\frac{\dot{\Omega}}{2\Omega} &= -\frac{1}{2}\varepsilon\mu_1\Omega\cos\psi_2 - \frac{1}{2}\varepsilon\zeta_{y1}a_2,\\
a_2\dot{\psi}_2 &= \frac{\omega_y^2-\Omega^2}{2\Omega}a_2 + \frac{1}{2}\varepsilon\mu_1\Omega\sin\psi_2,\\
\dot{\Omega} &= \varepsilon\mathcal{M}(\Omega) + \frac{1}{2}\varepsilon\eta_1\left(a_2\omega_y^2\cos\psi_2 + a_1\omega_x^2\sin\psi_1\right).
\end{aligned}
\tag{4.31}
$$

### *4.2.1 Steady-State Motion*

It is of interest to analyze the steady-state motion when $\dot{a}_1 = 0$, $\dot{a}_2 = 0$ and $\dot{\Omega} = 0$ and the corresponding Eq. (4.31) are

$$
\begin{aligned}
0 &= \mu\Omega_S\sin\psi_{1S} + \zeta_x a_{1S},\\
0 &= \frac{\omega_x^2-\Omega_S^2}{\Omega_S}a_{1S} - \mu\Omega_S\cos\psi_{1S},
\end{aligned}
\tag{4.32}
$$

$$0 = \mu\Omega_S \cos\psi_{2S} + \zeta_y a_{2S},$$
$$0 = \frac{\omega_y^2 - \Omega_S^2}{\Omega_S} a_2 + \mu\Omega_S \sin\psi_{2S}, \tag{4.33}$$

$$0 = \varepsilon\mathcal{M}(\Omega) + \frac{1}{2}\eta\left(a_{2S}\omega_y^2 \cos\psi_{2S} + a_{1S}\omega_x^2 \sin\psi_{1S}\right). \tag{4.34}$$

Eliminating $\psi_{1S}$ in Eq. (4.32) the steady-state amplitude $a_{1S}$ as the function of $\Omega$ is obtained

$$a_{1S} = \frac{\mu\Omega_S}{\sqrt{(\zeta_x)^2 + (\frac{\omega_x^2 - \Omega_S^2}{\Omega_S})^2}}. \tag{4.35}$$

Using the same procedure and eliminating $\psi_{2S}$ in Eq. (4.33) the steady-state amplitude $a_{2S}$ as the function of $\Omega$ yields

$$a_{2S} = \frac{\mu\Omega_S}{\sqrt{(\zeta_y)^2 + (\frac{\omega_y^2 - \Omega_S^2}{\Omega_S})^2}}. \tag{4.36}$$

Dividing equations in (4.32) and (4.33) the phase angles in the both directions of motion are obtained

$$\tan\psi_{1S} = \frac{\zeta_x\Omega_S}{\Omega_S^2 - \omega_x^2}, \quad \tan\psi_{2S} = \frac{\omega_y^2 - \Omega_S^2}{\zeta_y\Omega_S}. \tag{4.37}$$

Comparing the amplitudes $a_{1S}$ and $a_{2S}$ we obtain the condition for which the motion is out of resonance and also when the resonance occurs in one or both directions of motion. For $\omega_x \neq \Omega_S$ and $\omega_y \neq \Omega_S$ the motion is out of resonance regime. The amplitude of vibration in $x$ and $y$ direction depends on the damping properties of the system and on the difference between the angular velocity of the motor and the frequency of the system $\omega_x$ and $\omega_y$, respectively. However, it is of interest to analyze the motion when resonances appear.

Substituting (4.35)–(4.37) into (4.34), it is

$$0 = \varepsilon\mathcal{M}(\Omega) + \frac{\zeta_y\mu\Omega_S\eta\omega_y^2}{2(\zeta_y^2 + (\frac{\omega_y^2 - \Omega_S^2}{\Omega_S})^2} + \frac{\zeta_x\mu\Omega_S\eta\omega_x^2}{2(\zeta_x^2 + (\frac{\omega_x^2 - \Omega_S^2}{\Omega_S})^2}. \tag{4.38}$$

For the torque property (4.27) the relation (4.38) transforms into

$$\varepsilon M_0(1 - \frac{\Omega_S}{\Omega_0}) = -\frac{\eta\mu\Omega_S^3}{2}\left[\frac{\zeta_x\omega_x^2}{(\zeta_x\Omega_S)^2 + (\omega_x^2 - \Omega_S^2)^2} + \frac{\zeta_y\omega_y^2}{(\zeta_y\Omega_S)^2 + (\omega_y^2 - \Omega_S^2)^2}\right]. \tag{4.39}$$

The Eq. (4.39) gives the relation between the angular velocity $\Omega$ and parameter $\Omega_0$. Let us rewrite (4.39) as

$$\Omega_0 = f(\Omega_S) = \frac{\Omega_S}{1 + \frac{\eta\mu\Omega_S^3}{2\varepsilon M_0}\left[\frac{\zeta_x\omega_x^2}{(\zeta_x\Omega_S)^2 + (\omega_x^2 - \Omega_S^2)^2} + \frac{\zeta_y\omega_y^2}{(\zeta_y\Omega_S)^2 + (\omega_y^2 - \Omega_S^2)^2}\right]}. \tag{4.40}$$

Calculating the first derivative $d\Omega_0/d\Omega_S$ and equating with zero, we obtain the values of $\Omega_S$ which give the extreme values of $\Omega_0$, i.e., $\Omega_{0\min}(\Omega_S)$ and $\Omega_{0\max}(\Omega_S)$.

### 4.2.2 *Stability Analysis*

Using the results of steady-state motion (4.32)–(4.34) the perturbed amplitudes, phases and angular velocities are

$$\begin{aligned} a_1 &= a_{1S} + \xi_1, \quad a_2 = a_{2S} + \xi_2, \\ \psi_1 &= \psi_{1S} + \theta_1, \quad \psi_2 = \psi_{2S} + \theta_2, \\ \Omega &= \Omega_S + \varpi. \end{aligned} \tag{4.41}$$

where $\xi_1$, $\xi_2$, $\theta_1$, $\theta_2$ and $\varpi$ are small perturbation functions. Substituting (4.41) into (4.31) and after linearization the system of coupled first order differential equations follows

$$\begin{aligned} 2\dot{\xi}_1\Omega_S + a_{1S}\dot{\varpi} &= -(\zeta_x a_{1S} + 2\mu\Omega_S \sin\psi_{1S})\varpi \\ &\quad - \zeta_x\Omega_S\xi_1 - (\mu\Omega_S^2\cos\psi_{1S})\theta_1, \\ 2a_{1S}\dot{\theta}_1\Omega_S &= -2\Omega_S(a_{1S} - \mu\cos\psi_{1S})\varpi \\ &\quad + (\omega_x^2 - \Omega_S^2)\xi_1 + (\mu\Omega_S^2\sin\psi_{1S})\theta_1, \\ 2\dot{\xi}_2\Omega_S + a_{2S}\dot{\varpi} &= -(\zeta_y a_{2S} + 2\mu\Omega_S\cos\psi_{2S})\varpi \\ &\quad - (\zeta_y\Omega_S)\xi_2 + (\mu\Omega_S^2\sin\psi_{2S})\theta_2, \\ 2a_{2S}\dot{\theta}_2\Omega_S &= -2\Omega_S(a_{2S} - \varepsilon\mu_1\sin\psi_{2S})\varpi \\ &\quad + (\omega_y^2 - \Omega_S^2)\xi_2 + (\mu\Omega_S^2\cos\psi_{2S})\theta_2, \\ 2\dot{\varpi} &= -\frac{2\varepsilon M_0}{\Omega_0}\varpi \\ &\quad - (\eta a_{2S}\omega_y^2\sin\psi_{2S})\theta_2 + (\eta\omega_y^2\cos\psi_{2S})\xi_2 \\ &\quad + (\eta\omega_x^2\sin\psi_{1S})\xi_1 + (\eta a_{1S}\omega_x^2\cos\psi_{1S})\theta_1. \end{aligned} \tag{4.42}$$

Assuming the solution of (4.42) in the form

$$\begin{aligned}
\xi_1 &= A_1 \exp(\lambda t), \quad \xi_2 = A_2 \exp(\lambda t), \\
\theta_1 &= A_3 \exp(\lambda t), \quad \theta_2 = A_4 \exp(\lambda t), \\
\varpi &= A_5 \exp(\lambda t),
\end{aligned} \tag{4.43}$$

and substituting into (4.42), the system of linear algebraic equations is obtained

$$\begin{aligned}
0 &= -(2\lambda\Omega_S + \zeta_x\Omega_S)A_1 - (\mu\Omega_S^2 \cos\psi_{1S})A_3 \\
&\quad -(a_{1S}\lambda + \zeta_x a_{1S} + 2\mu\Omega_S \sin\psi_{1S})A_5 \\
0 &= (\omega_x^2 - \Omega_S^2)A_1 + (\mu\Omega_S^2 \sin\psi_{1S} - 2a_{1S}\lambda\Omega_S)A_3 \\
&\quad -2\Omega_S(a_{1S} - \mu\cos\psi_{1S})A_5, \\
0 &= -(2\Omega_S\lambda + \varepsilon\zeta_{y1}\Omega_S)A_2 + (\mu\Omega_S^2 \sin\psi_{2S})A_4 \\
&\quad -(a_{2S}\lambda + \zeta_y a_{2S} + 2\mu\Omega_S \cos\psi_{2S})A_5, \\
0 &= (\omega_y^2 - \Omega_S^2)A_2 + (\mu\Omega_S^2 \cos\psi_{2S} - 2a_{2S}\lambda\Omega_S)A_4 \\
&\quad -2\Omega_S(a_{2S} - \mu\sin\psi_{2S})A_5 \\
0 &= (\eta\omega_x^2 \sin\psi_{1S})A_1 + (\eta\omega_y^2 \cos\psi_2)A_2 \\
&\quad +(\eta a_1\omega_x^2 \cos\psi_{1S})A_3 - (\eta a_2\omega_y^2 \sin\psi_{2S})A_4 \\
&\quad -(2\lambda + \frac{2\varepsilon M_0}{\Omega_0})A_5.
\end{aligned} \tag{4.44}$$

The system has the nontrivial solution if the determinant is zero. The determinant of the system is a fifth order algebraic equation. Solving the equation and applying the Routh-Hurwitz criteria, the stability of the solutions is determined.

## 4.3 Special Cases

Two special cases are considered: one, when the resonance frequencies in both orthogonal directions are equal, and the second, when the resonance frequency in one direction is defined by half of the resonance frequency in the other direction.

### *4.3.1 Resonance Frequencies in Orthogonal Directions Are Equal*

For the special case when the frequencies in both direction are equal, i.e., $\omega_x = \omega_y = \omega$, the steady state amplitudes of vibration are

$$a_{1S} = \frac{\mu\Omega_S}{\sqrt{\zeta_x^2 + (\frac{\omega^2-\Omega_S^2}{\Omega_S})^2}}, \quad a_{2S} = \frac{\mu\Omega_S}{\sqrt{\zeta_y^2 + (\frac{\omega^2-\Omega_S^2}{\Omega_S})^2}},$$

while the corresponding phases are

$$\tan\psi_{1S} = \frac{\zeta_x\Omega_S}{\Omega_S^2 - \omega^2}, \quad \tan\psi_{2S} = \frac{\omega^2 - \Omega_S^2}{\zeta_y\Omega_S}.$$

For the resonant case when

$$\Omega_S = \omega - \Delta, \tag{4.45}$$

and the detuning function is $\Delta = \varepsilon\sigma$, the equations transform into

$$a_{1S} = \frac{\mu\Omega_S}{\sqrt{(\zeta_x^2 + 4(\varepsilon\sigma)^2}}, \quad a_{2S} = \frac{\mu\Omega_S}{\sqrt{\zeta_y^2 + 4(\varepsilon\sigma)^2}},$$

$$\tan\psi_{1S} = -\frac{\zeta_x}{2\varepsilon\sigma}, \quad \tan\psi_{2S} = \frac{2\varepsilon\sigma}{\zeta_y},$$

i.e.,

$$a_{1S} = \frac{\mu\Omega_S}{\sqrt{\zeta_x^2 + 4(\omega - \Omega_S)^2}}, \tag{4.46}$$

$$a_{2S} = \frac{\mu\Omega_S}{\sqrt{\zeta_y^2 + 4(\omega - \Omega_S)^2}}, \tag{4.47}$$

while the phase angles in the both directions of motion are

$$\tan\psi_{1S} = \frac{\zeta_x}{2(\omega - \Omega_S)}, \quad \tan\psi_{2S} = \frac{2(\omega - \Omega_S)}{\zeta_y}. \tag{4.48}$$

Substituting (4.46)–(4.48) into (4.34) it is

$$\varepsilon\mathcal{M}(\Omega) = -\frac{1}{2}\eta\mu\Omega_S^3 \left[\frac{\zeta_y}{\zeta_y^2 + 4(\varepsilon\sigma)^2} + \frac{\zeta_x}{\zeta_x^2 + 4(\varepsilon\sigma)^2}\right], \tag{4.49}$$

i.e.,

$$\varepsilon\mathcal{M}(\Omega_S) = -\frac{1}{2}\eta\mu\Omega_S^3 \left[\frac{\zeta_y}{\zeta_y^2 + 4(\omega - \Omega_S)^2} + \frac{\zeta_x}{\varepsilon\zeta_x^2 + 4(\omega - \Omega_S)^2}\right]. \tag{4.50}$$

The influence of the detuning parameter on the motor torque is evident. For the torque property (4.2) the relation (4.50) transforms into

$$\varepsilon M_0\left(1 - \frac{\Omega_S}{\Omega_0}\right) = -\frac{1}{2}\eta\mu\Omega_S^3 \left[\frac{\zeta_y}{\zeta_y^2 + 4(\omega - \Omega_S)^2} + \frac{\zeta_x}{\zeta_x^2 + 4(\omega - \Omega_S)^2}\right]. \tag{4.51}$$

If the damping properties in both direction are equal, i.e., for $\zeta_x = \zeta_y = \zeta_1$, the resonance occurs in both directions and the amplitude in both directions are the same

$$a_{1S} = a_{2S} = \frac{\mu\Omega_S}{\sqrt{\zeta_1^2 + 4(\varepsilon\sigma)^2}} = \frac{\mu\Omega_S}{\sqrt{\zeta_1^2 + 4(\omega - \Omega_S)^2}} = a_S. \tag{4.52}$$

For this assumption the angular velocity as the function of frequency $\Omega_0$ is obtained as

$$\varepsilon M_0\left(1 - \frac{\Omega_S}{\Omega_0}\right) = -\frac{\eta\mu\omega^3\zeta_1}{\zeta_1{}^2 + 4(\omega - \Omega_S)^2}. \tag{4.53}$$

In Fig. 4.2 according to (4.53) the evolution of the angular velocity is plotted. The parameter values are: $M = 0.064$, $m = 0.0021$, $J = 10^{-7}$, $d = 0.005$ and

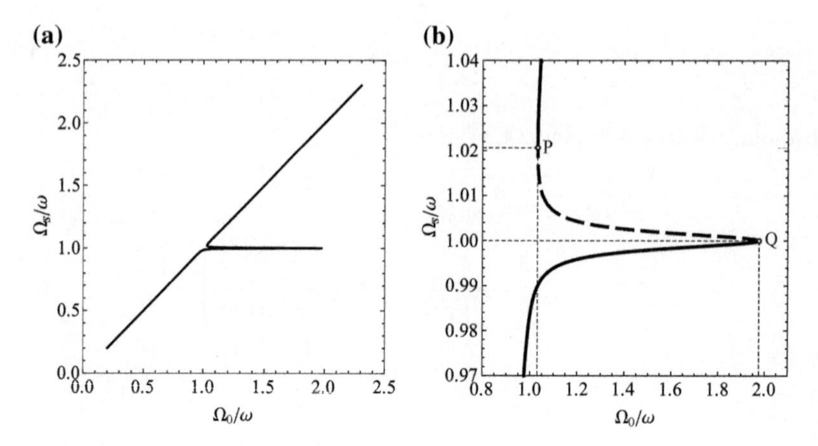


**Fig. 4.2** Angular velocity as the function of the parameter $\Omega_0$: **a** resonant case, **b** extrem angular velocities

$M_0 = 0.005$, the damping coefficient is $\zeta_1 = 0.006\omega$ where $\omega = 30\pi$. The motor is accelerated from rest to a fixed velocity by changing the parameter $\Omega_0$ and the $(\Omega_0/\omega) - (\Omega_S/\omega)$ curve is shown. As it can be seen only one resonant case appears.

Let us rewrite the relation (4.53) into

$$\Omega_0 = \Omega_S \left[ 1 + \frac{\eta\mu\omega^3\zeta_1}{\zeta_1{}^2 + 4(\omega - \Omega_S)^2} \frac{1}{\varepsilon M_0} \right]^{-1}. \tag{4.54}$$

Equating the first time derivative $\frac{d\Omega_0}{d\Omega_S}$ with 0, the condition for existence of extreme values for $\Omega_0$ is obtained

$$0 = 16(\omega - \Omega_S)^4 + 4\zeta_1 \left[ 2\zeta_1 - \frac{\eta\mu\omega^3}{\varepsilon M_0} \right] (\omega - \Omega_S)^2 - \frac{8\eta\mu\omega^4\zeta_1}{\varepsilon M_0}(\omega - \Omega_S) + \zeta_1^4 + \eta\mu\omega^3 \frac{1}{\varepsilon M_0}\zeta_1^3. \tag{4.55}$$

Solving the algebraic equation (4.55) for $\Omega_S$ two real values are obtained for which the extreme angular velocities exist (see Fig. 4.2b). For the assumed parameter values the extreme values are:

$$\left( \frac{\Omega_{0\,\min}}{\omega}, \frac{\Omega_S}{\omega} \right)_P = (1.03117,\ 1.0206400),$$
$$\left( \frac{\Omega_{0\,\max}}{\omega}, \frac{\Omega_S}{\omega} \right)_Q = (1.97570,\ 1.0000046). \tag{4.56}$$

From the Fig. 4.2b it is obvious that the number of solutions is one, two or three. Between P and Q three solutions exist. To examine the stability properties of the solutions the procedure suggested in previous section is applied. It is obvious that the stability of the solution depends on the torque characteristics $M_0$ and $\Omega_0$, characteristics of the system $\mu$, $\eta$ and $\varsigma$. In Fig. 4.2 it is shown that two solutions between P and Q are stable and one is unstable. The stable solutions are shown with the full-line, while the unstable solution is given with dotted-line.

Based on (4.52) and (4.53) the steady state amplitude is rewritten as

$$a_S^2 = \frac{\varepsilon M_0}{\eta\omega^3\zeta_1} \left( 1 - \frac{\Omega_S}{\Omega_0} \right) \mu\Omega_S^2. \tag{4.57}$$

Using (4.53) and (4.57) the amplitude as the function of parameter $\Omega_0$ is plotted in Fig. 4.3a, while in Fig. 4.3b the amplitude-frequency diagram (4.57) and the curves which depend on the motor torque for various values of parameter $\Omega_0$ are shown. In Fig. 4.3b) the curves are obtained for three various values of $\Omega_0$: 1. $\Omega_{0\,\min}/\omega = 1.03117$, 2. $\Omega_0/\omega = 1.5$,3. $\Omega_{0\,\max}/\omega = 1.97570$. motor properties for three values of angular velocity are presented.

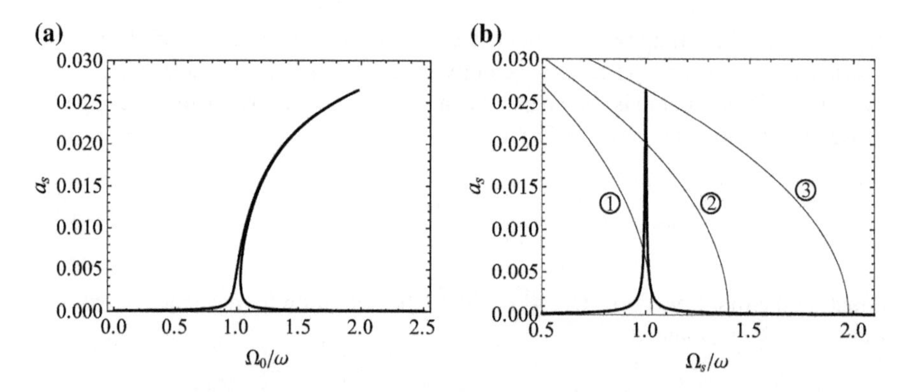


**Fig. 4.3** **a** Amplitude as the function of parameter $\Omega_0$; **b** Amplitude-frequency diagram and curves dependent on motor torque for: 1. $\Omega_{0\,\min}/\omega = 1.03117$, 2. $\Omega_0/\omega = 1.5$,3. $\Omega_{0\,\max}/\omega = 1.97570$

The curves 1 and 3 are boundary ones which satisfy the extreme conditions (4.56). The intersection of these curves and of the amplitude-frequency diagram gives the steady-state solutions. For boundary conditions (1) and (3) two steady-state solutions exist, while inside this interval there are three solutions (see intersection of (2) and the amplitude-frequency curve). In the regions outside these boundary ones, only one steady state solution exists.

### 4.3.2 *Resonance Frequency in One Direction Is Half of the Resonance frequency in Other Direction*

If the resonance frequency $\omega_x$ is defined by a half of the resonance frequency in the $y$ direction, i.e., for $\omega_x = \omega$ it is $\omega_y = 2\omega_x = 2\omega$, the two resonance frequencies are separated and two resonance features occur (Fig. 4.4).

For that case we obtain the steady state amplitudes (4.35) and (4.36) as

$$a_{1S} = \frac{\mu\Omega_S^2}{\sqrt{(\zeta_1\Omega_S)^2 + (\omega^2 - \Omega_S^2)^2}},$$

$$a_{2S} = \frac{\mu\Omega_S^2}{\sqrt{(\zeta_y\Omega_S)^2 + (4\omega^2 - \Omega_S^2)^2}}, \tag{4.58}$$

while the steady state phases (4.37) are

$$\tan\psi_{1S} = \frac{\zeta_x\Omega_S}{\Omega_S^2 - \omega^2}, \quad \tan\psi_{2S} = \frac{4\omega^2 - \Omega_S^2}{\zeta_y\Omega_S}. \tag{4.59}$$

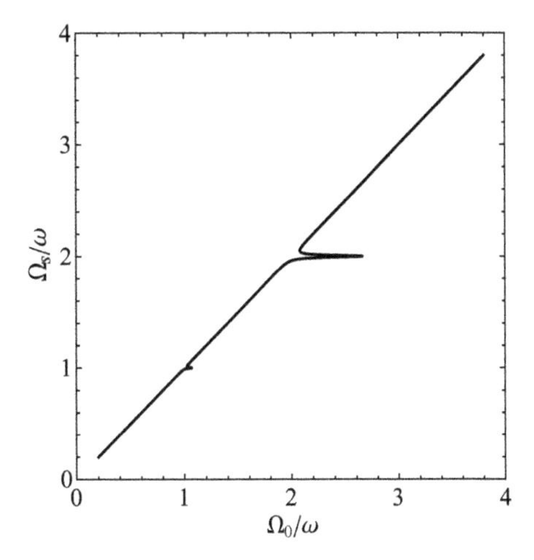


**Fig. 4.4** Frequency as a function of the motor torque parameter $\Omega_0$

The corresponding angular velocity - frequency relation (4.38) is

$$0 = \varepsilon \mathcal{M}(\Omega_S) + \frac{2\eta\zeta_y\mu\Omega_S^3\omega^2}{(\zeta_y\Omega_S)^2 + (4\omega^2 - \Omega_S^2)^2} + \frac{1}{2}\frac{\eta\zeta_x\mu\Omega_S^3\omega^2}{(\zeta_x\Omega_S)^2 + (\omega^2 - \Omega_S^2)^2}. \tag{4.60}$$

i.e.,

$$0 = \varepsilon M_0\left(1 - \frac{\Omega_S}{\Omega_0}\right) + \frac{2\eta\zeta_y\mu\Omega_S^3\omega^2}{(\zeta_y\Omega_S)^2 + (4\omega^2 - \Omega_S^2)^2}. \tag{4.61}$$

For numerical calculation the following numerical values are applied: $M = 0.064$, $m = 0.0021$, $J = 10^{-7}$, $d = 0.005$ and $M_0 = 0.005$, damping coefficients $\zeta_x = 0.012\omega$ and $\zeta_y = 0.024\omega$ where $\omega = 30\pi$.

Due to Fig. 4.4 it is evident that two resonances appear. In Fig. 4.5a, b the first and the second resonances with extreme values are plotted.

The extreme values for the first resonance are

$$\left(\frac{\Omega_{0\,\min}}{\omega}, \frac{\Omega_S}{\omega}\right)_P = (1.02397,\ 1.01507),$$
$$\left(\frac{\Omega_{0\,\max}}{\omega}, \frac{\Omega_S}{\omega}\right)_Q = (1.06596,\ 1.00029), \tag{4.62}$$

and for the second

$$\left(\frac{\Omega_{0\,\min}}{\omega}, \frac{\Omega_S}{\omega}\right)_R = (2.07929,\ 2.05142),$$

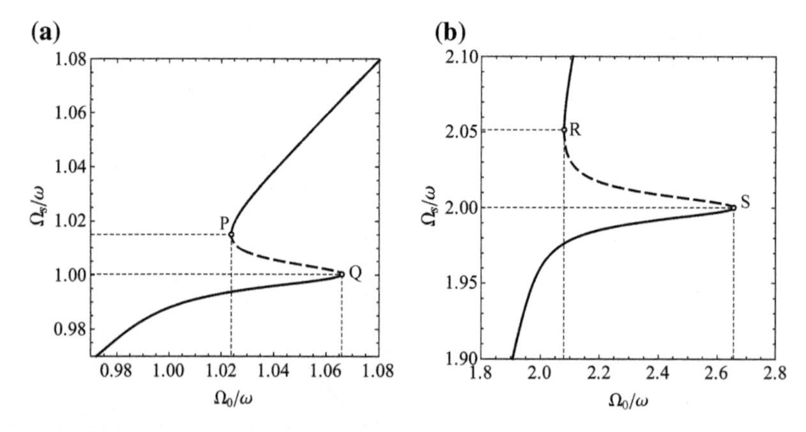


**Fig. 4.5** **a** First resonance, **b** Second resonance

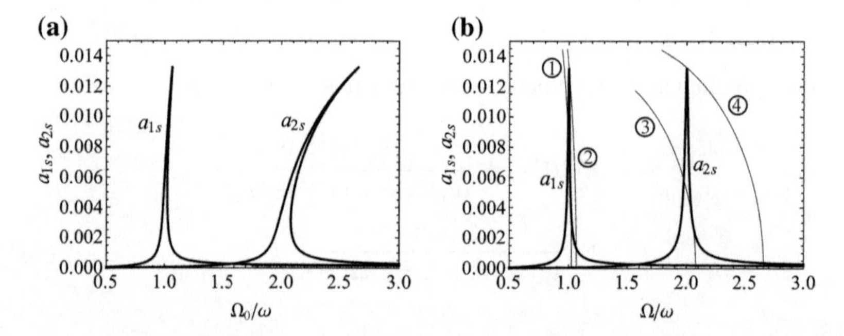


**Fig. 4.6** **a** Amplitudes as functions of the parameter $\Omega_0$; **b** Amplitude-frequency curves with motor characteristics for the extremal values of $\Omega_0$: *1* $(\Omega_{0\,\min}/\omega)_P = 1.02397$, *2* $(\Omega_{0\,\max}/\omega)_Q = 1.06596$, *3* $(\Omega_{0\,\min}/\omega)_R = 2.07929$, *4* $(\Omega_{0\,\max}/\omega)_S = 2.65594$

$$\left(\frac{\Omega_{0\,\max}}{\omega}, \frac{\Omega_S}{\omega}\right)_S = (2.65594,\ 2.00014)\,. \tag{4.63}$$

The second resonant is more significant. In Fig. 4.6a the amplitudes of vibration as functions of the parameter $\Omega_0$ of the motor torque and in Fig. 4.6b the amplitude-frequency curve with motor characteristics for the extreme values of $\Omega_0$ are plotted.

For the both resonant regimes the Sommerfeld effect occurs. The amplitude solution between P and Q (Fig. 4.7a) and S and R (Fig. 4.7b) is unstable and the jump phenomena occurs. From Fig. 4.7 it is evident that there are three steady state solutions in the interval of curves (1) and (2) for the first resonance and in the interval (3) and (4) for the second resonance. The solutions between P and Q and also R and S are unstable.

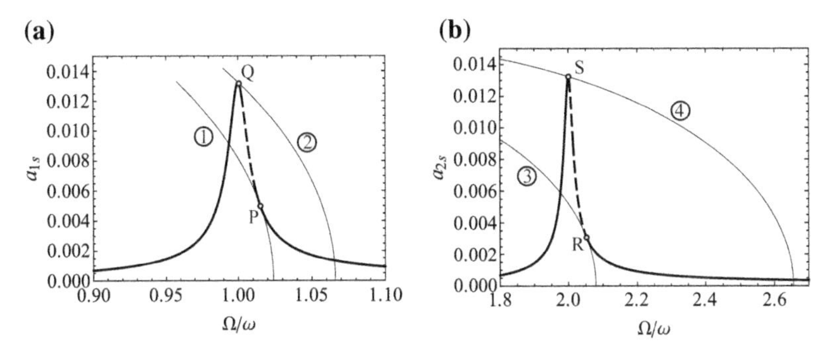


**Fig. 4.7** Amplitude-frequency diagram with motor characteristics for extrem angular velocities for: **a** first resonance, **b** second resonance

## 4.4 Numerical Simulation

Let us rewrite the equations of motion (4.14)–(4.16) into six first order differential equations

$$\dot{x} = x_1, \quad \dot{y} = y_1, \quad \dot{\varphi} = \Omega, \tag{4.64}$$

$$\dot{x}_1 = -\frac{\varepsilon\mu\mathcal{M}(\varphi_1)\sin\varphi}{\mu\eta - 1} + \frac{\mu\eta\sin(2\varphi)(\mu\varphi_1^2\sin\varphi - y\omega_y^2 - \zeta_y y_1)}{2(\mu\eta - 1)} + \frac{(\mu\eta\cos^2\varphi - 1)(\mu\Omega^2\cos\varphi - x\omega_x^2 - \zeta_x x_1)}{\mu\eta - 1}, \tag{4.65}$$

$$\dot{y}_1 = \frac{\varepsilon\mathcal{M}(\Omega)\mu\cos\varphi}{\mu\eta - 1} + \frac{\mu\eta\sin(2\varphi)(\mu\Omega^2\cos\varphi - x\omega_x^2 - \zeta_x x_1)}{2(\mu\eta - 1)} + \frac{(\mu\eta\sin^2\varphi - 1)(\mu\Omega^2\sin\varphi - y\omega_y^2 - \zeta_y y_1)}{\mu\eta - 1}, \tag{4.66}$$

$$\dot{\Omega} = -\frac{\varepsilon\mathcal{M}(\Omega)}{\mu\eta - 1} + \frac{\eta\cos\varphi(\mu\Omega^2\sin\varphi - y\omega_y^2 - \zeta_y y_1)}{\mu\eta - 1} - \frac{\eta\sin\varphi(\mu\Omega^2\cos\varphi - x\omega_x^2 - \zeta_x x_1)}{\mu\eta - 1}. \tag{4.67}$$

with motor torque function (4.27). Applying the fourth order Runge–Kutta procedure the equations are solved numerically.

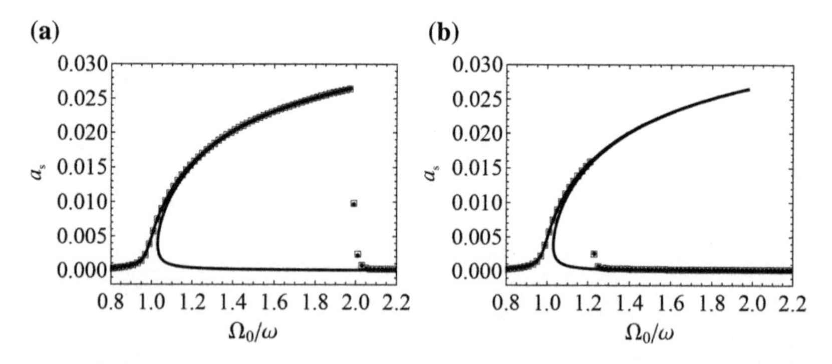


**Fig. 4.8** Amplitude as the function of motor torque parameter $\Omega_0$ for: **a** speeding up, **b** slowing down (analytical solution - *full line*, numerical solution in x direction - *squares*, numerical solution in y direction - *circles*)

Numerical solution is obtained for $\omega_x = \omega_y = \omega = 30\pi$ and parameter values $M = 0.064,\ m = 0.0021,\ J = 10^{-7},\ d = 0.005,\ M_0 = 0.005,\ \zeta = 0.006\omega$ and plotted in Fig. 4.8.

The procedure to obtain the result shown in Fig. 4.8a is performed by slowly increasing the parameter $\Omega_0$ of the DC motor and in Fig. 4.8b by slowly decreasing the parameter $\Omega_0$ applied to the motor. The numerically obtained results are given with squares in $x$ direction and with circles in $y$ direction, while the analytical solution with the full line. The numerical solution is compared with analytical one and shows good agreement.

Let us consider the case when $\omega_x = \omega$ and $\omega_y = 2\omega$ and the parameters of the system are $M = 0.064$, $m = 0.0021$, $J = 10^{-7}$, $d = 0.005$, $M_0 = 0.005$, $\zeta_x = 0.012\omega$ and $\zeta_y = 0.024\omega$ where $\omega = 30\pi$. Using the suggested procedure the solution for slow increase of the parameter $\Omega_0$ is plotted in Fig. 4.9. Two resonances occur: one in

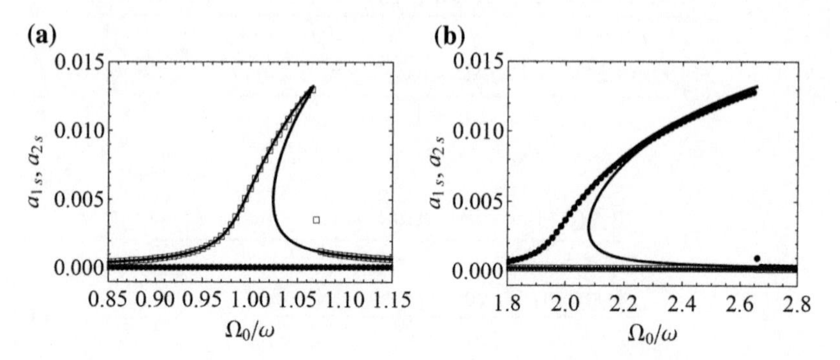


**Fig. 4.9** Amplitude as the function of motor torque parameter $\Omega_0$ for slow increase of $\Omega_0$: **a** first resonance, **b** second resonance. Analytical solution - *full line*, numerical solution in x direction - *squares*, numerical solution in y direction - *circles*

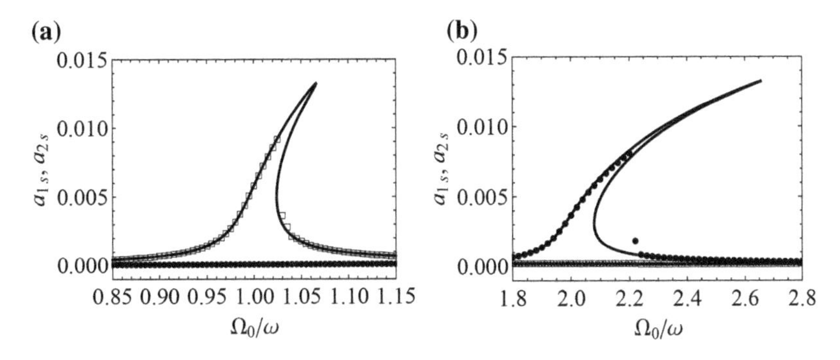


**Fig. 4.10** Amplitude as the function of motor torque parameter $\Omega_0$ for slow decrease of $\Omega_0$: **a** first resonance, **b** second resonance. Analytical solution - *full line*, numerical solution in x direction - *squares*, numerical solution in y direction - *circles*

$x$ direction (Fig. 4.9a) and the other in $y$ direction (Fig. 4.9b). The same procedure is applied for obtaining solutions during decreasing of the parameter $\Omega_0$. We obtain the first resonance in x direction (Fig. 4.10a) and the second in y direction (Fig. 4.10b). In the diagrams the jumps are observed when the system escapes resonance. Numerical results given with squares in x direction and circles in y direction are compared with analytical solution shown with full line. The numerical results are in good agreement with the analytical results.

## 4.5 Conclusions

In this chapter we considered a discrete parameter spring-mass-damper which moves in two orthogonal directions which is attached to a non-ideal rotating machine. Based on the analysis the following is concluded:

1. The system behaves different according to the values of resonance frequencies in the two orthogonal directions. Depending on the values of these frequencies the resonance can occur in both directions, only in one direction or can not occur. In the paper the limits between these cases are determined.
2. For the case when the frequencies of system in $x$ and $y$ direction are equal only one resonance regime appears. If the damping in both directions are also equal, the amplitudes of vibration in $x$ and $y$ direction are the same.
3. If the frequency of the system in $y$ direction is two times higher than in $x$ direction two resonances occur: one, in $x$ direction and other, in $y$ direction.
4. In the mechanical two degrees-of-freedom system with non-ideal excitation the Sommerfeld effect occurs. There is the jump in amplitude when the resonance regime is escaped. For the case when the frequency of the system in $x$ and $y$ directions are equal the Sommerfeld effect appears once time. If the frequency in $y$ direction is two times higher than that in $x$ direction the jump phenomena

occurs for two times for two different values of torque parameter. The procedure for analytical calculation of angular velocities and the corresponding amplitudes for which the jump occurs is suggested and the values are calculated for certain numerical data. The analytical procedure predicts the appearance of Sommerfeld effect for other relations between frequencies in two orthogonal directions, too.
5. Analytically obtained solutions are in good agreement with numerically obtained ones.
6. The analytical solutions are compared with experimental results given in Goncalves et al. (2016) and show good agreement.
7. The results published in this paper would be of special interest for engineers and technicians in prediction of resonances and their elimination.

## References

Balthazar, J. M., Chesankov, B. I., Rushev, D. T., Barbanti, L., & Weber, H. I. (2001). Remarks on the passage through resonance of a vibrating system with two degrees of freedom, excited by a non-ideal energy source. *Journal of Sound and Vibration*, *239*(5), 1075–1085.

Cveticanin, L., Zukovic, M., & Cveticanin, D. (2017). Two degree-of-freedom oscillator coupled to a non-ideal source. International Journal of Non-Linear Mechanics, Accessed on 6th March 2017.

Dantas, M. J. H., & Balthazar, J. M. (2006). A comment on a non-ideal centrifugal vibrator machine behavior with soft and hard springs. *International Journal of Bifurcation and Chaos*, *16*(4), 1083–1088.

Felix, J. L. P., Balthazar, J. M., & Brasil, R. M. L. R. F. (2005a). On tuned liquid column dampers mounted on a structural frame under a non-ideal excitation. *Journal of Sound and Vibration*, *282*, 1285–1292.

Felix, J. L. P., Balthazar, J. M., & Brasil, R. M. L. R. F. (2005b). On saturation control of a non-ideal vibrating portal frame foundation type shear-building. *Journal of Vibration and Control*, *11*, 121–136.

Felix, J. L. P., Balthazar, J. M., & Dantas, M. J. H. (2009). On energy pumping, synchronization and beat phenomenon in a nonideal structure coupled to an essentially nonlinear oscillator. *Nonlinear Dynamics*, *56*, 1–11.

Felix, J. L. P., Balthazar, J. M., & Brasil, R. M. L. R. F. (2013). On an energy exchange process and appearance of chaos in a non-ideal portal frame dynamical system. *Difeerential Equations and Dynamical of Systems*, *21*(4), 373–385.

Goncalves, P. J. P., Silveira, M., Petrocino, E. A., & Balthazar, J. M. (2016). Double resonance capture of a two-degree-of-freedom oscillator coupled to a non-ideal motor. *Meccanica*, *51*(9), 2203–2214.

Quinn, D. D. (1997). Resonance capture in a three degree-of-freedom mechanical system. *Nonlinear Dynamics*, *14*, 309–333.

Tsuchida, M., Guilherme, K. L., & Balthazar, J. M. (2005). On chaotic vibrations of a non-ideal system with two-degrees of freedom: 1:2 resonance and Sommerfeld effect. *Journal of Sound and Vibration*, *282*, 1201–1207.

Zniber, A., & Quinn, D. D. (2006). Resonance capture in a damped three-degree-of-freedom system: Experimental and analytical comparison. *International Journal of Non-Linear Mechanics*, *41*, 1128–1142.

# Chapter 5
# Dynamics of Polymer Sheets Cutting Mechanism

In this Chapter the theory of a non-ideal mechanical system, presented in previous chapters, is applied for solving of the problem of dynamics of polymer sheets cutting mechanism. Great variety of mechanisms, tools and devices are made for cutting throughout of materials based on specific requirements connected with the properties of the cutting object, its dimensions and form, strength and elasticity, etc., but also on the characteristics of the cutting tool and driving motor (Artobolevskij 1971). Most of these tools are analyzed and discussed and shown in the textbooks for mechanical engineers and technicians. For all of them it is common that have a simple construction. For example, for cutting of the parts of strings, rods or bands, which represent the continual cutting object, the cutting mechanism may be based on the four-bar one (see Cveticanin and Maretic 2000).

In this chapter a mechanism for throughout cutting of the polymer sheet, which represents the discontinual cutting object, is considered. Due to elastic properties of the polymer sheet and its tendency to crumple, and also to sheet dimensions, it was required the cutting to be done with an one direction cutting force. It was possible to be realized by a translator motion of the cutting tool. As the driving was with an electro motor, the mechanism had to transform the rotating motion of the leading element into the translator motion of the leaded element. Mechanism which transforms the rotation into straight motion is the slider-crank mechanism. This mechanism and its modification have been widely analyzed and applied for the internal combustion engines and other various purposes (see for example Metallidis and Natsiavas 2003; Koser 2004; Ha et al. 2006; Erkaya et al. 2007). There are a significant number of investigations done on a simple slider-crank mechanism as it is the basic element of the internal combustion motor. The most of investigation refers to mechanisms with rigid members whose crankshaft is supported rigidly and rotates with a constant angular velocity (see for example Metallidis and Natsiavas 2003). Most of these studies are analytical or numerical and investigate various mechanical aspects of the dynamic response and stability of slider-crank mechanism (Goudas and Natsiavas 2004). In the papers of Wauer and Buhrle (1997) and Goudas et al. (2004) the extension of the problem is done by including the non-ideal forcing and flexible supporting of the crankshaft. The two factors in conjunction with the kinematics

© Springer International Publishing AG 2018

L. Cveticanin et al., *Dynamics of Mechanical Systems with Non-Ideal Excitation*,
Mathematical Engineering, DOI 10.1007/978-3-319-54169-3_5

nonlinearity, lead to the model which is close to real mechanism. Due to its simplicity the slider-crank mechanism is assumed as a basic one for the cutting device. Joining together two slider-crank mechanisms an appropriate device is obtained which also transforms the rotating motion of the leading element into translator motion of the slider which is connected with a cutting tool. The idea of joining of two slider-crank mechanisms is not a new one. The double-slider crank mechanisms are already used in air compressors (Ogura and Daidoji 1982), two piston pumps (Wang et al. 2012), in the cutting machine for elliptical cylinder (Komatsubara et al. 2007), in the two-side piston engine (Kazimierski and Wojewoda 2011), in the haptic devices to generate pulling or pushing motion (Amemiya et al. 2007) and (Amemiya and Maeda 2009), in robotics (Masia et al. 2007; Xu and Wang 2008; Kim et al. 2008; Xu et al. 2011), and also as a continuous casting mold oscillation device (Ren et al. 2009).

In Cveticanin et al. (2012) the dynamics of the double-slider mechanism with rigid elements and with non-ideal forcing is considered. It is assumed that the motion of the leading element is with variable angle velocity which is caused by the constant cutting force on the output slider. The cutting mechanism is settled on the rigid support. Investigation of dynamics and vibration of the double-slider mechanism is difficult to perform, due to inherent nonlinearity associated with its kinematics, which is characterized by large rigid body rotations, but is necessary due to the practical significance of the subject. A more realistic system is considered in Zukovic et al. (2012): the mechanism is settled on the elastic support and the influence of the non-ideal forcing on the system and cutting process is analyzed. Forcing is of non-ideal type due to the requirement of the cutting force to be constant. As a consequence, the history of the angular rotation of the crank is included in the set of coordinates governing the dynamics of the mechanism. Direct integration technique gives the description of the run-up and close-down response of the mechanism. Steady-state periodic motions are obtained numerically, but also analytically for some special cases. Motion is compared with that for the case of ideal forcing when the angular velocity of the input element is constant. This result provides a basis for checking the accuracy of results obtained for non-ideal forcing.

The Chapter is divided into five sections. In Sect. 5.1 the structural synthesis of the cutting mechanism is considered. The advantages and disadvantages of the cutting mechanism based on the two slider-crank mechanism in comparison to the slider-crank mechanisms (simple and eccentric) are discussed. In Sect. 5.2 kinematic properties of the cutting mechanism are analyzed. In Sect. 5.3 the dynamics of the cutting mechanism settled on the rigid body is considered. Motion of the given mechanism is mathematically described. An analytical approximate procedure is developed for solving equations. Obtained solutions are compared with numerical ones. In Sect. 5.4 dynamics of the cutting mechanism i.e. double-slider mechanism with elastic support and non-ideal forcing is considered. Due to non-ideal forcing the motion is described with a system of two coupled differential equations. Namely, for ideal system only one differential equation is convenient to describe the motion. If the forcing is non-ideal the system is extended with an additional differential equation. The steady-state motion of the system with ideal and non-ideal forcing is also analyzed. The chapter ends with conclusion.

## 5.1 Structural Synthesis of the Cutting Mechanism

The structure of cutting mechanism is required to satisfy the following:

- the mechanism has to transform the input rotating motion into the translator one
- the cutting element has to move translatory
- the cutting process has to be during motion of the cutting element from up to down.

To fulfil these requirements, in this Chapter a device which contains two slider-crank mechanisms is suggested (see Fig. 5.1). The system is designed to have an eccentric $O_1AB$ and a simple $O_2DE$ slider-crank mechanism which are connected with a rod $BC$. The leading element of the mechanism is the crankshaft $O_1A$, while the slider is the cutting tool at the point $E$. The suggested mechanism converts the rotating motion of the crankshaft $O_1A$ into a straight-line motion of the slider $E$.

Mechanism has the following elements: $O_1A = a$, $AB = b$, $BC = c$, $O_2C = r$, $O_2D = g$, $DE = h$.

The position of the slider $B$ of the eccentric slider-crank mechanism $O_1AB$ (see Fig. 5.1) is given with the coordinates

$$x_B = a\cos\varphi + b\cos\theta = l, \tag{5.1}$$

$$y_B = -a\sin\varphi + b\sin\theta. \tag{5.2}$$

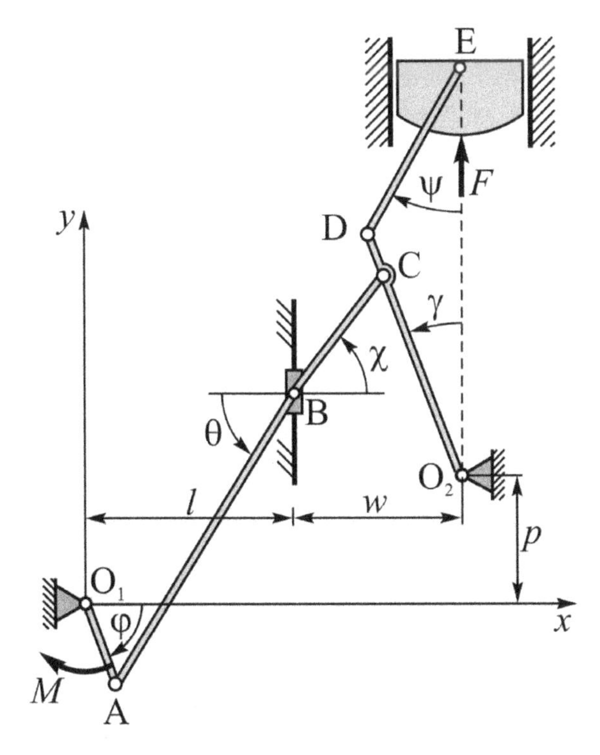


**Fig. 5.1** Model of the cutting mechanism

Eliminating $\theta$ in Eqs. (5.1) and (5.2) we obtain the position of the slider $B$ as a function of the leading angle $\varphi$

$$y_B = -a \sin\varphi + b\sqrt{1 - \left(\frac{l - a\cos\varphi}{b}\right)^2}. \tag{5.3}$$

For the simple slider-crank mechanism $O_2DE$ (see Fig. 5.1) the translatory motion of the slider is described as

$$O_2E = g\cos\gamma + h\cos\psi, \tag{5.4}$$

where the relation between the angles $\gamma$ and $\psi$ is given with the expression

$$g\sin\gamma = h\sin\psi. \tag{5.5}$$

Substituting Eq. (5.5) into Eq. (5.4) we have

$$O_2E = g\cos\gamma + h\sqrt{\left(1 - \frac{g^2}{h^2}\right) + \frac{g^2}{h^2}\cos^2\gamma}. \tag{5.6}$$

which describes the position of the slider $E$ as a function of the leading angle $\gamma$ of the slider-crank mechanism $O_2DE$.

Let us make the connection between these two slider-crank mechanisms. Due to the fact that after connection with the rod $BC$ the two slider-crank mechanism remains an one-degree-of-freedom system (as it was the case for the simple and eccentric slider-crank mechanisms), we have to determine the relation between the position of the slider $E$ and leading angle $\varphi$ of the crankshaft $O_1A$.

From Fig. 5.1 it is evident that the position of the slider $E$ in the coordinate system $xO_1y$ is

$$y_E = p + O_2E. \tag{5.7}$$

Moreover,

$$w = c\cos\chi + r\sin\gamma, \tag{5.8}$$

$$y_B + c\sin\chi = p + r\cos\gamma. \tag{5.9}$$

Eliminating $\chi$ in Eqs. (5.8) and (5.9) the $y_B - \gamma$ i.e., $\varphi - \gamma$ expression is obtained as

$$\left(c^2 - w^2 - r^2 - (p - y_B)^2 - 2r(p - y_B)\cos\gamma\right)^2 = 4w^2r^2(1 - \cos^2\gamma), \tag{5.10}$$

i.e.,

$$A_2\cos^2\gamma - A_1\cos\gamma + A_0 = 0, \tag{5.11}$$

where

$$A = c^2 - w^2 - r^2 - (p - y_B)^2,$$
$$A_0 = A^2 - 4w^2r^2, \quad A_1 = 4Ar(p - y_B),$$
$$A_2 = 4r^2\left((p - y_B)^2 + w^2\right), \tag{5.12}$$

and $p$ is a constant distance between fixed points $O_1$ and $O_2$ in $y$ direction. Solving the quadratic equation (5.11) for $\cos\gamma$ and substituting into Eqs. (5.7) with (5.6), the $y - \varphi$ relation follows.

### *5.1.1 Comparison of the Simple, Eccentric and Two Slider-Crank mechanisms*

In Fig. 5.2a the simple and in Fig. 5.2b the eccentric slider-crank mechanisms are plotted. The mechanisms differ as the distance between the fixed point O and the piston position is different.

In Fig. 5.3 the displacement-angle relations for: (a) simple (5.6), (b) eccentric (5.3) and (c) two slider-crank (5.7) mechanisms are plotted. It is assumed that for the simple and eccentric slider-crank mechanism the length of the leading shaft and of the connecting rod are equal for the both mechanisms. The dimensions of the two joined slider-crank mechanisms are: $a = 0.08$, $b = 0.32$, $c = 0.14$, $r = 0.20$, $g = 0.24$, $h = 0.18$, $l = 0.20$, $p = 0.12$, $w = 0.16$ and the cutting depth is $\delta = 0.12$. In our consideration the common assumption used for comparing the three mechanisms is that the cutting depth has to be equal and the cutting angle is calculated from the lowest position of the slider. In Fig. 5.2 the full line indicates the motion of the slider in the sheet (where the shaded area is for cutting) and the dotted line shows the

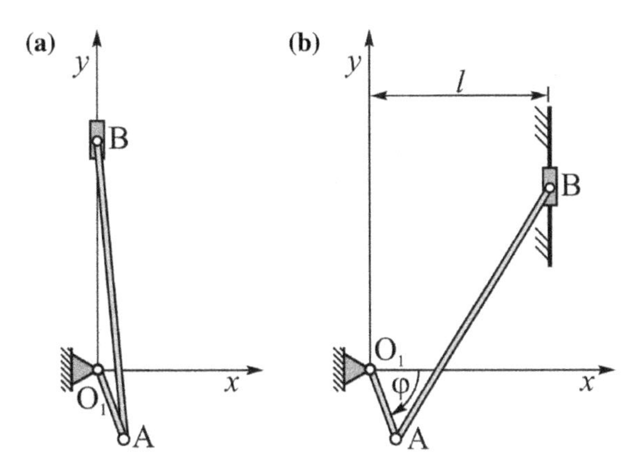


**Fig. 5.2** Slider-crank mechanisms: **a** simple, **b** eccentric

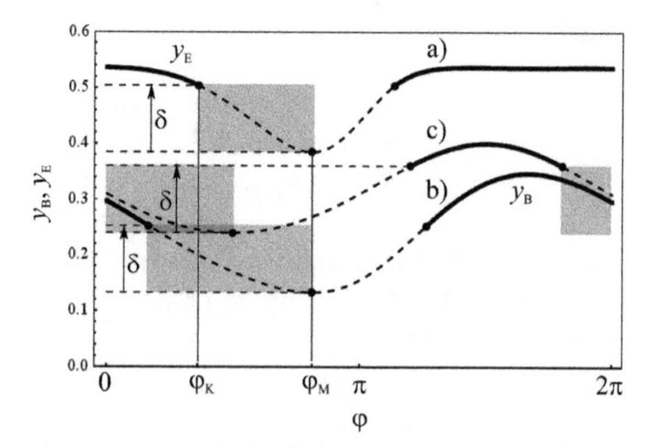


**Fig. 5.3** $y_B - \varphi$ diagrams for **a** simple slider-crank mechanism (Fig. 5.2a), **b** eccentric slider-crank mechanism (Fig. 5.2b), and **c** $y_E - \varphi$ diagram for two-joined slider-crank mechanism (Fig.5.1) with following notation: *shaded area* - cutting, *dotted line* - slider in the sheet, full line-slider out of sheet

motion of the slider out of the sheet. Comparing the diagrams in Fig. 5.3, it can be concluded:

1. Cutting lasts more longer with the simple and eccentric slider-crank mechanism than with the two joined slider-crank mechanism.
2. The interval in which the slider (cutting tool) is above the cutting object is much longer for the two joined slider-crank mechanism than for the simple and eccentric one. During this period the manipulation with the cutting sheet is possible to be finished. It is not the case for the simple and eccentric slider-crank mechanisms. Namely, the 'resting' period for the simple and eccentric slider-crank mechanisms is extremely short and does not give the opportunity to finish the manipulation with the sheet: setting and its removing from the machine.

It is the reason that the joined two-slider-crank mechanism is introduced and assumed for the cutting process. During one period of motion of the two-joined slider-crank mechanism the manipulation with the polymer sheet and also the cutting process is possible to be finished.

## 5.2 Kinematics of the Cutting Mechanism

Let us determine the velocity $v_E$ of the cutting tool as a function of the angular velocity $\dot{\varphi}$ of the leading crankshaft. Using the relations (5.6) and (5.7) the velocity of the cutting tool is

$$v_E = \dot{y}_E = -\dot{\gamma}\frac{gv_E \sin\gamma}{y_E - g\cos\gamma}. \tag{5.13}$$

The time derivative of (5.10) gives $\dot{\gamma}(\dot{y}_B)$ as

$$\dot{\gamma}r(w\cos\gamma - (p - y_B)\sin\gamma) = (p - y_B + r\cos\gamma)\dot{y}_B, \tag{5.14}$$

where according to (5.3)

$$\dot{y}_B = -a\dot{\varphi}\frac{y_B \cos\varphi + l \sin\varphi}{y_B + a\sin\varphi}. \tag{5.15}$$

Substituting (5.14) with (5.15) into (5.13) the velocity of the slider as the function of the angular velocity of the leading crankshaft is obtained

$$v_E = a\dot{\varphi} f(\varphi), \tag{5.16}$$

$$f(\varphi) = \frac{g}{r}\frac{y_E \sin\gamma}{y_E - g\cos\gamma}\frac{p - y_B + r\cos\gamma}{w\cos\gamma - (p - y_B)\sin\gamma} \frac{y_B \cos\varphi + l\sin\varphi}{y_B + a\sin\varphi}. \tag{5.17}$$

Function $f(\varphi)$ is a periodical with a period of $2\pi$.

## 5.3 Dynamic Analysis of the Mechanism with Rigid Support

Dynamics of a cutting mechanism with rigid support is analyzed. Mechanism contains two slider-crank mechanisms described in previous section. The cutting force in the cutting tool is required to be constant. Due to this assumption the input angular velocity of the driven element has to be controlled. Variation of the angular velocity of the driving motor are necessary to be calculated. The results are widely discussed in Cveticanin et al. (2012).

### *5.3.1 Mathematical Model of the Mechanism*

The considered two slider-crank mechanism has one degree of freedom and the generalized coordinate is the angle $\varphi$ of the leading crank $O_1A$. The Lagrange differential equation of motion of the mechanism for the generalized coordinate $\varphi$ is in general

$$\frac{d}{dt}\frac{\partial T}{\partial \dot{\varphi}} - \frac{\partial T}{\partial \varphi} + \frac{\partial \Phi}{\partial \dot{\varphi}} = Q_\varphi, \tag{5.18}$$

where $T$ is the kinetic energy of the mechanism, $\Phi$ is the dissipative function and $Q_\varphi$ is the generalized force.

It is assumed that the mass of the cutting tool is $m$ and the moment of inertia of the leading element is $J$. The inertial properties of other elements in mechanism can be omitted in comparison to the previous. Then, the kinetic energy of the mechanism is a sum of the kinetic energy of the cutting tool and of the leading element

$$T = \frac{1}{2}J\dot{\varphi}^2 + \frac{1}{2}mv_E^2, \tag{5.19}$$

where $v_E$ is the velocity of cutting tool given with (5.16). Substituting (5.16) into (5.19) we obtain

$$T = \frac{1}{2}J\dot{\varphi}^2 + \frac{1}{2}ma^2\dot{\varphi}^2 f^2, \tag{5.20}$$

where the kinetic energy is the function of the angular velocity $\dot{\varphi}$ and $f = f(\varphi)$ given with (5.17). Since all the mechanism members are rigid, the elastic energy of the system is zero.

The mechanism is driven by an electro motor whose characteristics is that the driving torque $M$ is the function of the velocity $\dot{\varphi}$, (Sandier 1999),

$$M = M_0\left(1 - \frac{\dot{\varphi}}{\omega_0}\right), \tag{5.21}$$

where $M_0 = const.$ and $\omega_0$ is the synchronal angular velocity of the motor. Thereby, the driving load is expressed as a function of the angular coordinate describing the crank rotation. Physically it means that the motion of the mechanism has an influence on the motor torque. Such mechanism is subjected to non-ideal forcing (see Nayfeh and Mook 1979; Zukovic and Cveticanin 2007; Zukovic and Cveticanin 2009).

The cutting process is required to be during motion of the cutting tool from up to down in the angle interval $[\varphi_K, \varphi_M]$ where $\varphi_M$ corresponds to the lowest position of the cutting tool which satisfies the relation $dy(\varphi_M)/d\varphi = 0$ and $\varphi_K$ is the angle position for which the cutting starts and has to be adopted to the thickness of the sheet $\delta$: $y(\varphi_K) = y(\varphi_M) + \delta$. In this interval the cutting force is required to be constant and sufficiently strong to provide the cutting without folding of the sheet. Otherwise, the cutting force is assumed to be zero. Mathematically, for $\varphi \in [\varphi_K, \varphi_M]$ the constant force is $F = F_0$ and for $\varphi \in [0, \varphi_K) \cup (\varphi_M, 2\pi]$ it is $F = 0$.

As the cutting process is periodical, the cutting force is modeled as a UnitStep function

$$\begin{aligned} F = F(\varphi) = F_0\bar{F}(\varphi) = F_0(&UnitStep(\text{mod}(\varphi, 2\pi) - \varphi_K) \\ &-UnitStep(\text{mod}(\varphi, 2\pi) - \varphi_M)), \end{aligned} \tag{5.22}$$

where the unit step function is defined as $UnitStep(x) = \begin{cases} 1, & x \geq 0 \\ 0, & x < 0 \end{cases}$. The force distribution is plotted in Fig. 5.4 ($\varphi_K = 2.06379$, $\varphi_M = 2.55591$, $\delta = 0.03$).The driving torque $M$ and the cutting force $F$ give the virtual works for virtual angle and displacement variations, respectively, i.e.,

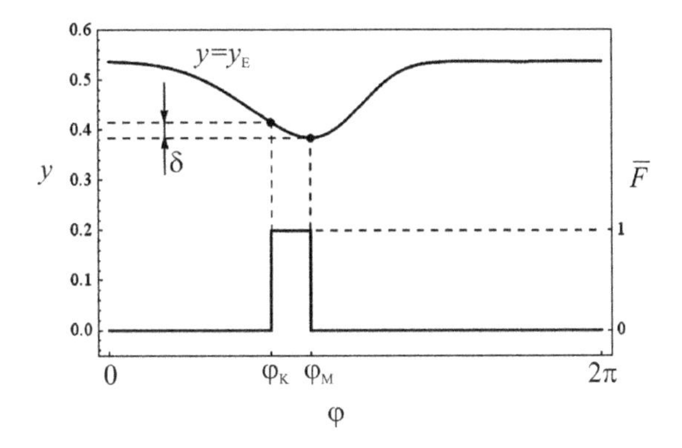


**Fig. 5.4** $y - \varphi$ and $\bar{F} - \varphi$ diagrams of the cutting tool

$$\delta A = M\delta\varphi + F\delta y. \tag{5.23}$$

According to (5.16) the variation of the variable $y$ is

$$\delta y = af\delta\varphi. \tag{5.24}$$

Substituting (5.24) into (5.23) we obtain $\delta A = Q_\varphi \delta\varphi$ where the generalized force is

$$Q_\varphi = M + afF. \tag{5.25}$$

During the cutting the damping force acts. For energy dissipation during the slider motion through various materials of the polymer sheet, the damping force is assumed to be proportional to the velocity of the cutting tool, i.e.,

$$\mathbf{F}_w = -q\mathbf{v}_E. \tag{5.26}$$

The corresponding dissipative function is

$$\Phi = \frac{1}{2}qv_E^2, \tag{5.27}$$

where $q$ is the damping coefficient. According to (5.16), the dissipative function (5.24) is

$$\Phi = \frac{1}{2}qa^2f^2\dot{\varphi}^2. \tag{5.28}$$

The derivatives of kinetic energy function (5.20) suitable for Lagrange equation (5.18) are

$$\frac{\partial T}{\partial \varphi} = a^2mf\frac{df}{d\varphi}\dot{\varphi}^2, \tag{5.29}$$

$$\frac{\partial T}{\partial \dot{\varphi}} = J\dot{\varphi} + a^2 f^2 \dot{\varphi}, \tag{5.30}$$

$$\frac{d}{dt}\frac{\partial T}{\partial \dot{\varphi}} = J\ddot{\varphi} + a^2(2f\dot{f}\dot{\varphi} + f^2\ddot{\varphi}), \tag{5.31}$$

where $(\dot{\ }) = d/dt$. The time derivative of the function $f$ expressed with (5.17) is

$$\dot{f} = \frac{df}{dt} = \dot{\varphi}\frac{df}{d\varphi}. \tag{5.32}$$

As $f$ explicitly and implicitly depends on the angle $\varphi$ the total derivative of $f$ is

$$\frac{df}{d\varphi} = \frac{\partial f}{\partial \varphi} + \frac{\partial f}{\partial y_B}\frac{dy_B}{d\varphi} + \left(\frac{\partial f}{\partial y_E}\frac{dy_E}{d\gamma} + \frac{\partial f}{\partial \gamma}\right)\frac{d\gamma}{d\varphi}. \tag{5.33}$$

Introducing the notation

$$\begin{aligned} &s_1 = y_E \sin\gamma, \quad s_2 = y_E - g\cos\gamma, \quad s_3 = y_B \cos\varphi, \\ &s_4 = y_B + a\sin\varphi, \quad s_5 = p - y_B + r\cos\gamma, \\ &s_6 = w\cos\gamma - (p - y_B)\sin\gamma, \end{aligned} \tag{5.34}$$

and substituting into (5.17), the function $f$ is

$$f(\varphi) = \frac{g}{r}\frac{s_1}{s_2}\frac{s_3}{s_4}\frac{s_5}{s_6}. \tag{5.35}$$

The corresponding derivatives of (5.35) according to (5.33) are

$$\begin{aligned} &\frac{\partial f}{\partial \varphi} = \frac{g}{r}\frac{s_1}{s_2}\frac{s_5}{s_6}\frac{\partial}{\partial \varphi}\left(\frac{s_3}{s_4}\right), \quad \frac{\partial f}{\partial y_E} = \frac{g}{r}\frac{s_2\sin\gamma - s_1}{s_2^2}\frac{s_3 s_5}{s_4 s_6}, \\ &\frac{\partial f}{\partial \gamma} = \frac{g}{r}\frac{s_3}{s_4}\frac{\partial}{\partial \gamma}\left(\frac{s_1}{s_2}\frac{s_5}{s_6}\right), \quad \frac{\partial f}{\partial y_B} = \frac{g}{r}\frac{s_1}{s_2}\frac{\partial}{\partial y_B}\left(\frac{s_3}{s_4}\frac{s_5}{s_6}\right), \end{aligned} \tag{5.36}$$

$$\begin{aligned} &\frac{\partial s_3}{\partial \varphi} = -y_B\sin\varphi + l\cos\varphi, \quad \frac{\partial s_6}{\partial y_B} = \sin\varphi, \\ &\frac{\partial s_1}{\partial \gamma} = y_E\cos\gamma, \quad \frac{\partial s_2}{\partial \gamma} = \sin\gamma, \quad \frac{\partial s_5}{\partial \gamma} = -r\sin\gamma, \\ &\frac{\partial s_4}{\partial \varphi} = a\cos\varphi, \quad \frac{\partial s_6}{\partial \gamma} = -(p - y_B)\cos\gamma - w\sin\gamma, \\ &\frac{\partial s_3}{\partial y_B} = \cos\varphi, \quad \frac{\partial s_4}{\partial y_B} = -\frac{\partial s_5}{\partial y_B} = 1. \end{aligned} \tag{5.37}$$

For (5.3), (5.6) and (5.10) the derivatives in angle $\varphi$ are

$$\frac{\partial y_B}{\partial \varphi} = -a\frac{s_5}{s_6}, \qquad \frac{\partial y_B}{\partial \gamma} = -g\frac{s_1}{s_2}, \qquad \frac{\partial \gamma}{\partial \varphi} = \frac{a}{r}\frac{s_3}{s_4}\frac{s_5}{s_6}. \tag{5.38}$$

Substituting (5.38) and the also (5.3), (5.6) and (5.10) into (5.33) the $(df/d\varphi)$ relation is calculated. Substituting into (5.18) the relations (5.15), (5.20) and (5.28) and the corresponding derivatives calculated in Appendix, the differential equation of motion is obtained

$$(J + ma^2 f^2)\ddot{\varphi} + ma^2 f\frac{df}{d\varphi}\dot{\varphi}^2 + qa^2 f^2\dot{\varphi} = M(\dot{\varphi}) + afF(\varphi), \tag{5.39}$$

where $f$ and $(df/d\varphi)$ are $\varphi$ - periodical functions with period of $2\pi$. (see Eqs. (5.17) and (5.33)).

According to (5.17) and (5.33), the functions $f(\varphi) - \varphi$, $(df/d\varphi) - \varphi$ and $f(\varphi)(df(\varphi)/d\varphi) - \varphi$ are plotted in Fig. 5.5.

Introducing the dimensionless values

$$\tau = \omega_0 t, \quad I = \frac{J\omega_0^2}{M_0}, \quad \lambda = \frac{F_0 a}{M_0}, \quad Q = \frac{qa^2\omega_0}{M_0}, \quad \mu = \frac{ma^2\omega_0^2}{M_0}, \tag{5.40}$$

the differential equation (5.39) transforms into

$$(I + \mu f^2)\varphi'' + \mu f\frac{df}{d\varphi}(\varphi')^2 + Qf^2\varphi' = (1 - \varphi') + \lambda f\bar{F}(\varphi), \tag{5.41}$$

where $\varphi' = d\varphi/d\tau$ and $\varphi'' = d^2\varphi/d\tau^2$, $\mu$ is dimensionless mass of the cutting tool, $I$ is dimensionless moment of inertia of the leading crank $O_1A$, $Q$ is the dimensionless damping coefficient, $\lambda$ is dimensionless cutting force and $\tau$ is the dimensionless time. Equation (5.41) is strong nonlinear. Only for some special parameter values the closed

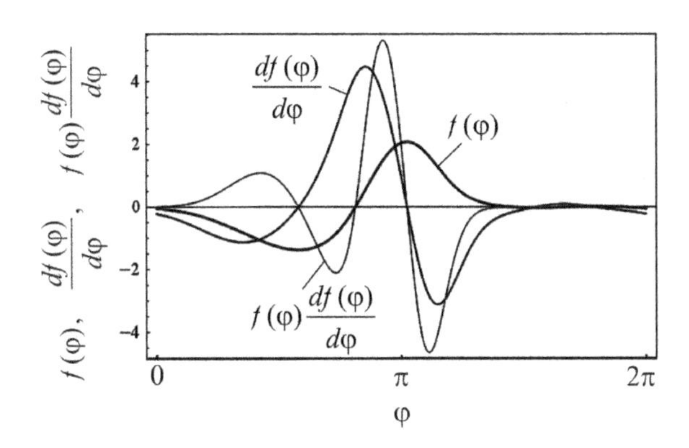


**Fig. 5.5** $f(\varphi) - \varphi$, $(df/d\varphi) - \varphi$ and $f(\varphi)(df(\varphi)/d\varphi) - \varphi$ curves

form analytical solution is possible to be obtained. Otherwise, (5.41) has to be solved numerically using the Runge–Kutta procedure.

### 5.3.2 Numerical Simulation

Solving (5.41) for various values of $\bar{F}(\varphi)$ the influence of the cutting force on the angular velocity of the motor is obtained. In Fig. 5.6 the $\varphi' - \tau$ curves for various values of $\bar{F}(\varphi)$ are plotted. (Dimensionless parameters are $\lambda = 0.033$, $I = 1.4557\ 10^{-4}$, $\mu = 1.051\ 10^{-3}$, $Q = 0.00134$ and the initial conditions $\varphi(0) = 0$ and $\varphi'(0) = 1$).

Analyzing the curves in Fig. 5.5 it can be concluded:

1. For the case when the cutting force is zero, $\bar{F}(\varphi) = 0$, and the motion of the mechanism is without loading, the angular velocity of the leading crank $O_1A$ varies as it is shown in Fig. 5.5 (curve I). Variation of the angular velocity is periodical.
2. If it is assumed that the mechanism is loaded with a force $\bar{F}(\varphi) = 1$ for all positions of the leading crank, the influence of the force on the angular velocity of the motor motion is extremely high (see curve II in Fig. 5.5).
3. For the case when the cutting process is discontinual and the cutting force has the form (5.22) there is a jump in the angular velocity curve (see curve III, Fig. 5.5). For this case the influence of the cutting parameter $\lambda$ on the $\varphi' - \tau$ is evident (see Fig. 5.6). The higher the cutting force the higher the velocity variation.

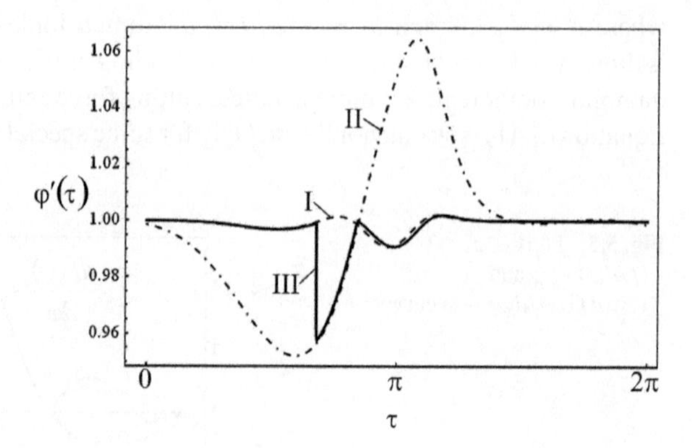


**Fig. 5.6** $\varphi' - \tau$ curves for various values of $\bar{F}(\varphi)$: $I - \bar{F}(\varphi) = 0$, $II - \bar{F}(\varphi) = 1$, $III - UnitStep$ function

### 5.3.3 Analytical Consideration

Let us consider the case when $M_0$ is significantly larger than the parameters $I$, $\mu$, $Q$ and $\lambda$ which for the small parameter $\varepsilon << 1$ have the form

$$I = \varepsilon I_1,\ \mu = \varepsilon \mu_1,\ Q = \varepsilon Q_1,\ \lambda = \varepsilon \lambda_1. \tag{5.42}$$

Substituting (5.42) into (5.41) we have

$$(1-\varphi') = \varepsilon(I_1+\mu_1 f^2(\varphi))\varphi'' + \varepsilon\mu_1 f(\varphi)\frac{df(\varphi)}{d\varphi}\varphi'^2 + \varepsilon Q_1 f^2(\varphi)\varphi' - \varepsilon f(\varphi)\lambda_1 \bar{F}(\varphi). \tag{5.43}$$

Using the series expansion of the variable $\varphi$ and its time derivatives up to the first order of the small parameter, we obtain

$$\begin{aligned}
\varphi &= \varphi_0 + \varepsilon\varphi_1 + \dots, \quad \varphi' = \varphi_0' + \varepsilon\varphi_1' + \dots, \\
\varphi'' &= \varphi_0'' + \varepsilon\varphi_1'' + \dots, \quad \bar{F}(\varphi) \approx \bar{F}(\varphi_0), \\
f(\varphi) &= f(\varphi_0 + \varepsilon\varphi_1) \approx f(\varphi_0) + \varepsilon f'(\varphi_0)\varphi_1, \\
\frac{df(\varphi)}{d\varphi} &= \frac{df(\varphi_0 + \varepsilon\varphi_1)}{d\varphi} \approx \left(\frac{df(\varphi)}{d\varphi}\right)_{\varphi 0} + \varepsilon\varphi_1 \left(\frac{d^2 f(\varphi)}{d\varphi^2}\right)_{\varphi 0}.
\end{aligned} \tag{5.44}$$

Substituting (5.44) into (5.43) and separating the terms with the same order of small parameter $\varepsilon$ up to the small value of second order, the system of equations follows

$$\varepsilon^0: \quad 0 = 1 - \varphi', \tag{5.45}$$

$$\begin{aligned}
\varepsilon^1: \quad \varphi_1' &= f(\varphi_0)\lambda_1 \bar{F}(\varphi_0) - Q_1 f^2(\varphi_0)\varphi_0' \\
&- (I_1 + \mu_1 f^2(\varphi_0))\varphi_0'' - \mu_1 f(\varphi_0)\left(\frac{df(\varphi)}{d\varphi}\right)_{\varphi 0} \varphi_0'^2.
\end{aligned} \tag{5.46}$$

Solution of (5.45) is $\varphi_0' = 1 = const.$ which after integration gives

$$\varphi_0 = \tau. \tag{5.47}$$

Substituting (5.47) into (5.46) we obtain

$$\varphi_1' = -\mu_1 f(\varphi_0)\left(\frac{df(\varphi)}{d\varphi}\right)_{\varphi 0} - Q_1 f^2(\varphi_0) + f(\varphi_0)\lambda_1 \bar{F}(\varphi_0). \tag{5.48}$$

According to (5.47), (5.48) and (5.44) the first order approximate analytical solution is

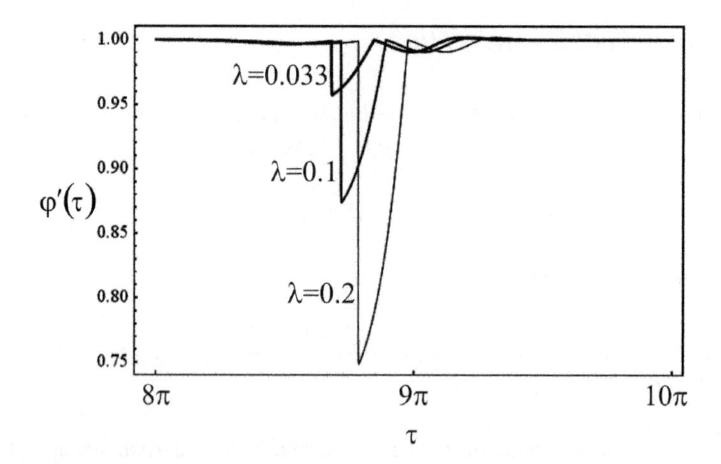


**Fig. 5.7** $\varphi' - \tau$ curves for various values of $\lambda$

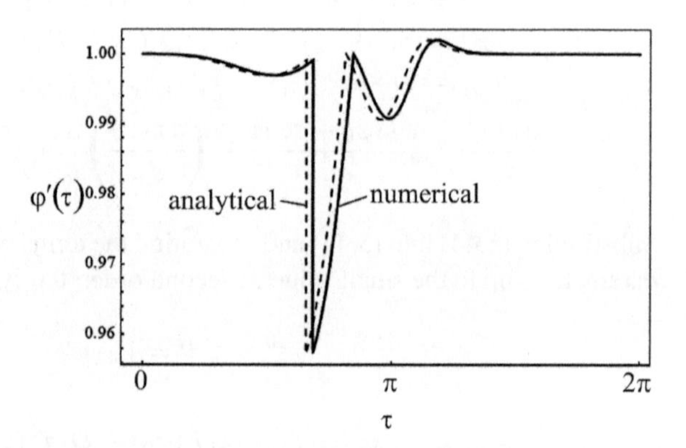


**Fig. 5.8** Comparison of the analytical and numerical $\varphi' - \tau$ functions

$$\varphi'(\tau) = 1 + \left( -\mu f(\tau) \left( \frac{df(\varphi)}{d\varphi} \right)_{\tau} - Q_1 f^2(\tau) + f(\tau)\lambda_1 \bar{F}(\tau) \right). \tag{5.49}$$

Influence of mass and damping parameters, and also of the cutting force on the angular velocity of the leading element is obtained.

In Fig. 5.7 the analytical result (5.49) is compared with numerical one which is valid for differential equation (5.43). The difference between the results is negligible (Fig. 5.8).

### 5.3.4 Comparison of Analytical and Numerical Results

Equation (5.43) and the analytically obtained solution (5.49) yield:

1. For the mechanism with omitted mass of the leading crank and of the cutting tool, the angular velocity variation is $\varphi' = (1+\lambda f \bar{F}(\varphi))/(1+Qf^2)$. For higher values of coefficient of damping the angular velocity is smaller. The influence of the cutting force $\lambda$ on the angular velocity $\varphi'$ is significant: the higher the cutting force, the larger the angular velocity variation.
2. If the mass of the cutting tool and the damping coefficient during cutting and are omitted, the differential equation is $I\varphi" = (1-\varphi') + \lambda f \bar{F}(\varphi)$. It depends on the moment of inertia $I$ of the leading crankshaft and on the cutting force $\lambda$.
3. For the case when damping is neglected and the cutting force is zero, for the initial angular velocity $\varphi_0'$ the angular velocity of the leading element varies as $\varphi' = 1 + (\varphi_0' - 1)\exp(-\tau/I)$. For the steady state motion when time $\tau$ tends to infinity, the angular velocity of the leading element tends to a constant value: $\varphi' = 1 = const.$
4. If the dimensionless driving torque $M_0$ is significant in comparison to other parameters of the mechanism, the angular velocity in the first approximation is obtained as $\varphi' \approx 1 + \varepsilon\varphi_1'$, where $\varphi_1' = -\mu f(df/d\varphi) - Q_1 f^2 + \lambda f \bar{F}$. For certain parameter values the analytically obtained result is compared with exact numerical one (see Fig. 5.7). The difference between the results is negligible.

Based on the previous results, the following is concluded:

1. The damping during cutting has a significant influence on the angular velocity of the leading element of the cutting mechanism. If mass of the leading crank and of the cutting tool are quite small it is obvious that for higher values of the damping, the angular velocity of the leading crank is smaller.
2. The influence of the cutting force on the angular velocity is also significant: the higher the cutting force, the larger is the angular velocity variation.
3. The angular velocity variation effects the stability of motion and also the quality of cutting process. Namely, for high values of angular velocity variation of the leading element, the motor can get from the steady state stable motion into the unstable one. Besides, the higher the cutting force, the cutting process is retard due to the fact that the averaged velocity is smaller.

## 5.4 Dynamics of the Cutting Mechanism with Flexible Support and Non-ideal Forcing

Dynamics of a cutting mechanism, which is a double-slider one where one of the sliders represents the cutting tool, is investigated in Zukovic et al. (2012). Mechanism is subjected to non-ideal forcing. The constant cutting force requires the driving torque to be the function of the angular velocity of the driving crank. Mechanism examined involves rigid elements but elastic support. The cutting mechanism with the elastic support is modelled as a two-degree-of-freedom system whose motion is described with two coupled second order nonlinear differential equations. For the non-ideal case the steady-state dynamics of the system is examined by introducing the approximate analytical solution of the averaged differential equations of motion for the case of primary resonance. Parameters of the system are varied and their influence on the motion is tested. The analytically obtained solution is compared with exact numerical one and shows a good agreement. The ideal forcing case, when the motion of the mechanism is with constant angular velocity, is also analyzed. For this case the steady-state motion is obtained analytically, too.

### 5.4.1 Mathematical Model of Motion of the Cutting Mechanism

Cutting mechanism connected with the support is shown in Fig. 5.9.

The considered mechanism contains two slider-crank mechanisms $O_1AB$ and $O_2DE$ connected with a rod $BC$. Rotation of the driving element $O_1A$ is transformed into the straightforward motion of the slider $E$ which represents the cutting tool. During the contact between the slider $E$ and the object in $G$, the cutting tool cuts

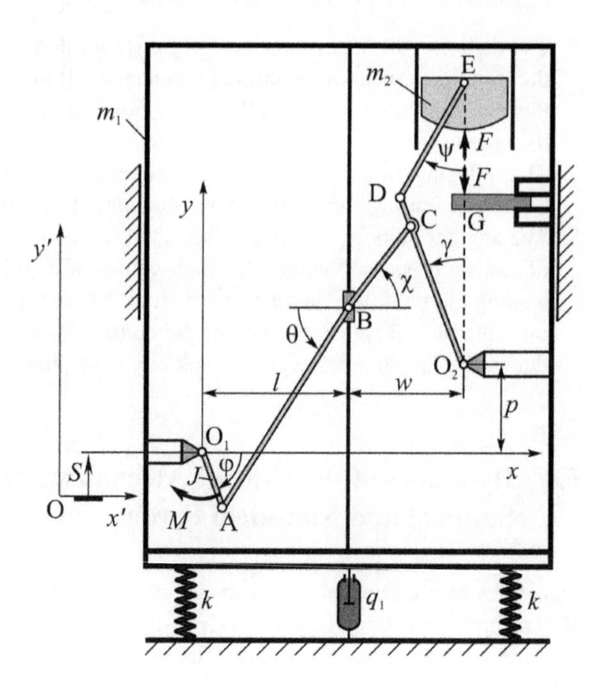


**Fig. 5.9** Model of the cutting mechanism on the elastic support

the sheet in $G$. The kinematic properties of the mechanism are widely discussed in previous section. In order to obtain the more realistic description of the dynamics of the mechanism it is necessary to include its interaction with the support. Namely, the mechanism has an influence on the support, but also the support influences the dynamics of the cutting process. The mechanism-support system has two-degrees-of-freedom and the two generalized coordinates are the angle position of the driving element due to the rotation, given with generalized coordinate $\varphi$, and the displacement of the support due to its straightforward motion, given with the other independent generalized coordinate $S$. Then, the Lagrange's differential equations of motion are in general

$$\begin{aligned}\frac{d}{dt}\frac{\partial T}{\partial \dot{S}} - \frac{\partial T}{\partial S} &= -\frac{\partial V}{\partial S} - \frac{\partial \Phi}{\partial \dot{S}} + Q_S, \\ \frac{d}{dt}\frac{\partial T}{\partial \dot{\varphi}} - \frac{\partial T}{\partial \varphi} &= -\frac{\partial V}{\partial \varphi} - \frac{\partial \Phi}{\partial \dot{\varphi}} + Q_\varphi,\end{aligned} \tag{5.50}$$

where $T$ is the kinetic energy, $V$ is the potential energy, $\Phi$ is the dissipative function of the system, while $Q_S$ and $Q\varphi$ are the generalized forces.

We assume that the mass $m_2$ of the slider E and of the support $m_1$ are significantly higher than the masses of the other elements in the mechanism. If the mass moment of inertia of motor is $J$, the kinetic energy of the mechanism has three terms: kinetic energy of the motor (due to rotation), $T_m$, kinetic energy of the support, $T_1$, and of the slider, $T_2$, respectively,

$$T = T_m + T_1 + T_2, \tag{5.51}$$

i.e.,

$$T = \frac{1}{2}J\dot{\varphi}^2 + \frac{1}{2}m_1\dot{S}^2 + \frac{1}{2}m_2 v_E^2, \tag{5.52}$$

where $v_E$ is the velocity of the slider $E$. To determine the velocity $v_E$ some geometric properties of the mechanism have to be considered.

Let us determine the coordinate $y_E$ as a function of the generalized coordinates $\varphi$ and $S$. From Fig. 5.9. the coordinate $y_E$ in the fixed coordinate system $x'Oy'$ is

$$y_E = S + p + g\cos\gamma + h\cos\psi, \tag{5.53}$$

where the lengths of the elements are $g = O_2D$ and $h = DE$. Using the geometric properties of the mechanism $O_2DE$ we have

$$g\sin\gamma = h\sin\psi. \tag{5.54}$$

The relation (5.53) transforms into

$$y_E = S + p + g\cos\gamma + h\sqrt{1 - \left(\frac{g\sin\gamma}{h}\right)^2}. \tag{5.55}$$

For the known distances $l$ and $w$ we have

$$\cos\theta = \frac{l - a\cos\varphi}{b} \leq 1, \tag{5.56}$$

and

$$\cos\chi = \frac{w - r\sin\gamma}{c} \leq 1, \tag{5.57}$$

where $a = O_1A, b = AB, r = O_2C, c = BC$. Using the relation

$$b\sin\theta - a\sin\varphi = p + r\cos\gamma - c\sin\chi, \tag{5.58}$$

with (5.56) and (5.57), we obtain

$$b\sqrt{1 - \left(\frac{l - a\cos\varphi}{b}\right)^2} - a\sin\varphi = p + r\cos\gamma - c\sqrt{1 - \left(\frac{w - r\sin\gamma}{c}\right)^2}. \tag{5.59}$$

After some transformation we obtain the equation

$$4r^2(A^2 + w^2)\cos^2\gamma + 4ABr\cos\gamma + (B^2 - 4w^2r^2) = 0, \tag{5.60}$$

whose solutions are

$$\cos\gamma = \frac{-K_1(K_1^2 + K_2) \pm w\sqrt{4r^2(w^2 + K_1^2) - (K_1^2 + K_2)^2}}{2r(w^2 + K_1^2)}, \tag{5.61}$$

where $|\cos\gamma| \leq 1$ and $4r^2(w^2 + K_1^2) - (K_1^2 + K_2)^2 \geq 0$ for $K_2 = r^2 - c^2 + w^2$ and $K_1(\varphi) = K_1 = a\sin\varphi + p - \sqrt{b^2 - (l - a\cos\varphi)^2}$.

Substituting the solution (5.61) into (5.55) we obtain

$$y_E = S + p + af_1(\varphi), \tag{5.62}$$

where

$$f_1(\varphi) = \sqrt{\frac{h^2 - g^2}{a^2} + \frac{g^2}{a^2}\left(\frac{-K_1(K_1^2 + K_2) \pm w\sqrt{4r^2(w^2 + K_1^2) - (K_1^2 + K_2)^2}}{2r(w^2 + K_1^2)}\right)^2}$$
$$+ \frac{g}{a}\frac{-K_1(K_1^2 + K_2) \pm w\sqrt{4r^2(w^2 + K_1^2) - (K_1^2 + K_2)^2}}{2r(w^2 + K_1^2)}. \tag{5.63}$$

Relation (5.62) with (5.63) gives the coordinate of $E$ as a function of $S$ and $\varphi$.
Using the $y_E(\varphi)$ relation (5.62) given in Appendix, the velocity of the slider is

$$v_E = \frac{dy_E}{dt} = \dot{S} + af(\varphi)\dot{\varphi}, \tag{5.64}$$

where $f(\varphi) = f = \frac{df_1(\varphi)}{d\varphi}$ and $\dot{\varphi} = \frac{d\varphi}{dt}$. Substituting (5.64) into (5.52) the kinetic energy of the mechanism is

$$T = \frac{1}{2}(J + m_2 a^2 f^2)\dot{\varphi}^2 + \frac{1}{2}(m_1 + m_2)\dot{S}^2 + m_2 af\dot{\varphi}\dot{S}. \tag{5.65}$$

Since only the support has elastic property, the potential energy of the system is

$$V = \frac{1}{2}k_1 S^2, \tag{5.66}$$

where $k_1$ is the coefficient of rigidity. In the system the energy dissipation is due to cutting and due to damping properties of the support. Then, the energy dissipation is expressed as

$$\Phi = \frac{1}{2}q_1\dot{S}^2 + \frac{1}{2}q_2 v_E^2, \tag{5.67}$$

where $q_1$ and $q_2$ are the damping coefficients of the support and of the cutting tool, respectively. Substituting (5.64) into (5.67), it yields

$$\Phi = \frac{1}{2}(q_1 + q_2)\dot{S}^2 + q_2 af\dot{S}\dot{\varphi} + \frac{1}{2}q_2 a^2 f^2\dot{\varphi}^2. \tag{5.68}$$

The mechanism is driven with a motor torque M and produces the cutting force $F$. For the purpose of the suggested mechanism the torque appears in the form

$$M = M(\dot{\varphi}) = M_0\left(1 - \frac{\dot{\varphi}}{\Omega_0}\right), \tag{5.69}$$

where $M_0$ is the fixed moment and $\Omega_0$ is the synchronous angular velocity of the motor. As the force $F$ is required to be constant during the cutting ($F = F_0 = const.$ for $\varphi \in [\varphi_K, \varphi_M]$) and otherwise to be zero ($F = 0$ for $\varphi \in [0, \varphi_K) \cup (\varphi_M, 2\pi]$), it is modeled as a UnitStep function:

$$F = F(\varphi) = F_0 \bar{F}(\varphi), \tag{5.70}$$

where

$$\bar{F}(\varphi) = F_0(UnitStep(\text{mod}(\varphi, 2\pi) - \varphi_K) - UnitStep(\text{mod}(\varphi, 2\pi) - \varphi_M)). \tag{5.71}$$

Using (5.69) and (5.70), the expression of the virtual work of the force and the torque in the system is

$$\delta A = M\delta\varphi + F(\delta y_E - \delta y_G), \tag{5.72}$$

where $\delta y_E$ is the variation of the coordinate $y_E$, (see Eq. (A.10)), and $\delta y_G$ is the variation of the coordinate $y_G$. From Fig. 5.9 it is evident that $y_G = S + p + (O_2G)$ and $(O_2G)$ is a fixed distance. Substituting the variation of the coordinates $\delta y_E$ and $\delta y_G$:

$$\delta y_E = \delta S + af\delta\varphi, \quad \delta y_G = \delta S,$$

into (5.72) we have

$$\delta A = (M + Faf(\varphi))\delta\varphi. \tag{5.73}$$

The generalized forces of the system are according to (5.73)

$$Q_\varphi = M + Faf(\varphi), \tag{5.74}$$

$$Q_S = 0. \tag{5.75}$$

Employing the relations (5.65)–(5.67), (5.74) and (5.75) the Lagrange's equations (5.50) yield the equations of motion of the mechanism in the form

$$\begin{aligned} 0 = {} & (m_1 + m_2)\ddot{S} + m_2af\ddot{\varphi} + (q_1 + q_2)\dot{S} + m_2af'\dot{\varphi}^2 + q_2af\dot{\varphi} \\ & + k_1S, \end{aligned} \tag{5.76}$$

$$\begin{aligned} & (J + m_2a^2f^2)\ddot{\varphi} + m_2af\ddot{S} + m_2a^2ff'\dot{\varphi}^2 + q_2af\dot{S} + q_2a^2f^2\dot{\varphi} \\ & = M(\dot{\varphi}) + afF_0\bar{F}(\varphi), \end{aligned} \tag{5.77}$$

where

$$f' = \frac{df}{d\tau}, \quad \tau = \omega_0 t, \quad \omega_0 = \sqrt{\frac{k_1}{m_1 + m_2}}. \tag{5.78}$$

For computational reason, let us introduce beside (5.78) the following non-dimensional variables

$$s = \frac{S}{a}, \quad \mu = \frac{m_2}{m_1 + m_2}, \quad Q_1 = \frac{q_1}{\sqrt{k_1}\sqrt{m_1 + m_2}}, \quad Q_2 = \frac{q_2}{\sqrt{k_1}\sqrt{m_1 + m_2}},$$
$$\kappa = \frac{m_2 a^2}{J}, \quad \lambda_M = \frac{M_0}{J\omega_0^2}, \quad \lambda_F = \frac{aF_0}{J\omega_0^2}, \quad \nu_0 = \frac{\Omega_0}{\omega_0}, \quad \bar{M}(\varphi') = 1 - \frac{\varphi'}{\nu_0}. \tag{5.79}$$

Substituting them into (5.76) and (5.77) the following system of two coupled non-linear differential equations is obtained

$$s'' + (Q_1 + Q_2)s' + s = -\mu \frac{df}{d\varphi}\varphi'^2 - \mu f \varphi'' - Q_2 f \varphi', \tag{5.80}$$

$$(1 + \kappa f^2)\varphi'' = \lambda_M \bar{M}(\varphi') + \lambda_F f \bar{F}(\varphi) - \kappa f s''$$
$$- \frac{\kappa Q_2}{\mu} f(s' + f\varphi') - \kappa f \frac{df}{d\varphi}\varphi'^2, \tag{5.81}$$

where $(') = \frac{d}{d\tau}$, $('') = \frac{d^2}{d\tau^2}$.

The Eqs. (5.80) and (5.81) are nonlinear. Besides, as the driving torque is a function of the crank rotation, the mechanical system examined with the equations is put into the class of non-ideal dynamic systems (Kononenko 1969; Nayfeh and Mook 1979; Balthazar et al. 2002, 2003; Zukovic and Cveticanin 2007, 2009), with consequences that will be encountered and discussed in the following sections. Before solving the complete set of equations of motion (5.80) and (5.81) of the mechanism models, some typical results are first presented for the special case where the angular velocity $\Omega$ of the crank-shaft $O_1A$ is constant. In Sect. 5.3, the ideal forcing conditions are discussed. In Sect. 5.4. the more realistic model of the system examined is based on expressing both the external torque and cutting force as a function of the motion of the mechanism.

### 5.4.2 *Ideal Forcing Conditions*

For the special case when angular velocity $\Omega$ of the crank-shaft is constant, that is when $\varphi'' = 0$, $\varphi' = \Omega$, $\varphi = \Omega\tau$, the Eqs. (5.80) and (5.81) simplify to

$$s" + (Q_1 + Q_2)s' + s = -\mu\Omega^2 \left(\frac{df}{d\varphi}\right)_{\varphi=\Omega t} - Q_2 f \Omega, \tag{5.82}$$

$$\lambda_F \bar{F}(\Omega\tau) = \kappa s" + \frac{\kappa Q_2}{\mu} s' - \frac{\lambda_M \bar{M}(\Omega)}{f} + \frac{\kappa \Omega f Q_2}{\mu} + \kappa\Omega^2 \left(\frac{df}{d\varphi}\right)_{\varphi=\Omega t}. \tag{5.83}$$

Equation (5.82) can be solved independently to yield the support lateral displacement $s$. The equation describes the forced vibrations of the support. For the linear system the exact analytical solution of (5.82) has the form

$$s = s_h + s_p = C\exp(-\delta\tau)\cos\psi + f^*(\Omega\tau), \tag{5.84}$$

where $\delta = \frac{Q_1+Q_2}{2}$, $\psi = \tau\sqrt{1-\delta^2} + \theta$, and $C$ and $\theta$ are arbitrary constants which satisfy the initial conditions $s(0) = s_0$, $\theta(0) = \theta_0$.

Besides, $f^*(\Omega\tau) = s_p$ is the known particular solution of (5.82). Substituting the obtained solution (5.84) into (5.83) the external forcing necessary for creating and sustaining this type of motion is obtained

$$\lambda_F \bar{F}(\Omega\tau) = \kappa C\exp(-\delta\tau)\left(\left(\delta^2 - \sqrt{1-\delta^2} - \frac{Q_2\delta}{\mu}\right)\cos\psi + \left(2\delta - \frac{Q_2}{\mu}\right)\sqrt{1-\delta^2}\sin\psi\right) + \frac{\kappa f Q_2 \Omega}{\mu} - \frac{\lambda_M \bar{M}(\Omega)}{f} + \kappa\left(\Omega^2 + \frac{Q_2}{\mu}\right)\left(\frac{df}{d\varphi}\right)_{\varphi=\Omega\tau} + \kappa\left(\frac{d^2 f}{d\varphi^2}\right)_{\varphi=\Omega\tau}. \tag{5.85}$$

For the steady-state motion, when the transient terms disappear, the cutting force is

$$\lambda_F \bar{F}(\Omega\tau) = \frac{\kappa f Q_2 \Omega}{\mu} - \frac{\lambda_M \bar{M}(\Omega)}{f} + \kappa\left(\Omega^2 + \frac{Q_2}{\mu}\right)\left(\frac{df}{d\varphi}\right)_{\varphi=\Omega\tau} + \kappa\left(\frac{d^2 f}{d\varphi^2}\right)_{\varphi=\Omega\tau}, \tag{5.86}$$

which depends not only on the angular velocity $\Omega$ but also on the input torque and damping properties of the system. For the case when the support of the machine is rigid, Eq. (5.86) transforms into

$$\lambda_F \bar{F}(\Omega\tau) = \frac{\kappa f Q_2 \Omega}{\mu} - \frac{\lambda_M \bar{M}(\Omega)}{f} + \kappa\Omega^2 \left(\frac{df}{d\varphi}\right)_{\varphi=\Omega\tau}. \tag{5.87}$$

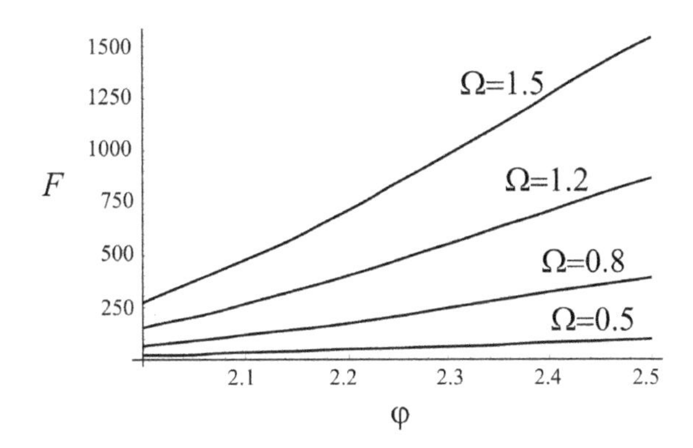


**Fig. 5.10** Cutting force – angle position curves for various values of angular velocity

The dimensional cutting force is

$$F = m_2 a\Omega^2 \left(\frac{df}{d\varphi}\right)_{\varphi=\Omega\tau} + q_2 af\Omega - \frac{M}{af}. \tag{5.88}$$

In Fig. 5.10 the cutting force as a function of the angle position of the driving element for various values of the velocity $\Omega$ is plotted.

It can be seen that for the case of constant angular velocity of the driving element the cutting force is not constant, as it is required: the cutting force increases with increase of the angle of the driving element. This phenomenon is more expressed in the systems with higher values of the angular velocity of the driving element: the higher the angular velocity, the increase of the force is higher.

To satisfy the requirement for the constant cutting force, the driving torque of the motor has to be varied during the cutting process. Then the cutting force affects the motor properties i.e., the motor angular velocity and torque. Such a mechanical system is of non-ideal type.

### *5.4.3 Non-ideal Forcing Conditions*

If the non-linearity is small, the system of Eqs. (5.80) and (5.81) can be rewritten in the form

$$s'' + s = \varepsilon f_s(s, s', \varphi, \varphi', \varphi''), \tag{5.89}$$

$$\left(1 + \kappa f^2(\varphi)\right) \varphi'' = \varepsilon f_\varphi(s', s'', \varphi, \varphi'), \tag{5.90}$$

where

$$\varepsilon f_s(s, s', \varphi, \varphi', \varphi'') = -(Q_1 + Q_2)s' - \mu \frac{df}{d\varphi}\varphi'^2 - \mu f \varphi'' - Q_2 f \varphi', \tag{5.91}$$

$$\varepsilon f_{\varphi}(s', s", \varphi, \varphi') = \lambda_M \bar{M}(\varphi') + \lambda_F f \bar{F}(\varphi) - \kappa f s" \\ - \frac{\kappa Q_2}{\mu} f(s' + f\varphi') - \kappa f \frac{df}{d\varphi} \varphi'^2. \tag{5.92}$$

We seek approximate solutions which are uniformly valid for small $\varepsilon$. Thus we are considering the case in which $m_2$ is small compared with $m_1$. If the damping force is small in comparison with the restoring force of the support, the damping term appears to be of the same order as the nonlinearity and the excitation.

Let us concern with a relative narrow band of frequencies which encloses the natural frequency of the system (unity in the dimensionless variables). Accordingly we let

$$\varphi' = \Omega \approx 1 + \varepsilon\Delta, \tag{5.93}$$

where $\varepsilon\Delta$ is used to distinguish between the speed of the driving element and the unit natural frequency of the rectilinear motion. Frequency of the rectilinear motion is expected to differ for the value $\xi'$ on the angular speed of the driving element $\varphi'$. Accordingly, we let

$$s = A\cos(\varphi + \xi). \tag{5.94}$$

First and the second time derivatives of (5.94) are, respectively,

$$s' = -A\sin(\varphi + \xi), \tag{5.95}$$

with

$$A'\cos(\varphi + \xi) - A(\varepsilon\Delta' + \xi') = 0, \tag{5.96}$$

and

$$s" = -A'\sin(\varphi + \xi) - A(1 + \varepsilon\Delta + \xi')\cos(\varphi + \xi). \tag{5.97}$$

Substituting (5.94) and (5.97) into (5.90) and (5.91), and using the relation (5.93), leads to

$$-A'\sin(\varphi+\xi) - A(\varepsilon\Delta+\xi')\cos(\varphi+\xi) = \varepsilon f_s(A\cos(\varphi+\xi), -A\sin(\varphi+\xi), \varphi, \Omega, 0), \tag{5.98}$$

$$\left(1 + \kappa f^2(\varphi)\right)\varepsilon\Delta' \equiv \left(1 + \kappa f^2(\varphi)\right)\Omega' = \varepsilon f_{\varphi}(-A\sin(\varphi+\xi), -A\Omega\cos(\varphi+\xi), \varphi, \Omega). \tag{5.99}$$

Solving (5.98) and (5.96), and using the relation (5.93) we obtain

$$A' = -\varepsilon f_s(A\cos(\varphi + \xi), -A\sin(\varphi + \xi), \varphi, \Omega, 0)\sin(\varphi + \xi), \tag{5.100}$$

$$A\xi' = (1 - \Omega)A - \varepsilon f_s(A\cos(\varphi + \xi), -A\sin(\varphi + \xi), \varphi, \Omega, 0)\cos(\varphi + \xi). \tag{5.101}$$

Adding the relations (5.93) and (5.99) to (5.100) and (5.101), the system of four coupled first order differential equations is obtained, where $A$, $\varphi$, $\Omega$ and $\xi$ are the new functions dependent on the variable $\tau$. The system of equations (5.93), (5.99)–(5.101), represents the transformed version of the differential equations (5.80), (5.81) into the new variables. To solve this system of differential equations is not an easy task. It is at this point when the averaging procedure is introduced. Using the averaging procedure of (5.99)–(5.101) and integrating the equations in the interval from 0 to $2\pi$, it follows

$$A' = -\frac{1}{2\pi}(\pi A(Q_1 + Q_2) - \cos\xi(-\mu\Omega^2 I_1 + Q_2\Omega I_2) - \sin\xi(\mu\Omega^2 I_2 + Q_2\Omega I_1), \tag{5.102}$$

$$A\xi' = (1 - \Omega)A - \frac{1}{2\pi}(-\cos\xi(\mu\Omega^2 I_2 + Q_2\Omega I_1) - \sin\xi(\mu\Omega^2 I_1 - Q_2\Omega I_2), \tag{5.103}$$

$$\begin{aligned}\left(1 + \kappa\frac{I_3}{2\pi}\right)\Omega' = {} & \lambda_M \bar{M}(\Omega) + \lambda_F \frac{I_4}{2\pi} - \frac{\kappa Q_2}{\mu}\Omega f \frac{I_3}{2\pi} \\ & + \frac{1}{2\pi}\cos\xi\left(\kappa A\Omega I_1 + \frac{\kappa Q_2}{\mu} A I_2\right) \\ & + \frac{1}{2\pi}\sin\xi\left(-\kappa A\Omega I_2 + \frac{\kappa Q_2}{\mu} A I_1\right),\end{aligned} \tag{5.104}$$

where

$$\begin{aligned} I_1 &= \int_0^{2\pi} f(\varphi)\cos\varphi d\varphi = -\int_0^{2\pi} f'(\varphi)\sin\varphi d\varphi, \\ I_2 &= \int_0^{2\pi} f(\varphi)\sin\varphi d\varphi = \int_0^{2\pi} f'(\varphi)\cos\varphi d\varphi, \\ I_3 &= \int_0^{2\pi} f^2(\varphi)d\varphi, \quad I_4 = \int_0^{2\pi} f(\varphi)\bar{F}(\varphi)d\varphi. \end{aligned} \tag{5.105}$$

It is of special interest to analyze the steady-state motion.

**Steady-state motion**

For the steady-state motion, when $A' = \xi' = 0$ and $\Omega' = 0$, the relations (5.102)–(5.104) transform into

$$\pi A(Q_1 + Q_2) = \cos\xi(-\mu\Omega^2 I_1 + Q_2\Omega I_2) + \sin\xi(\mu\Omega^2 I_2 + Q_2\Omega I_1), \quad (5.106)$$

$$-2\pi(1-\Omega)A = \cos\xi(\mu\Omega^2 I_2 + Q_2\Omega I_1) + \sin\xi(\mu\Omega^2 I_1 - Q_2\Omega I_2), \quad (5.107)$$

$$\lambda_M \bar{M}(\Omega) + \lambda_F \frac{I_4}{2\pi} - \frac{\kappa Q_2}{\mu}\Omega f \frac{I_3}{2\pi})$$
$$= -\frac{1}{2\pi}\cos\xi\left(\kappa A\Omega I_1 + \frac{\kappa Q_2}{\mu} A I_2\right) - \frac{1}{2\pi}\sin\xi\left(-\kappa A\Omega I_2 + \frac{\kappa Q_2}{\mu} A I_1\right). \quad (5.108)$$

Using (5.106) and (5.107) we eliminate the variable $\xi$

$$\tan\xi = \frac{I_1(4Q_1Q_2 + 4Q_2^2 - 8\mu\Omega(1-\Omega) + I_2(Q_2(8 + 4\Omega(\mu-2)) + 4Q_1\mu\Omega)}{I_2(4Q_1Q_2 + 4Q_2^2 - 8\mu\Omega(1-\Omega) - I_1(Q_2(8 + 4\Omega(\mu-2)) + 4Q_1\mu\Omega)},$$

and the amplitude-frequency relation

$$A = \frac{\mu\Omega^2\left(\frac{\sqrt{I_1^2+I_2^2}}{\pi}\sqrt{1 + \left(\frac{Q_2}{\mu\Omega}\right)^2}\right)}{\sqrt{(Q_1+Q_2)^2 + 4(1-\Omega)^2}},$$

i.e., the amplitude-frequency curve is obtained

$$f_{A\Omega 1}(A,\Omega) = A - \frac{\mu\Omega^2\left(\frac{\sqrt{I_1^2+I_2^2}}{\pi}\sqrt{1 + \left(\frac{Q_2}{\mu\Omega}\right)^2}\right)}{\sqrt{(Q_1+Q_2)^2 + 4(1-\Omega)^2}} = 0. \quad (5.109)$$

Eliminating $\xi$ in (5.108) the characteristic curve $f_{A\Omega 2}(A,\Omega) = 0$ is obtained. Along this curve the control or regulator parameter $\nu_0$ is constant. The steady-state position $(A,\Omega)$ is obtained as the intersection of the curves $f_{A\Omega 1}(A,\Omega) = 0$ and $f_{A\Omega 2}(A,\Omega) = 0$. Dependently on the value of the parameter $\nu_0$, one, two ore three real solutions exist.

In Figs. 5.11 and 5.12, the amplitude-frequency (thick line) and the characteristic curves (thin lines) for various values of the control parameter and cutting properties of the mechanism without and with cutting force, respectively, are plotted. The parameters of the system are: $a = 0.08$, $b = 0.32$, $c = 0.14$, $r = 0.20$, $g = 0.24$, $h = 0.18$, $l = 0.20$, $p = 0.12$, $w = 0.16$, $\mu = 0.15$, $Q_1 = 0.1$, $Q_2 = 0.01$, $\lambda_F = 0.8$, $\lambda_M = 0.5$, $\kappa = 0.1$. The cutting interval is $\varphi_K = 2.063785512$ to $\varphi_M = 2.55590711$. As the special case the characteristic curves which are tangents to the amplitude-frequency curve are plotted (see Figs. 5.11b and 5.12b). It can be concluded that for the fixed driving properties of the motor, frequency of vibration is smaller if the cutting is included. Besides, for the values where the characteristic

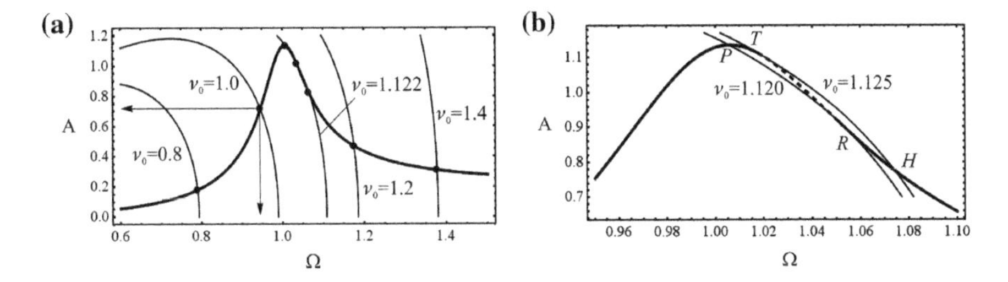


**Fig. 5.11** Amplitude-frequency curve for the case without cutting with: **a** characteristic curves for various values of $\nu_0$ and **b** tangents of curves

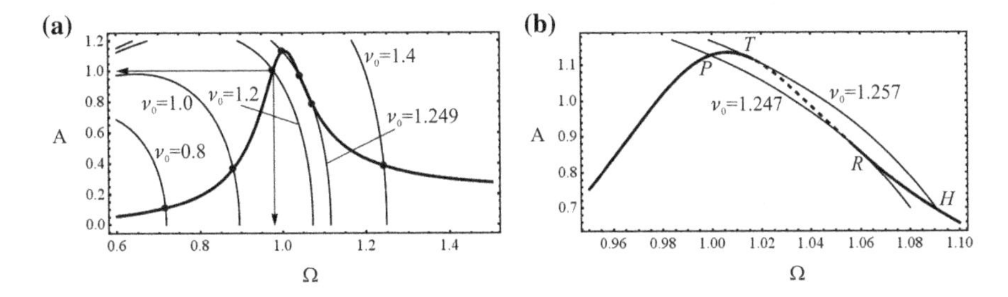


**Fig. 5.12** Amplitude-frequency curve for the case of cutting with: **a** characteristic curves for various values of $\nu_0$ and **b** tangents of curves

curve touches the amplitude-frequency curve the stable motion (solid line) changes into unstable (dashed line). In this paper the values for which the stability change occurs are obtained numerically and the unstable region is plotted with the dashed-line.

The characteristic curve which is the tangent of the amplitude-frequency diagram gives us the points $P$ and $R$ and the other the points $T$ and $H$ during slow increasing and decreasing of the control parameter. The system cannot be made to respond at a frequency between $\Omega_T$ and $\Omega_H$ by simple increasing the control setting from a low value. At $T$ the characteristic of the motion suddenly changes. An increase in the input power causes the amplitude to decrease considerably and the frequency to increase. The phenomenon is called the Sommerfeld effect. The same phenomena is registered if the control setting is continually decreased, and there is no response for the interval $\Omega_R$ and $\Omega_p$. In other words the right side of the resonance spike between $T$ and $R$ cannot be reached by any continual change of control system. The non-ideal source causes a jump phenomenon to occur.To prove the correctness of the analytically obtained solutions the numerical integration of the differential equations (5.99)–(5.101) is done. For the previously parameter values, the steady-state curves for the case of cutting and without cutting are plotted in Fig. 5.13. Control parameter $\nu_0$ is varied.

The solid lines show the response predicting by the theory and the black and grey circles the results obtained numerically for the case of quasi-static close-down

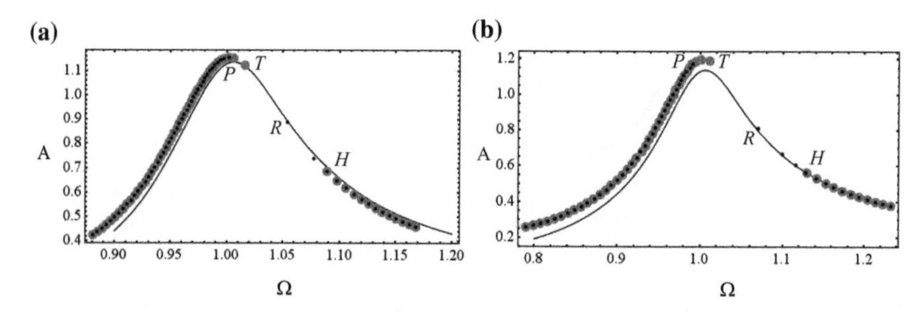


**Fig. 5.13** Amplitude-frequency digrams obtained analytically (*black line*) and also numerically for run-up (*grey big circles*) and for close-down (*black small circles*): **a** cutting force is zero, **b** constant cutting force exists

and run-up, respectively. Figure 5.13 shows that for the both cases (without and with cutting force) there are gaps where no steady-state response exists. The gaps are not the same for increasing or decreasing of $\Omega$ but there is some overlap. The points labeled $P$, $R$, $T$ and $H$ correspond to those in Figs. 5.11 and 5.12.

### 5.4.4 Non-stationary Motion

If the non-stationary motion has to be analyzed the solution of the equations (5.99)–(5.101) have to be determined for the initial conditions $s(0) = \dot{s}(0) = 0$, $\varphi(0) = \dot{\varphi}(0) = 0$.

In Fig. 5.14 the amplitude of vibration of the support and of the driving angular velocity for the case with and without cutting are plotted. Namely, in Fig. 5.14a the amplitude-frequency curve and the characteristic curves for the both mentioned cases and the constant control parameter $\nu_0$ are plotted. In Fig. 5.14b, c the time history diagrams for the support vibration and for the driving speed are shown.

It can be seen that the interaction between the control parameter $\nu_0$ and cutting process is significant. For the same value of control parameter the amplitude of vibration of the support increases or decreases if the cutting is on. Namely, for extremely small value of the control parameter the amplitude of vibration is smaller during cutting than for the case without cutting. For extremely high values of the control parameter the situation is opposite in comparison to the previous: the amplitude of vibration is higher for cutting than without cutting. The most significant difference in the amplitude of vibration is for the case when the characteristic curve is the tangent of the amplitude-frequency one. Independently on the value of the control parameter, the angular velocity of the mechanism decreases during cutting.

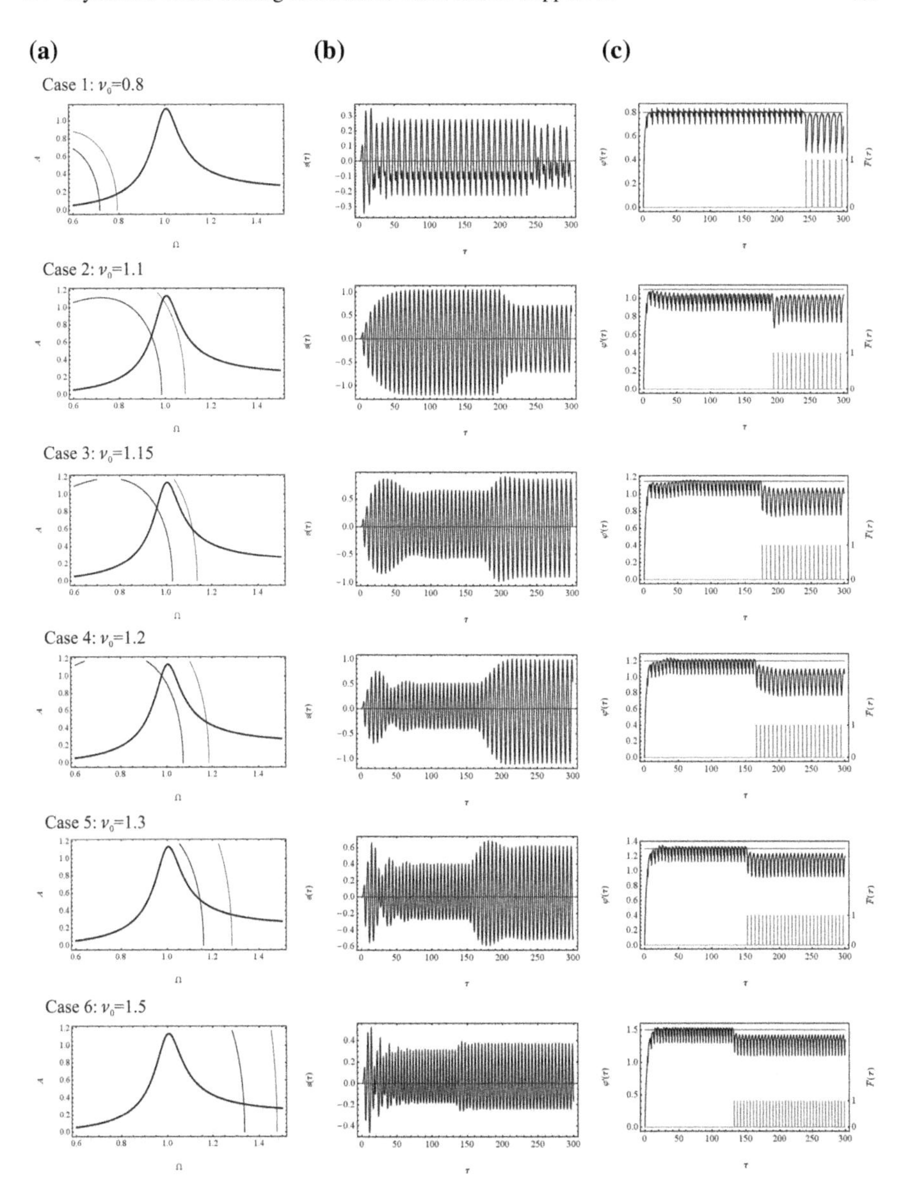


**Fig. 5.14** **a** Amplitude-frequency (*solid black line*) and characteristic curves for the case without cutting (*thin line*) and with cutting (*thick line*), **b** $s - \tau$ diagram, **c** $\varphi' - \tau$ (*black line*) and $\bar{F} - \tau$ (*grey line*) diagrams

## 5.5 Conclusion

The following can be concluded:

1. Introducing the constant cutting force into consideration, the input torque variation has to be calculated. Cutting mechanism-support system can be considered as a non-ideal mechanical system where the cutting process has an influence on the input torque. Mathematical model of such a system is given with two coupled differential equations whose solution gives us the variation of the driving torque and also of the driving angular velocity for the constant cutting force.
2. Amplitude of vibration of the system and also the angular velocity of the driving element strongly depends on the synchronous angular velocity of the driving motor. For small angular velocity of the motor, the amplitude of vibration of the support is smaller during cutting than for the case of the free motion of the mechanism. It is evident for the pre-resonant case. In contrary, for high angular velocity of the motor, i.e., for the post resonance case, the amplitude of vibration of the support is larger during cutting than for the free motion of the mechanism. The higher the angular velocity of the motor in the post-resonance region, the motion of the support becomes independent on the working conditions i.e., cutting force. The conclusion is that the cutting has to be done in the regime far from the resonant one.
3. In the near-resonant regime the following cases may appear: a) the amplitude of vibration of the support increases during the cutting, b) decreases during the cutting or c) remains unchangeable. The transient from one to another case is based on the small change of the angular velocity. It is suggested to avoid cutting in the near-resonant regime.
4. Angular velocity of the driving element depends on the cutting force. For all values of the cutting force the angular velocity decreases in comparison to the case without cutting.
5. In the cutting mechanism-support system the Sommerfeld effect appears. For certain values of forcing during run-up or close-down the unstable motion appears where in spite of energy adding only the frequency of vibration varies, but the amplitude of vibration is approximately constant. Region of unstable motion depends on the driving motor properties and on the cutting force, too. Elimination of the unstable motion is realized by increasing of the motor power and decreasing of the cutting force.
6. The analytically obtained solutions are in a good agreement with 'exact' numerical one, not only qualitatively, but also quantitatively. It means that the assumed mathematical model is convenient to be applied for such problems.

# References

Amemiya, T., & Maeda, T. (2009). Directional force sensation by asymmetric oscillation from a double-layer slider-crank mechanism. *Journal of Computing and Information Science in Engineering*, *9*, 1–8. art. no. 011001.

Amemiya, T., Kawabuchi, I., Ando, H., & Maeda, T. (2007). Double-layer slider-crank mechanism to generate pulling and pushing ground. In *IEEE/RSJ International Conference on Intelligent Robots and Systems*, art. no. 4399211 (pp. 2101–2106).

Artobolevskij, I. I. (1971). *Mechanisms in contemporary technique*. Moscow: Nauka.

Balthazar, J. M., Mook, D. T., Weber, H. I., Brasil, R. M. L. R. F., Fenili, A., Belato, D., et al. (2002). An overview of non-ideal vibrations. *Meccanica*, *330*, 1–9.

Balthazar, J. M., Mook, D. T., Weber, H. I., Brasil, R. M. I. R. F., Fenili, A., Beltano, D., et al. (2003). An overview on non-ideal vibrations. *Meccanica*, *38*, 613–621.

Cveticanin, L., & Maretic, R. (2000). Dynamic analysis of a cutting mechanism. *Mechanism and Machine Theory*, *35*, 1391–1411.

Cveticanin, L., Maretic, R., & Zukovic, M. (2012). Dynamics of polymer sheets cutting mechanism. *Journal of Mechanical Engineering (Strojniski Vestnik)*, *58*(5), 354–361.

Erkaya, S., Su, S., & Uzmay, I. (2007). Dynamic analysis of a slider-crank mechanism with eccentric connector and planetary gears. *Mechanism and Machine Theory*, *42*, 393–408.

Goudas, I., & Natsiavas, S. (2004). Nonlinear dynamics of engine mechanisms with a flexible connection rod. *Proceedings of the Institution of Mechanical Engineers, Part K: Journal of Multi-body Dynamics*, *218*(2), 67–80.

Goudas, I., Stavrakis, I., & Natsiavas, S. (2004). Dynamics of slider-crank mechanisms with flexible supports and non-ideal forcing. *Nonlinear Dynamics*, *35*(3), 205–227.

Ha, J. L., Fung, R. F., Chen, K. Y., & Hsien, S. C. (2006). Dynamic modeling and identification of a slider-crank mechanism. *Journal of Sound and Vibration*, *289*, 1019–1044.

Kazimierski, Z., & Wojewoda, J. (2011). Double internal combustion piston engine. *Applied Energy*, *88*, 1983–1985.

Kim, H. S., Park, J. J., & Song, J. B. (2008). Safe joint mechanism using double slider mechanism and spring force. In *8th IEEE-RAS International Conference on Humanoid Robots 2008, Humanoids 2008*, art. no. 4755934 (pp. 73–78).

Komatsubara, H., Mitome, K.-I., & Sasaki, Y. (2007). A new cutting machine for elliptical cylinder. *JSME (Japan Society of Mechanical Engineers) International Journal, Series C: Mechanical Systems, Machine Elements and Manufacturing*, *73*, 891–896.

Kononenko, V. O. (1969). *Vibrating systems with a limited power supply*. London: Iliffe Books Ltd.

Koser, K. (2004). A slider crank mechanism based robot arm performance and dynamic analysis. *Mechanism and Machine Theory*, *39*, 169–182.

Masia, L., Krebs, H. I., Cappa, P., & Hogen, N. (2007). Design and characterization of hand module for whole-arm rehabilitation following stroke. *IEEE/ASME Transactions on Mechatronics*, *12*, 399–407.

Metallidis, P., & Natsiavas, S. (2003). Linear and nonlinear dynamics of reciprocating engines. *International Journal of Non-Linear Mechanics*, *38*(5), 723–738.

Nayfeh, A. H., & Mook, D. T. (1979). *Nonlinear Oscillations*. New York: Wiley-Interscience.

Ogura, M., & Daidoji, S. (1982). Application and practice of the double slider crank mechanism for the air compressor. *JSME (Japan Society of Mechanical Engineers) International Journal, Series B: Fluids and Thermal Engineering*, *48*, 1483–1491.

Ren, T., He, B., Chen, J., & Jin, X. (2009). Non-sinusoidal waveform and parameters of distance changeable double-slider crank mechanism for mold. *Chinese Journal of Mechanical Engineering*, *45*, 269–273.

Sandier, B. Z. (1999). *Designing the mechanisms for automated machinery*. New York: Academic Press.

Wang, M., Song, Q., & Zhong, K. (2012). The double role piston pump based on the symmetrical tears and crank-link-slider mechanism driven by servo motor. *Applied Mechanics and Materials, 121–126*, 2308–2312.

Wauer, J., & Buhrle, P. (1997). Dynamics of a flexible slider-crank mechanism driven by a non-ideal source of energy. *Nonlinear Dynamics, 13*, 221–242.

Xu, F., & Wang, X. (2008). Design and experiments on a new wheel-based cable climbing robot. In *IEEE/ASME International Conference on Advanced Intelligent Mechatronics*, art. no. 4601697 (pp. 418–423). AIM.

Xu, F., Wang, X., & Wang, L. (2011). Cable inspection robot for cable-stayed bridges: Design, analysis, and application. *Journal of Field Robotics, 28*, 441–459.

Zukovic, M., & Cveticanin, L. (2007). Chaotic responses in a stable Duffing system of non-ideal type. *Journal of Vibration and Control, 13*(6), 751–767.

Zukovic, M., & Cveticanin, L. (2009). Chaos in non-ideal mechanical system with clearance. *Journal of Vibration and Control, 15*(8), 1229–1246.

Zukovic, M., Cveticanin, L., & Maretic, R. (2012). Dynamics of the cutting mechanism with flexible support and non-ideal forcing. *Mechanism and Machine Theory, 58*, 1–12.

# Chapter 6
# Non-ideal Energy Harvester with Piezoelectric Coupling

Advances in silicon electronics and MEMS technology reduced significantly the power consumption of devices such as wireless sensors, portable and wearable electronics. A large number of the locations where those devices are used are either remote or inaccessible. Most of these low-power devices rely heavily on electromechanical batteries as a source of power. However, batteries have a limited life span and number of recharging cycles. They are also constantly in need for recharging or replacement. For application such as wireless sensing and remote monitoring, battery replacement or recharging can be expensive, challenging or impossible in some cases. Another serious problem with batteries is the fact that they contain hazardous chemical materials that are harmful to the environment if not recycled.

New technologies have triggered the needs to new energy sources, smaller and more efficient, so the research about energy harvesting has increased substantially. (The energy harvesting is the process of converting energy into electric energy.) The low power design trends combined with self-sustainability needs presented an opportunity for researches to find alternative ways to power such devices and eliminate or reduce dependency on batteries. Several energy harvesting approaches have been proposed using solar, thermoelectric, electromagnetic, piezoelectric and capacitive schemes. In the research about the energy harvesting devices, many researchers have concentrated their efforts on finding the best configuration and optimization of its power output. As the kinetic energy is a source of energy easily found in the environment, devices that convert kinetic energy into electrical energy has been widely studied. So, one promising avenue to achieve the goal of harvesting is to exploit ambient vibration energy source. Vibration energy harvesting technology has been making significant strides over the last few years as it aims to provide a continuous and uninterrupted source of power for low-power electronic devices and wireless sensors. While the idea of converting environmental vibration energy into electrical energy has been used before, advances in micro-electronics and low power consumption of silicon-based electronics and wireless sensors have given it an added

© Springer International Publishing AG 2018

L. Cveticanin et al., *Dynamics of Mechanical Systems with Non-Ideal Excitation*, Mathematical Engineering, DOI 10.1007/978-3-319-54169-3_6

significance. Finally, energy harvesting is an alternative to traditional power sources such as batteries.

Nowadays, a special attention has been devoted to devices that use piezoelectric elements as means of energy transduction. The conversion of the wasted mechanical energy to electric energy is done using piezoelectric materials as a transducer. The wasted mechanical energy is conversed into electrical energy using piezoelectric materials.

Several different devices for energy harvesting have been developed. In all these devices, a new way of harvesting energy is the use of piezoelectric material as a transducer To harvest energy from ambient mechanical vibrations many researchers have recently applied the piezoceramics which is used as piezomagnetoelastic structure (Stephen 2006; Erturk et al. 2009; Litak et al. 2012; Bendame et al. 2015). The reuse of the wasted energy is explored that is very important nowadays to some applications, including renewable energy. Usually, it is assumed that the relation between the strain and the electric field in the piezoceramic material is constant and independent on strain. Then, the model of the energy harvesting is linear. Unfortunately, such assumption gives us not a correct prediction of energy harvester performance. Crawley and Anderson (1990) published the experimental results for the piezoelectric constant on induced strain in the material. They claimed that the constant exhibits a significant dependence on the induced strain. Such strain dependence introduces nonlinear behavior in the energy harvesting. Using the constitutive laws of piezoelectric materials, the role of nonlinearity in the electromechanical coupling in the design of energy harvesting system is taken into account (Triplett and Quinn 2009). The nonlinear coupling incorporates the more realistic effect of the piezoelements, because of the constitutive laws of piezoelectric materials specifically the nonlinear relationship between the strain and the electric field in the piezoceramic material. In the papers of Tereshko et al. (2004); Iliuk et al. (2013c, 2014a, b); Daqaq et al. (2014); Rocha et al. (2016) and Litak et al. (2016) energy harvesting using nonlinear piezoelectric material is considered. The energy harvesting including nonlinearity in the piezoelectric coupling and a non-ideal forces of excitation are considered by Cveticanin et al. (2017). It is investigated how the power harvested was influenced by the nonlinear vibrations of the structure as well as by the influence of the nonlinearities in the piezoelectric coupling.

The chapter is divided into 5 sections. In Sect. 6.1 the constitutive equation for the piezoceramic material is given. In Sect. 6.2 the ideally excited harvesting system is considered. Parameters of the constitutive equation of the piezomaterial are obtained numerically. For these parameter values the averaged harvested energy is calculated. Analytical approximative solution to equations which describe the harvester system are obtained. Two special cases are analyzed: one, when the piezomaterial is with linear properties and the other, when the properties are nonlinear. It has to be mentioned that the considered oscillator is of the Duffing type. In Sect. 6.3, harvester with non-ideal excitation is considered. First the model of the system is formed. An analytical procedure for solving equations is developed and the steady state solution is determined. One of the most important value in harvesting is the power which calculated. Based on the power distribution the averaging of the value is done. Finally, the analytically obtained results are compared with those obtained numerically. Two types of piezoelectric material for harvesting are applied: the linear and the nonlinear one. The influence of nonlinearity is discussed. In Sect. 6.4 harvesting in the non-ideal system where the torque function is exponential is considered. Numerical simulation is introduced. Both, the linear and nonlinear energy harvesting is analyzed. For certain values of parameters of the system, chaos appears. Various methods for chaos control are introduced. The passive control in the harvesting system is done with the pendulum. In Sect. 6.5 dynamics of the system with control part is shown. The calculation is numerical.

## 6.1 Constitutive Equation of the Piezoceramic Material

Let us consider the energy harvesting system which consists of a mass which is attached to an excitation which is of ideal or non-ideal type and has a piezoelectric coupling. On the mass a force, arising from the mechanical elasticity and damping, as well as the electro-mechanical force from the piezoelectric coupling, acts. If the displacement of the mass is represented by $z(t)$ the piezoelectric coupling to the mechanical component is given as (see Triplett and Quinn 2009)

$$\frac{\Theta(z)}{C}Q, \tag{6.1}$$

where $\Theta(z)$ is a strain dependent coupling coefficient $C$ is the piezoelectric capacitance and $Q$ is the electric charge developed in the coupled circuit. The voltage $V$ across the piezoelectric material is described by an electromechanical constitutive relation

$$V = -\frac{\Theta(z)}{C}z + \frac{Q}{C}. \tag{6.2}$$

For $V = -R\dot{Q}$ the Eq. (6.2) transforms into

$$R\dot{Q} = \frac{\Theta(z)}{C}z - \frac{Q}{C}, \tag{6.3}$$

where $R$ is electrical resistance. Based on the investigation given by Crawley and Anderson (1990) it is concluded that the piezoelectric coupling coefficient is nonlinear and is modeled as

$$\Theta(z) = \Theta_{lin}(1 + \Theta_{nl}|z|), \tag{6.4}$$

where $\Theta_{lin}$ is a constant, while $\Theta_{nl}|z|$ depends on $z$. In the following text the ideal and non-ideal harvesting systems will be considered.

## 6.2 Harvesting System with Ideal Excitation

In Fig. 6.1 the model of the vibration-based energy harvesting system is shown Triplett and Quinn (2009).

Mass $m$ is connected to the excited supporting base. The elastic force in the connection has a linear and a nonlinear part and is described with a strong nonlinear displacement function

$$F_e = kz(1 + a\,|z|^{s-1}), \tag{6.5}$$

where $s \geq 1$ is the order of nonlinearity (a real integer of noninteger), $k$ is the coefficient of the linear stiffness and $a$ of the nonlinear component. Damping coefficient is $c$.

Using the momentum balance to the mass and the elastic (6.5), damping and electromechanical forces (6.1), assuming that the base excitation is harmonic, i.e., the excitation force is a periodical trigonometric time function, we have

$$m\ddot{z} + kz(1 + a\,|z|^{s-1}) + b\dot{z} = F_0 \sin \Omega t + \frac{\theta(z)}{C} Q, \tag{6.6}$$

where $F_0$ and $\Omega$ are the amplitude and the frequency of the excitation force. Introducing the dimensionless coordinates and parameters

$$x = \frac{z}{c_x}, \quad \gamma = \frac{c_x F_0 C}{c_q^2}, \quad \tau = t\sqrt{\frac{k}{m}}, \quad 2\varepsilon\zeta = \frac{b}{\sqrt{km}}, \quad \rho = RC\sqrt{\frac{k}{m}},$$
$$\varepsilon\alpha = a c_x^2, \quad \varepsilon = \left(\frac{c_x}{c_q}\right)^{s-1} \frac{1}{Ck}, \quad q = \frac{Q}{c_q}, \quad \omega = \Omega\sqrt{\frac{m}{k}}, \tag{6.7}$$

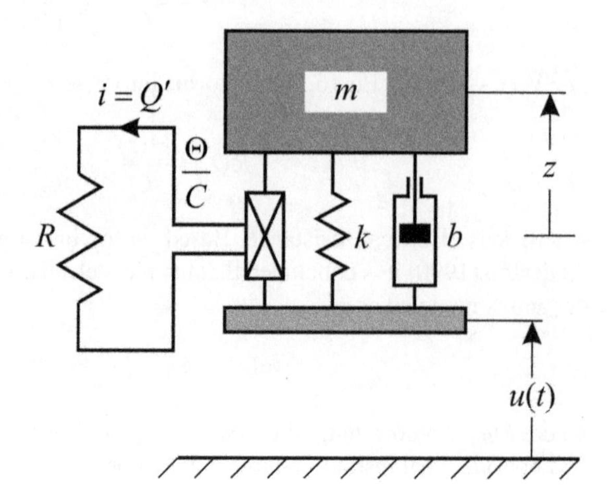


**Fig. 6.1** Model of the vibration-based energy harvesting system

the nondimensional equation of motion (6.6) is

$$\ddot{x} + 2\varepsilon\zeta_1\dot{x} + x(1 + \varepsilon\alpha\,|x|^{s-1}) - \varepsilon\hat{\Theta}(x)q = \varepsilon\gamma\sin(\omega\tau), \tag{6.8}$$

where $(\dot{\ })=\frac{d}{d\tau}$ and $\hat{\Theta}(x) = \left(\frac{c_x}{c_q}\right)\Theta(c_x x)$ is the non-dimensional piezoelectric coupling coefficient. For (6.4) we have

$$\hat{\Theta}(x) = \theta(1 + \beta\,|x|), \tag{6.9}$$

where the coupling coefficients are

$$\theta = \frac{c_x}{c_q}\Theta_{lin}, \quad \beta = c_x\Theta_{nl}. \tag{6.10}$$

Substituting (6.7) and (6.9) into (6.3) and (6.8), and assuming that the nonlinearity is cubic, the nondimensional equations of motion follow as

$$\ddot{x} + 2\varepsilon\zeta_1\dot{x} + x(1 + \varepsilon\alpha\,|x|^2) - \varepsilon\theta(1 + \beta\,|x|)q = \varepsilon\gamma\sin(\omega\tau), \tag{6.11}$$

$$\rho\dot{q} - \theta(1 + \beta\,|x|)x + q = 0. \tag{6.12}$$

Analyzing (6.11) and (6.12) it can be seen that the behavior of the system depends on a set of parameters $\varepsilon$, $\gamma$, $\alpha$, $\theta$, $\psi$ and $\beta$. In the relation (6.9) due to nondimensionalization two unspecified parameters $c_x$ and $c_q$ are introduced. Triplett and Quinn (2009) suggested to choose these values and to eliminate the direct influence of two non-dimensional parameters from the set ($\varepsilon$, $\gamma$, $\alpha$, $\theta$, $\psi$). For example, $\varepsilon$ and $\gamma$ are fixed and the influence of $\alpha$, $\theta$, $\beta$ are varied. Then, the influence of these parameters on the response of the system is obtained. Based on $\varepsilon$ and $\gamma$ the appropriate values of $c_x$ and $c_q$ are calculated. Using this conception, du Toit and Wardle (2007) calculated the dimensional and nondimensional parameters of system for $a = 0$ and $\Theta(x) = \Theta_{lin} = const$ (see Table 6.1).

**Table 6.1** System parameters for $c_q/c_x = 4.57 \times 10^{-3}$ N/V

| Parameter | Value |
|---|---|
| Equivalent stiffness, $k$ | $4.59 \times 10^3$ N/m |
| Mechanical damping, $c$ | 0.218 Ns/m |
| Equivalent mass, $m$ | $9.12 \times 10^{-3}$ kg |
| Electrical resistance, $R$ | $105 \times 10^3$ Ω |
| Piezoelectric capacitance, $C$ | $8.60 \times 10^{-8}$ F |
| Linear piezoelectric coefficient, $\Theta_{lin}$ | $4.57 \times 10^{-3}$ N/V |
| $\varepsilon$ | 0.0528 |
| $\xi$ | 0.3180 |
| $\rho$ | 6.41 |
| $\theta$ | 1 |

For these parameter values the energy harvesting in the system is calculated

The electrical power harvested from the mechanical component is $V = R\dot{Q}$. Rewriting the relation in the nondimensional form, it is

$$P = \rho \dot{q}^2. \tag{6.13}$$

It is of special interest to calculate the averaged value

$$P_{avg} = \frac{1}{T} \int_0^T P(t) dt. \tag{6.14}$$

Using the result of numerical integration of (6.11) and (6.12) for zero initial conditions and the parameter values

$$\begin{aligned} \gamma = 2.00, \quad \varepsilon = 0.10, \quad \zeta = 0.25, \quad \alpha = 0.25, \\ \rho = 2.00, \quad \theta = 1.00, \quad \beta = 1.00, \quad \omega = 1.00, \end{aligned} \tag{6.15}$$

we plot the $x - t$ curve in Fig. 6.2a and the $P - t$ curve in Fig. 6.2b.

In Fig. 6.3 the average power during a forcing cycle generated by the energy harvesting system as a function of the excitation frequency is shown.

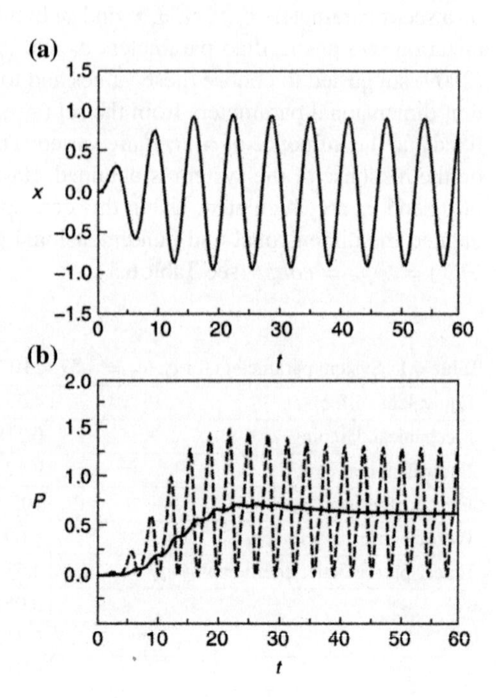


**Fig. 6.2** **a** $x - t$ diagram and **b** $P - t$ diagram with average power harvested (*solid line*) (Triplett and Quinn 2009)

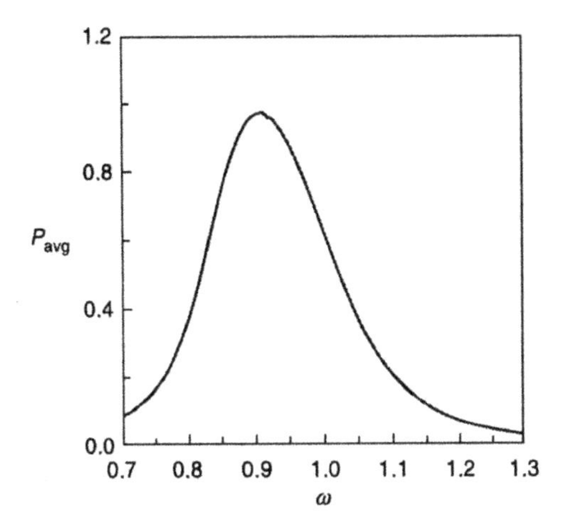


**Fig. 6.3** Average power harvested versus frequency (Triplett and Quinn 2009)

## 6.2.1 Analytical Procedure

For further analysis it is of interest to find the solution of the coupled equations (6.11) and (6.12) of the harvesting system. The approximate solving procedure based on the Lindstedt–Poincare perturbation method is developed (Triplett and Quinn 2009). For the new time variable

$$\tau^* = \omega\tau + \phi, \tag{6.16}$$

the corresponding derivatives are

$$\begin{aligned} \frac{d}{d\tau} &= \frac{d}{d\tau^*}\frac{d\tau^*}{d\tau} = \omega\frac{d}{d\tau^*}, \\ \frac{d^2}{d\tau^2} &= \frac{d}{d\tau}(\omega\frac{d}{d\tau^*}) = \omega^2\frac{d^2}{d\tau^{*2}}, \end{aligned} \tag{6.17}$$

where $\phi$ is a phase angle. Substituting (6.16) and (6.17) into (6.11) and (6.12) we have

$$\begin{aligned} \omega^2\frac{d^2x}{d\tau^{*2}} + x = \varepsilon\gamma\sin(\tau^* - \phi) \\ -2\varepsilon\zeta\omega\frac{dx}{d\tau^*} - \varepsilon\alpha x^3 + \varepsilon\theta(1 + \beta\,|x|), \end{aligned} \tag{6.18}$$

$$\rho\omega\frac{dq}{d\tau^*} - \theta(1 + \beta\,|x|)x + q = 0. \tag{6.19}$$

Let us assume the series expansion for $x$ and $q$ in the form

$$x = x_0 + \varepsilon x_1 + ..., \quad q = q_0 + \varepsilon q_1 + ... \tag{6.20}$$

and the near resonant excitation

$$\omega = 1 + \varepsilon\sigma. \tag{6.21}$$

Substituting (6.20) and (6.21) into (6.18) and separating the terms with the same order of the small parameter it is

$$\varepsilon^0: \quad \frac{d^2x_0}{d\tau^{*2}} + x_0 = 0, \tag{6.22}$$

$$\varepsilon^1: \quad \frac{d^2x_1}{d\tau^{*2}} + x_1 = \gamma\sin(\tau^* - \phi) - 2\sigma\frac{d^2x_0}{d\tau^{*2}} - 2\zeta_1\frac{dx_0}{d\tau^*} - \alpha x_0^3 + \theta(1 + \beta|x_0|)q_0, \tag{6.23}$$

$$\dots$$

while for (6.23), we have

$$\varepsilon^0: \quad \rho\frac{dq_0}{d\tau^*} - \theta(1 + \beta|x_0|)x_0 + q_0 = 0, \tag{6.24}$$

$$\dots$$

Solution of (6.22) is

$$x_0 = X_0\sin\tau^*, \tag{6.25}$$

in which the amplitude $X_0$ is unknown. Substituting (6.25) into (6.24), we have

$$F(\tau) = 0, \tag{6.26}$$

where

$$\rho\frac{dq_0}{d\tau^*} + q_0 = \theta(1 + \beta\left|X_0\sin\tau^*\right|)X_0\sin\tau^* \equiv F(\tau^*). \tag{6.27}$$

Equation (6.26) with (6.27) can be solved via convolution integral as

$$q_0(\tau^*) = q_0(0)\exp(-\tau^*/\rho) + \frac{\exp(-\tau^*/\rho)}{\rho}\int_0^T F(t)\exp(t/\rho)dt. \tag{6.28}$$

Requirement that the solution has to be periodic with $2\pi$ gives

$$q_0(\tau^*) = \frac{\exp(-\tau^*/\rho)}{\rho}\left(\int_0^{\tau^*} F(t)\exp(t/\rho)dt + \frac{1}{\exp(T/\rho) - 1}\int_0^T F(t)\exp(t/\rho)dt\right). \tag{6.29}$$

Substituting (6.29) into (6.23) we obtain

$$\begin{aligned}\frac{d^2x_1}{d\tau^{*2}} + x_1 = \gamma \sin(\tau^* - \phi) + 2\sigma X_0 \sin \tau^* \\ - 2\zeta X_0 \cos \tau^* - \frac{1}{4}\alpha X_0^3 \left(3 \sin \tau^* - \sin(3\tau^*)\right) \\ + \theta(1 + \beta |x_0|)q_0. \end{aligned} \tag{6.30}$$

Based on (6.25) and (6.29) the electromechanical coupling $\theta(1+\beta|x_0|)q_0$ represented in a Fourier expansion is

$$\theta(1 + \beta |x_0|)q_0 = \frac{\theta^2 X_0}{\pi}\left(\frac{a_0}{2} + \sum \left(a_k \cos(k\tau^*) + b_k \sin(k\tau^*)\right)\right), \tag{6.31}$$

with

$$\frac{\theta^2 X_0}{\pi} a_i = \frac{1}{\pi}\int_0^{2\pi} \theta(1 + \beta |x_0|)q_0 \cos(kt)dt, \tag{6.32}$$

$$\frac{\theta^2 X_0}{\pi} b_i = \frac{1}{\pi}\int_0^{2\pi} \theta(1 + \beta |x_0|)q_0 \sin(kt)dt. \tag{6.33}$$

For the special case when the piezoelectric nonlinearity is $x_0|x_0|$, using (6.31) the Eq. (6.30) is

$$\begin{aligned}\frac{d^2x_1}{d\tau^{*2}} + x_1 = \gamma \sin(\tau^* - \phi) + 2\sigma X_0 \sin \tau^* \\ - 2\zeta X_0 \cos \tau^* - \frac{1}{4}\alpha X_0^3 \left(3 \sin \tau^* - \sin(3\tau^*)\right) \\ + \frac{\theta^2 X_0}{\pi}\left(\frac{a_0}{2} + a_1 \cos \tau^* + b_1 \sin \tau^*\right), \end{aligned} \tag{6.34}$$

where $a_0$, $a_1$ and $b_1$ are constants derived from the Fourier expansion. Separating the secular terms from (6.34), we obtain

$$\begin{aligned}\gamma \sin \phi = \left(\frac{\theta^2 a_1(X_0)}{\pi} - 2\zeta\right) X_0, \\ \gamma \cos \phi = \frac{3}{4}\alpha X_0^3 - \left(\frac{\theta^2 b_1(X_0)}{\pi} + 2\sigma\right) X_0, \end{aligned} \tag{6.35}$$

where

$$a_1(X_0) = \frac{-X_0^2\beta^2}{2}\left[\frac{\pi\rho}{1+4\rho^2} + \frac{1-\exp(-\pi/\rho)}{\left(1+\frac{1}{4\rho^2}\right)^2(1+\exp(-\pi/\rho))}\right] - \frac{4X_0\beta\rho}{1+\rho^2} - \frac{\pi\rho}{1+\rho^2}, \tag{6.36}$$

$$b_1(X_0) = -X_0^2\beta^2\left[\frac{\pi(1+8\rho^2)}{4(1+4\rho^2)} - \frac{\rho(1-\exp(-\pi/\rho))}{\left(1+\frac{1}{4\rho^2}\right)^2(1+\exp(-\pi/\rho))}\right] + \frac{16X_0\beta}{3(1+\rho^2)} + \frac{\pi}{1+\rho^2}. \tag{6.37}$$

Based on (6.35) the amplitude and phase for the stationary state are, respectively

$$\gamma^2 = \left(\frac{\theta^2 a_1(X_0)}{\pi} - 2\zeta_1\right)^2 X_0^2 + \left[\frac{3}{4}\alpha X_0^3 - \left(\frac{\theta^2 b_1(X_0)}{\pi} + 2\sigma\right)X_0\right]^2, \tag{6.38}$$

and

$$\tan\phi = \frac{\theta^2 a_1(X_0) - 2\zeta_1\pi}{0.75\alpha X_0^2\pi - \theta^2 b_1(X_0) - 2\sigma\pi}. \tag{6.39}$$

Solving Eq. (6.38), we obtain $X_0$. Substituting into (6.29) the charge $q_0(\tau^*)$ follows.

Further, two types of energy harvester will be considered: one, with piezoelement with linear properties and the other, where the properties of the piezoelement are nonlinear.

### 6.2.2 *Harvester with Linear Piezoelectricity*

If the piezoelectric coupling is linear and $\beta = 0$, the relation for amplitude of the response (6.38) reduces to

$$\gamma^2 = \left(\frac{\theta^2\rho}{1+\rho^2} + 2\zeta_1\right)^2 X_0^2 + \left[\frac{3}{4}\alpha X_0^2 - \left(\frac{\theta^2}{1+\rho^2} + 2\sigma\right)\right]^2 X_0^2. \tag{6.40}$$

The relation (6.40) has the same form as the amplitude-frequency relation for the forced Duffing oscillator.

The averaged harvested power (6.14) is for amplitude $Q_0$ of the charge $q_0(\tau^*)$

$$P_{avg} = \frac{\rho}{2} Q_0^2 = \frac{X_0^2}{2} \frac{\theta^2 \rho}{1+\rho^2}. \tag{6.41}$$

For further analysis, let us introduce the parametric variable $s$ as

$$2s = \frac{3}{4} \alpha X_0^2 - \left( \frac{\theta^2}{1+\rho^2} + 2\sigma \right). \tag{6.42}$$

The detuning parameter is then

$$\sigma = \frac{3}{8} \alpha X_0^2 - s - \frac{\theta^2}{2(1+\rho^2)},$$

and the amplitude

$$X_0 = \frac{\gamma}{\sqrt{\left( \frac{\theta^2 \rho}{1+\rho^2} + 2\zeta_1 \right)^2 + (2s)^2}}. \tag{6.43}$$

For $s = 0$, the maximum amplitude is

$$X_{0\,\max} = \frac{\gamma}{2(\zeta_1 + \zeta_{eq})}, \tag{6.44}$$

while the maximum average power is

$$P_{\max} = \frac{1}{8\left(\zeta_1 + \zeta_{eq}\right)^2} \frac{\theta^2 \rho \gamma^2}{1+\rho^2}, \tag{6.45}$$

at

$$\sigma_{\max} = \frac{3}{32} \frac{\alpha \gamma^2}{\left(\zeta_1 + \zeta_{eq}\right)^2} - \frac{\theta^2}{2(1+\rho^2)}, \tag{6.46}$$

where the parameter of electromechanical coupling is given as an equivalent damping

$$\zeta_{eq} = \frac{1}{2} \frac{\theta^2 \rho}{1+\rho^2}. \tag{6.47}$$

In Fig. 6.4 the average nondimensional power harvested as a frequency function for $\beta = 0$ (linear piezoelectricity), $\gamma = 2.00$, $\zeta = 0.25$, $\rho = 1.00$ is plotted. In Fig. 6.4a the system with linear stiffness ($\alpha = 0$) and in Fig. 6.4b with nonlinear stiffness ($\alpha = 0$) and various values of $\theta$ is plotted.

From Fig. 6.4a is evident that increase of $\theta$ broadens the frequency-amplitude response and moves the peak to left. Besides, the peak value increases with increasing

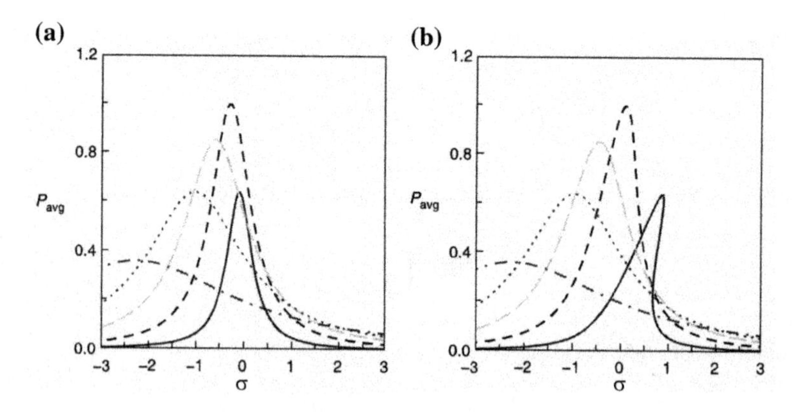


**Fig. 6.4** Average power harvested versus frequency for: **a** $\alpha = 0$, and **b** $\alpha = 0.25$, and $\theta = 0.50$ (____), $\theta = 1.0$ (- - - *black*), $\theta = 1.50$ (- - - *gray*), $\theta = 2.0$ (.... *gray*), $\theta = 3.00$ (-. -. .- *gray*) (Triplett and Quinn 2009)

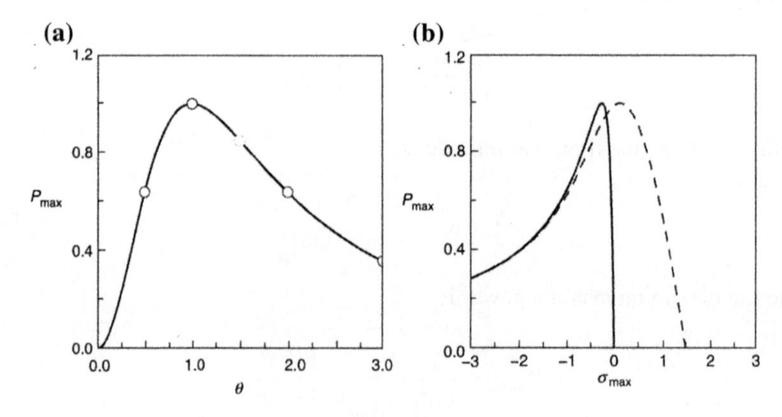


**Fig. 6.5** **a** Maximum power harvested versus $\theta$, **b** Maximum power harvested versus $\sigma_{\max}$ for $\alpha = 0.00$ (—) and $\alpha = 0.25$ (- - -) (Triplett and Quinn 2009)

of $\theta$ up to a maximal value and than decreases. The behavior is similar as for the oscillator with damping. In Fig. 6.4b it is shown that the nonlinearity with positive stiffness gives the effect of hardening and the curves are bending on right.

From (6.45) it is evident that the maximal power harvested is independent on the parameter of nonlinearity $\alpha$. In Fig. 6.5a the $P_{\max} - \theta$ curve and in Fig. 6.5b the $P_{\max} - \theta_{\max}$ curve, according to (6.45) and (6.46), for $\beta = 0$, $\gamma = 2.00$, $\zeta = 0.25$, $\rho = 1.00$ are plotted. The largest power is harvested for $\theta = \frac{2\zeta_1(1+\rho^2)}{\rho}$ and has the value

$$P = \frac{\gamma^2}{4\zeta_1}. \tag{6.48}$$

This value depends on the nondimensional electrical parameter $\rho$.

### 6.2.3 Harvester with Nonlinear Piezoelectricity

For $\beta \neq 0$ the amplitude equation (6.38) is

$$\gamma^2 = \left( \frac{\theta^2 a_1(X_0)}{\pi} - 2\zeta_1 \right)^2 X_0^2 + (2s)^2 X_0^2, \tag{6.49}$$

which occurs at detuning frequency

$$\sigma = \frac{3}{8}\alpha X_0^2 - s - \frac{\theta^2 b_1(X_0)}{2\pi}. \tag{6.50}$$

The relations depend on $a_1(X_0)$ and $b_1(X_0)$ given with (6.36) and (6.37).

For $s = 0$ the maximum amplitude is

$$X_{0\,\max} = \frac{\gamma}{\frac{\theta^2 a_1(X_0)}{\pi} - 2\zeta_1}, \tag{6.51}$$

for

$$\sigma = \frac{3}{8}\alpha X_0^2 - \frac{\theta^2 b_1(X_0)}{2\pi}. \tag{6.52}$$

The influence of the nonlinearity in the coupling coefficient is investigated through the harvested power. Solutions of (6.49) and (6.50) with (6.36) and (6.37) are determined numerically.

In Fig. 6.6 based on these values the averaged power harvested - frequency diagram for $\theta = 1.00$, $\omega = 1 + \varepsilon\sigma$ and various values of $\beta$ is plotted. It is seen that the increase of $\theta$ broadens the frequency-amplitude response and moves the peak to left. Besides, the peak value increases with increasing of $\theta$ up to a maximal value and than decreases.

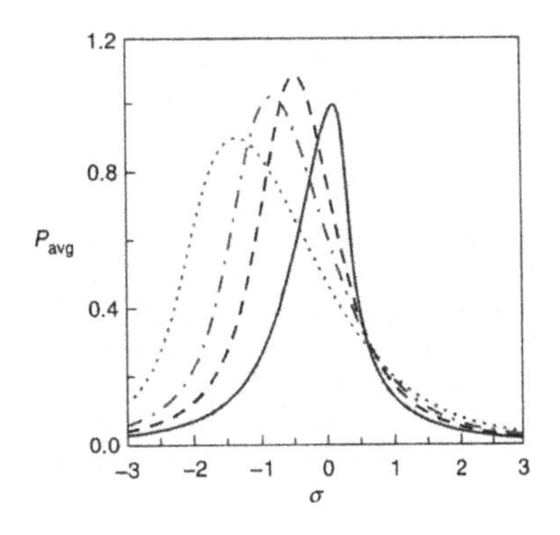


**Fig. 6.6** Average harvested versus frequency: $\beta = 0.00$ (___), $\beta = 0.50$ (- - -), $\beta = 1.00$ (-.-.-), $\beta = 2.00$ (. . .) (Triplett and Quinn 2009)

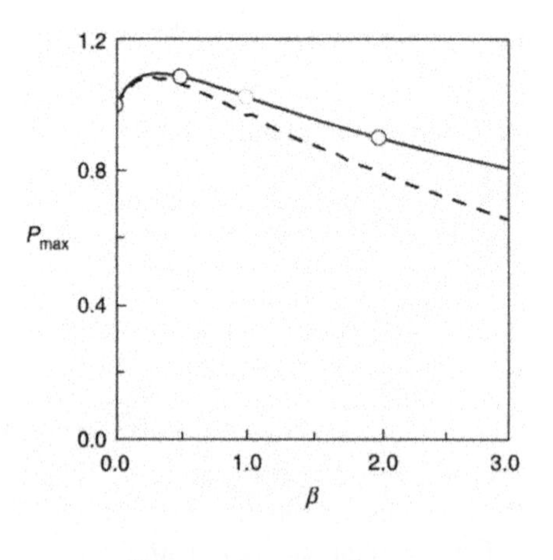


**Fig. 6.7** Maximum power harvester versus nonlinear coupling coefficient $\beta$: analytical (___), numerical (- - -). Marked points correspond to the power frequency curves shown in previous figure (Triplett and Quinn 2009)

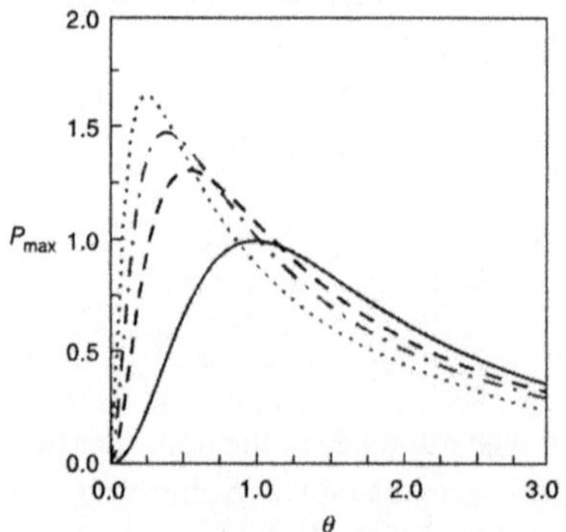


**Fig. 6.8** Maximum power harvested versus coupling coefficient $\theta$: $\beta = 0.00$ (___), $\beta = 0.50$ (- - -), $\beta = 1.00$ (-.-.-), $\beta = 2.00$ (. . .) (Triplett and Quinn 2009)

In Fig. 6.7 the trend in $P_{\max}$ as a function of the nonlinear piezoelectric coefficient $\beta$ is treated for $\theta = 1.00$, $\gamma = 2.00$, $\alpha = 0.25$, $\zeta = 0.25$, $\rho = 1.00$. It is shown that for $\beta < 0.5$ the maximum of power harvester increases with $\beta$, but for $\beta > 0.5$ the increase of $\beta$ causes the decrease of the maximum of the power harvester.

In Fig. 6.8 the maximum power harvested versus the coupling coefficient $\theta$ for various values of $\beta$ and $\gamma = 2.00$, $\alpha = 0.25$, $\zeta = 0.25$, $\rho = 1.00$ is plotted.

For certain value of $\beta$ the increase of $\theta$ causes the sharp increase of the maximum of the power harvester and after that the power harvester decreases with $\theta$. Power harvester reaches its optimum for relatively low values of $\theta$.

*Remark 2* The analytically obtained results are valid only for small nonlinear piezoelectricity. If we include the electromechanical coupling with strong nonlinearity we cannot predict the power harvester with high accuracy. Then, improvement of the model is necessary.

## 6.3 Harvesting System with Non-ideal Excitation

In this section the piezoelectric harvesting in a real mechanical system which contains an unbalanced motor settled on a clamped beam is considered (see Crawley and Anderson 1990). The system is non-ideal, i.e., the motion of the motor has an influence on the beam motion, but also the vibration of the beam affects the rotor rotation (Cveticanin 2010; Cveticanin and Zukovic 2015a; Cveticanin and Zukovic 2015b). The effect of the piezo element is incorporated as nonlinear, because the constitutive laws of piezoelectric materials specify the nonlinear relationship between the strain and the electric field in the piezoceramic material (Triplett and Quinn 2009). Due to nonlinear properties of the oscillator and of the piezoelectric material the model of the system is strong nonlinear.

In this section the influence of nonlinear properties of the oscillator and of the piezoelement on energy harvesting in a non-ideal system is investigated. The paper is divided into 5 subsections. In the Sect. 6.3.2, the model of the one degree-of-freedom non-ideal system with piezoelectric element is formed. The system is described with a system of three coupled differential equations where two of them correspond to the non-ideal motor-structure system and the third to the energy harvesting. In the Sect. 6.3.3, an analytical method for solving the equations for the resonant case is developed. The averaging procedure is modified for solving the problem. The steady-state solution is considered. The jump phenomena the so called 'Sommerfeld effect' are discussed. The special attention is given to calculation of the harvesting power. The influence of the nonlinear properties of the oscillator and of the piezoelectric material on the power harvesting is discussed. In Sect. 6.3.4 the analytically obtained solutions are compared with numerically obtained ones. The results are given in Cveticanin et al. (2017).

### 6.3.1 *Model of the Non-ideal Mechanical System with Harvesting Device*

The non-ideal energy harvester consists of a cantilever beam, covered with piezoelectric layers on the both sides, and a direct current (DC) electric motor attached at the free end of the beam. The rotor has an unbalanced mass. On the clamped end, piezoelectric elements are connected via electrodes to an electrical load, for example, a resistor (Fig. 6.9). Model of the system is shown in Fig. 6.10.

The cantilever beam whose mass is $m_1$ is covered with piezoelectric layers $P_1$ and $P_2$ on the both sides. A direct current (DC) electric motor is attached at the beam with nonlinear cubic rigidity and linear damping properties. The rotor has an unbalance mass $m_2$. The piezoelectric elements are connected via electrodes to an electrical load, for example, a resistor $R$ (Fig. 6.10).

Mass of the system ($M = m_1 + m_2$) is moving oscillatory in $x$ direction. On the mass a force arising from the mechanical elasticity and damping, as well as, the

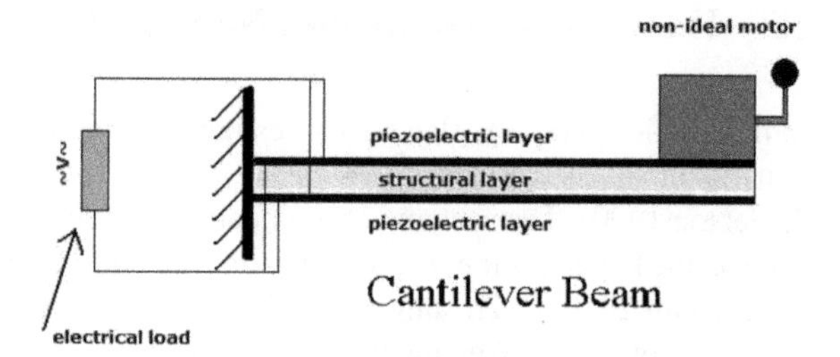


**Fig. 6.9** Non-ideal energy harvester (Iliuk et al. 2013a)

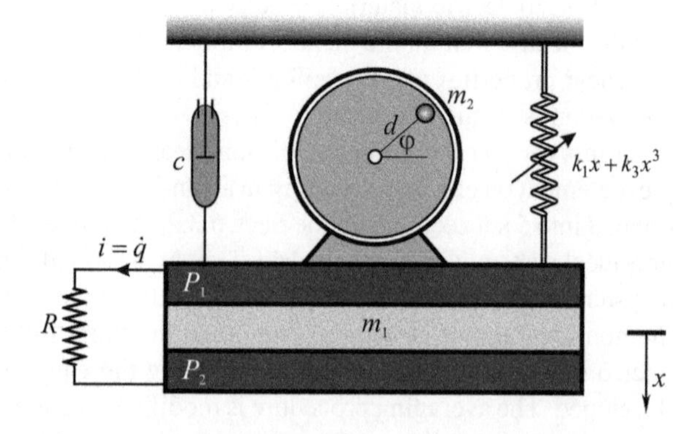


**Fig. 6.10** Model of the one degree-of-freedom non-ideal system with piezoelectric harvester

electromechanical force from the piezoelectric coupling act. The elastic force in the spring is assumed to be of stable hardening Duffing type with potential energy

$$U = \frac{1}{2}k_1x^2 + \frac{1}{4}k_3x^4, \tag{6.53}$$

where $k_1$ and $k_3$ are coefficients of linear and cubic rigidity. The dissipative function due to viscous damping is

$$\Phi = \frac{1}{2}c\dot{x}^2, \tag{6.54}$$

where $c$ is the damping coefficient.

As already mentioned in the Sect. 6.1, the piezoelectric coupling to the mechanical component is given as (see Triplett and Quinn 2009)

$$\frac{\Theta(x)}{C}Q, \tag{6.55}$$

where $\Theta(x)$ is a strain dependent coupling coefficient, $C$ is the piezoelectric capacitance and $Q$ is the electric charge developed in the coupled circuit. The voltage $V$ across the piezoelectric material is described by an electromechanical constitutive relation (du Toit and Wardle 2007)

$$V = -\frac{\Theta(x)}{C}x + \frac{Q}{C}. \tag{6.56}$$

For $V = -R\dot{Q}$ the Eq. (6.56) transforms into

$$R\dot{Q} = \frac{\Theta(x)}{C}x - \frac{Q}{C}, \tag{6.57}$$

where $R$ is electrical resistance. Based on the investigation given by Crawley and Anderson (1990) it is concluded that the piezoelectric coupling coefficient is nonlinear and is modeled as

$$\Theta(x) = \Theta_{lin}(1 + \Theta_{nl}\,|x|), \tag{6.58}$$

where $\Theta_{lin}$ is a constant, while $\Theta_{nl}\,|x|$ depends on $x$. In the following text the non-ideal harvesting systems will be considered. Namely, the rotor of the motor whose moment of inertia is $J$ has an unbalance $m_2$ which is settled on the distance $d$ to the center of the rotor shaft. Position of the unbalance mass is defined with the angle position $\varphi$. The motor-structure system has two degrees-of-freedom given with two generalized coordinates $x$ and $\varphi$. Lagrange's equations of motion in general form are

$$\begin{aligned} \frac{d}{dt}\frac{\partial T}{\partial \dot{x}} - \frac{\partial T}{\partial x} + \frac{\partial U}{\partial x} + \frac{\partial \Phi}{\partial \dot{x}} &= Q_x, \\ \frac{d}{dt}\frac{\partial T}{\partial \dot{\varphi}} - \frac{\partial T}{\partial \varphi} + \frac{\partial U}{\partial \varphi} + \frac{\partial \Phi}{\partial \dot{\varphi}} &= Q_\varphi, \end{aligned} \tag{6.59}$$

where $T$ is the kinetic energy, and $Q_x$ and $Q_\varphi$ are generalized forces. The kinetic energy of the system is

$$T = \frac{1}{2}(m_1 + m_2)\dot{x}^2 + \frac{1}{2}J\dot{\varphi}^2 + \frac{1}{2}m_2 v_2^2,$$

where $M = m_1 + m_2$ is the total mass of the system and $v_2$ is the velocity of unbalance. For position coordinates

$$x_2 = x + d\cos\varphi, \quad y_2 = d\sin\varphi,$$

the velocity follows as

$$v_2 = \sqrt{\dot{x}_2^2 + \dot{y}_2^2} = \sqrt{\dot{x}^2 + d^2\dot{\varphi}^2 - 2d\dot{x}\dot{\varphi}\sin\varphi}. \tag{6.60}$$

Substituting (6.60) into (6.53) we obtain

$$T = \frac{1}{2}(m_1 + m_2)\dot{x}^2 + \frac{1}{2}(J + m_2 d^2)\dot{\varphi}^2 - m_2 d\dot{x}\dot{\varphi}\sin\varphi. \tag{6.61}$$

The motion of the system is excited with the motor torque. The torque of the motor contains two terms: the characteristic of the motor $L(\dot{\varphi})$ and the resisting moment $H(\dot{\varphi})$ due primarily to windage of the rotating parts outside the motor

$$\mathcal{M}(\dot{\varphi}) = L(\dot{\varphi}) - H(\dot{\varphi}).$$

Generally, $L(\dot{\varphi})$ and $H(\dot{\varphi})$ are determined experimentally. Various types of mathematical description of the motor property are suggested. One of the most often applied and the simplest one is the linear mode which is a function of the angular velocity $\dot{\varphi}$

$$\mathcal{M}(\dot{\varphi}) = M_0 \left(1 - \frac{\dot{\varphi}}{\Omega_0}\right), \tag{6.62}$$

and depends on two constant parameters $M_0$ and $\Omega_0$ which define the limited source of power as the angular velocity increases. The expression (6.62) defines the characteristic curve of the motor, where for angular velocities greater than $\Omega_0$ the torque reduces to zero and when the angular velocity is zero the torque is maximum. The Eq. (6.62) is valid for positive values of torque and angular velocity. Based on (6.55) and for the suggested motor torque functions the generalized forces are

$$Q_x = \frac{\Theta(x)}{C} Q, \quad Q_{\varphi} = \mathcal{M}(\dot{\varphi}). \tag{6.63}$$

Substituting (6.53), (6.54), (6.61) and (6.63) into (6.59) and using (6.57) the following system of equations is obtained

$$(m_1 + m_2)\ddot{x} + k_1 x + k_3 x^3 + c\dot{x} - \frac{\Theta(x)}{C} Q \tag{6.64}$$

$$= m_2 d(\ddot{\varphi} \cos\varphi - \dot{\varphi}^2 \sin\varphi),$$

$$(J + m_2 d^2)\ddot{\varphi} = m_2 d\ddot{x} \cos\varphi + \mathcal{M}(\dot{\varphi}), \tag{6.65}$$

$$R\dot{Q} = \frac{\Theta(x)}{C} x - \frac{Q}{C}. \tag{6.66}$$

The equations are valid if the gravity potential energy is neglected. The Eqs. (6.64)–(6.66) are coupled and nonlinear. In this paper the calculation for the linear torque function (6.62) is done.

Introducing the dimensionless functions

$$z = \frac{x}{d}, \quad \tau = \omega t, \tag{6.67}$$

and parameters

$$\omega^2 = \frac{k_1}{m_1 + m_2}, \quad \varepsilon\zeta_1 = \frac{c}{(m_1 + m_2)\omega}, \quad \varepsilon\mu_1 = \frac{m_2}{m_1 + m_2}, \quad \rho = R\omega C,$$
$$q = \frac{Q}{C}, \quad \theta = d\Theta_{lin}, \quad \varepsilon\theta_n = d\Theta_{nl}, \quad \varepsilon = \frac{C}{d^2\omega^2},$$
$$\varepsilon k_z = \frac{k_3 d^2}{k_1}, \quad \varepsilon\eta_1 = \frac{m_2 d^2}{J + m_2 d^2}, \quad \varepsilon\mathcal{M}(\varphi') = \frac{\mathcal{M}(\dot{\varphi})}{(J + m_2 d^2)\omega^2}. \tag{6.68}$$

into Eqs. (6.64)–(6.66) we have

$$z'' + z = -\varepsilon k_z z^3 - \varepsilon\zeta_1 z' + \varepsilon\theta(1 + \varepsilon\theta_n |x|)q \tag{6.69}$$
$$+\varepsilon\mu_1(\varphi'' \cos\varphi - \varphi'^2 \sin\varphi),$$
$$\varphi'' = \varepsilon\eta_1 z'' \cos\varphi + \varepsilon\mathcal{M}(\varphi'), \tag{6.70}$$
$$\rho q' + q = \theta(1 + \varepsilon\theta_n |z|)z. \tag{6.71}$$

For (6.62) the dimensionless motor torque is

$$\varepsilon\mathcal{M}(\varphi') = \varepsilon M_0 \left(1 - \frac{\omega}{\Omega_0}\varphi'\right), \tag{6.72}$$

where $\varepsilon\mathcal{M}_0 = \frac{M_0}{(J+m_2d^2)\omega^2}$. Neglecting the terms with the second order small parameter $O(\varepsilon^2)$ the Eqs. (6.69)–(6.72) are transformed into

$$z'' + z = -\varepsilon k_z z^3 - \varepsilon\zeta_1 z' + \varepsilon\theta q - \varepsilon\mu_1\varphi'^2 \sin\varphi, \tag{6.73}$$
$$\varphi'' = \varepsilon\eta_1 z'' \cos\varphi + \varepsilon\mathcal{M}_0 \left(1 - \frac{\omega}{\Omega_0}\varphi'\right), \tag{6.74}$$
$$\rho q' + q = \theta(1 + \varepsilon\theta_n |z|)z. \tag{6.75}$$

Introducing the new variables

$$z_1 = z, \; z_2 = z', \; \varphi' = \Omega,$$

the Eqs. (6.73)–(6.75) are transformed into five coupled first order differential equations

$$\begin{aligned}
z_1' &= z_2, \\
z_2' &= -z_1 - \varepsilon k_z z_1^3 - \varepsilon\zeta_1 z_2 + \varepsilon\theta q - \varepsilon\mu_1\Omega^2 \sin\varphi, \\
\varphi' &= \Omega, \\
\Omega' &= \varepsilon\eta_1 z_2' \cos\varphi + \varepsilon\mathcal{M}_0 \left(1 - \frac{\omega}{\Omega_0}\Omega\right), \\
\rho q' + q &= \theta(1 + \varepsilon\theta_n |z_1|)z_1.
\end{aligned} \tag{6.76}$$

Our aim is to solve the system of Eq. (6.76).

### 6.3.2 Analytical Solving Procedure

For simplicity let us rewrite the Eq. (6.76) by introducing the new variables $a$ and $\psi$ which satisfy the relations

$$z_1 = a\cos(\varphi + \psi), \tag{6.77}$$

and

$$z_2 = -a\Omega\sin(\varphi + \psi), \tag{6.78}$$

where $a$ and $\psi$ are time dependent functions. Comparing the assumed relation (6.78) with the first time derivative of (6.77) the following constraint exists

$$a'\cos(\varphi + \psi) - a\psi'\sin(\varphi + \psi) = 0. \tag{6.79}$$

Substituting (6.77), (6.78) and its time derivative into $(6.76)_1$ it follows

$$\begin{aligned} &-a'\Omega\sin(\varphi + \psi) - a\Omega\psi'\cos(\varphi + \psi) - a\Omega^2\cos(\varphi - \psi) \\ &= -a\cos(\varphi + \psi) - \varepsilon\zeta_1 a\sin(\varphi + \psi) - \varepsilon k_z a^3\cos^3(\varphi + \psi) \\ &-\varepsilon\mu_1\Omega^2\sin\varphi + \varepsilon\theta q. \end{aligned} \tag{6.80}$$

After some modification of (6.79) and (6.80) the Eq. $(6.76)_{1,2}$ are transformed into two coupled first order equations

$$\begin{aligned} a' &= a\frac{1-\Omega^2}{\Omega}\sin(\varphi + \psi)\cos(\varphi + \psi) - \varepsilon\zeta_1 a\sin^2(\varphi + \psi) \\ &+\frac{\varepsilon k_z a^3}{\Omega}\cos^3(\varphi + \psi)\sin(\varphi + \psi) \\ &+\varepsilon\mu_1\Omega\sin\varphi\sin(\varphi + \psi) - \frac{\varepsilon\theta}{\Omega}q\sin(\varphi + \psi), \end{aligned} \tag{6.81}$$

$$\begin{aligned} \psi' &= \frac{1-\Omega^2}{\Omega}\cos^2(\varphi + \psi) - \varepsilon\zeta_1\sin(\varphi + \psi)\cos(\varphi + \psi) \\ &+\frac{\varepsilon k_z a^2}{\Omega}\cos^4(\varphi + \psi) \\ &+\frac{\varepsilon\mu_1\Omega}{a}\sin\varphi\cos(\varphi + \psi) - \frac{\varepsilon\theta}{a\Omega}q\cos(\varphi + \psi). \end{aligned} \tag{6.82}$$

Further, eliminating the terms with small values $O(\varepsilon^2)$ and higher, the Eq. $(6.76)_{3-5}$ are

$$\varphi' = \Omega,$$
$$\Omega' = -\varepsilon\eta_1 a\Omega^2 \cos\varphi\cos(\varphi+\psi) + \varepsilon\mathcal{M}_0\left(1 - \frac{\omega}{\Omega_0}\Omega\right),$$
$$q' = -\frac{1}{\rho}q + \frac{\theta}{\rho}(1 + \varepsilon\theta_n\,|a\cos(\varphi+\psi)|)a\cos(\varphi+\psi). \tag{6.83}$$

Analyzing the five Eqs. (6.81)–(6.83) it is evident that for the near resonant case when $\Omega \approx 1$ and $(1 - \Omega) = O(\varepsilon)$ the variables $a'$, $\psi'$ and $\Omega'$ are of the order $O(\varepsilon)$. For simplification, the averaging procedure over the period of the function $\varphi$ is introduced. The averaged Eqs. (6.81)–(6.83) are

$$a' = -\frac{1}{2}\varepsilon\zeta_1 a + \frac{1}{2}\varepsilon\mu_1\Omega\cos\psi - \frac{\varepsilon\theta}{\Omega}\langle q\sin(\varphi+\psi)\rangle, \tag{6.84}$$
$$\psi' = \frac{1-\Omega^2}{2\Omega} + \frac{3}{8}\frac{\varepsilon k_z a^2}{\Omega} - \frac{1}{2}\frac{\varepsilon\mu_1\Omega}{a}\sin\psi - \frac{\varepsilon\theta}{a\Omega}\langle q\cos(\varphi+\psi)\rangle, \tag{6.85}$$
$$\varphi' = \Omega, \tag{6.86}$$
$$\Omega' = -\frac{1}{2}\varepsilon\eta_1 a\Omega^2\cos\psi + \varepsilon\mathcal{M}_0\left(1 - \frac{\omega}{\Omega_0}\Omega\right), \tag{6.87}$$
$$q' = -\frac{1}{\rho}q + \frac{\theta}{\rho}(1 + \varepsilon\theta_n\,|a\cos(\varphi+\psi)|)a\cos(\varphi+\psi). \tag{6.88}$$

where $\langle\bullet\rangle = \frac{1}{2\pi}\int_0^{2\pi}(\bullet)\,d\varphi$.

### 6.3.3 Steady-State Motion

For $a' = 0$, $\psi' = 0$ and $\Omega' = 0$, the steady state motion is

$$a = a_S = const.,\ \psi = \psi_S = const.,\ \Omega = \Omega_S = const., \tag{6.89}$$

and

$$\varphi = \Omega_S t. \tag{6.90}$$

Substituting the conditions for the steady-state motion (6.89) and (6.90) into (6.88) we have

$$q' = -\frac{1}{\rho}q + \frac{\theta}{\rho}(1 + \varepsilon\theta_n\,|a_S\cos(\Omega_S t + \psi_S)|)a_S\cos(\Omega_S t + \psi_S). \tag{6.91}$$

Analyzing the relation (6.91) it is obvious that the harvesting function $q$ is periodical and depends on the amplitude of vibration of the system. Using the Fourier series expansion

$$|\cos(\Omega_S t + \psi_S)|)\cos(\Omega_S t + \psi_S) = \frac{8}{3\pi}\cos(\Omega_S t + \psi_S) + \frac{8}{15\pi}\cos(3(\Omega_S t + \psi_S)) + \ldots \tag{6.92}$$

and introducing the first term into (6.92), we obtain

$$q' = -\frac{1}{\rho}q + \left(\frac{\theta}{\rho}a_S + \varepsilon\theta_n \frac{8}{3\pi^2}\frac{\theta}{\rho}a_S^2\right)\cos(\Omega_S t + \psi_S). \tag{6.93}$$

The particular solution of (6.93) is

$$q = q_S = D\cos(\Omega_S t + \psi_S + \delta), \tag{6.94}$$

where

$$D = \frac{\theta a_S(3\pi + 8a_S\varepsilon\theta_n)}{3\pi\sqrt{1+\rho^2\Omega_S^2}}, \quad \tan\delta = -\rho\Omega_S. \tag{6.95}$$

Introducing (6.94) into the steady-state equations (6.84), (6.85) and (6.87)

$$-\frac{1}{2}\varepsilon\zeta_1 a_S + \frac{1}{2}\varepsilon\mu_1\Omega_S\cos\psi_S - \frac{\varepsilon\theta}{\Omega}\langle q\sin(\Omega_S t + \psi_S)\rangle = 0,$$

$$\frac{1-\Omega_S^2}{2\Omega_S} + \frac{3}{8}\frac{\varepsilon k_z a_S^2}{\Omega_S} - \frac{1}{2}\frac{\varepsilon\mu_1\Omega_S}{a_S}\sin\psi_S - \frac{\varepsilon\theta}{a_S\Omega_S}\langle q\cos(\Omega_S t + \psi_S)\rangle = 0,$$

$$-\frac{1}{2}\varepsilon\eta_1 a_S\Omega_S^2\cos\psi_S + \varepsilon\mathcal{M}_0\left(1 - \frac{\omega}{\Omega_0}\Omega_S\right) = 0,$$

it is

$$-\frac{1}{2}\varepsilon\zeta_1 a_S + \frac{1}{2}\varepsilon\mu_1\Omega_S\cos\psi_S - \frac{\varepsilon\theta}{2\Omega_S}D\sin\delta = 0,$$

$$\frac{1-\Omega_S^2}{2\Omega_S} + \frac{3}{8}\frac{\varepsilon k_z a_S^2}{\Omega_S} - \frac{1}{2}\frac{\varepsilon\mu_1\Omega_S}{a_S}\sin\psi_S - \frac{\varepsilon\theta}{a_S\Omega_S}D\cos\delta = 0,$$

$$-\frac{1}{2}\varepsilon\eta_1 a_S\Omega_S^2\cos\psi_S + \varepsilon\mathcal{M}_0\left(1 - \frac{\omega}{\Omega_0}\Omega_S\right) = 0. \tag{6.96}$$

From (6.96) it is evident that the harvester has an influence on the system through parameters $D$ and $\delta$. Vibrations depend on the linear and nonlinear coefficients of the piezoelectric element.

Using (6.95) the relations (6.96) yield

$$
\begin{aligned}
&-\varepsilon\zeta_1 a_S + \varepsilon\mu_1\Omega_S\cos\psi_S - \varepsilon\theta^2\rho\frac{a_S(3\pi + 8a_S\varepsilon\theta_n)}{3\pi\sqrt{1+\rho^2\Omega_S^2}} = 0,\\
&\frac{1-\Omega_S^2}{\Omega_S} + \frac{3}{4}\frac{\varepsilon k_z a_S^2}{\Omega_S} - \frac{\varepsilon\mu_1\Omega_S}{a_S}\sin\psi_S - \frac{\varepsilon\theta^2}{\Omega_S}\frac{(3\pi + 8a_S\varepsilon\theta_n)}{3\pi\sqrt{1+\rho^2\Omega_S^2}} = 0,\\
&-\frac{1}{2}\varepsilon\eta_1 a_S\Omega_S^2\cos\psi_S + \varepsilon\mathcal{M}_0\left(1 - \frac{\omega}{\Omega_0}\Omega_S\right) = 0.
\end{aligned}
\tag{6.97}
$$

Relations (6.97) are three coupled algebraic equations which give us the steady state amplitude $a_S$, phase $\psi_S$ and angular velocity $\Omega_S$. Eliminating the variable $\psi_S$ in $(6.97)_1$ and $(6.97)_2$, we have the amplitude -frequency characteristic

$$
\begin{aligned}
(\varepsilon\mu_1\Omega_S)^2 &= \left(\varepsilon\zeta_1 a_S + \varepsilon\rho\theta^2\frac{a_S(3\pi + 8\varepsilon a_S\theta_n)}{3\pi(1+\rho^2\Omega^2)}\right)^2\\
&+ \left(\frac{1-\Omega_S^2}{\Omega_S}a_S + \frac{3}{4}\frac{\varepsilon k_z a_S^3}{\Omega_S} - \frac{\varepsilon\theta^2}{\Omega_S}\frac{a_S(3\pi + 8a_S\varepsilon\theta_n)}{3\pi\sqrt{1+\rho^2\Omega_S^2}}\right)^2.
\end{aligned}
\tag{6.98}
$$

The phase angle is obtained according to $(6.97)_1$ and $(6.97)_2$ as

$$
\psi_S = \tan^{-1}\frac{\dfrac{1-\Omega_S^2}{\Omega_S}a_S + \dfrac{3}{4}\dfrac{\varepsilon k_z a_S^3}{\Omega_S} - \dfrac{\varepsilon\theta^2}{\Omega_S}\dfrac{a_S(3\pi + 8a_S\varepsilon\theta_n)}{3\pi\sqrt{1+\rho^2\Omega_S^2}}}{\varepsilon\theta^2\rho\dfrac{a_S(3\pi + 8a_S\varepsilon\theta_n)}{3\pi\sqrt{1+\rho^2\Omega_S^2}} + \varepsilon\zeta_1 a_S}.
\tag{6.99}
$$

From Eq. $(6.97)_1$ and $(6.97)_3$ it is

$$
\varepsilon\mathcal{M}_0(1 - \frac{\omega}{\Omega_0}\Omega_S) = \frac{\varepsilon\eta_1\Omega_S a_S^2}{2\varepsilon\mu_1}\left[\varepsilon\zeta_1 + \varepsilon\rho\theta^2\frac{(3\pi + 8\varepsilon a_S\theta_n)}{3\pi(1+\rho^2\Omega^2)}\right].
\tag{6.100}
$$

The Eq. (6.100) gives the relation between $a_S$, $\Omega_S$ and $\Omega_0$. For the certain value of $\Omega_0$ the Eqs. (6.98) and (6.100) give the $a_S - \Omega_S$ relation. In Fig. 6.11 the steady-state amplitude - frequency diagram (thick line) and the intersecting curves which take into consideration the properties of the motor (thin line) are plotted. The numerical data are:

$$
\varepsilon = 0.1,\quad k_z = 0.025,\quad \eta_1 = 1,\quad \mu_1 = 1,\quad \varsigma_1 = 1,\quad \rho = 1,\quad \omega = 1,\quad \mathcal{M}_0 = 2.
\tag{6.101}
$$

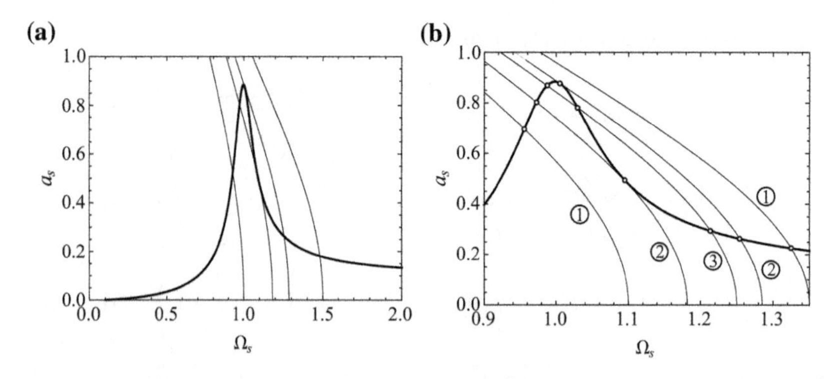


**Fig. 6.11** **a** Amplitude - frequency diagram (*thick line*) and motor property curves (*thin line*), **b** Notation of the number of intersections between curves (1, 2 and 3)

In Fig. 6.11b the number of intersections between the curves is signed. It can be seen that the Eqs. (6.98) and (6.100) have one, two or three solutions.

### 6.3.4 Harvested Energy

The harvested power from the mechanical component is $V^2/R$. Using the non-dimensional values the harvested power is given as

$$P = \rho q'^2. \tag{6.102}$$

According to (6.94) the steady-state power - time function is

$$P = \rho D^2 \Omega_S^2 \sin^2(\Omega_S t + \psi_S + \delta),$$

where $D$ and $\delta$ correspond to (6.95). After elimination of the transient of the system response the averaged harvested power is

$$\begin{aligned} P_{avg} &= \frac{1}{T}\int_0^T P(\tau)d\tau = \rho D^2 \Omega_S^2 \frac{1}{2\pi/\Omega_S}\int_0^{2\pi/\Omega_S} \sin^2(\Omega_S t + \psi_S + \delta)dt \\ &= \frac{1}{2}\rho D^2 \Omega_S^2, \end{aligned} \tag{6.103}$$

i.e., due to $(6.95)_1$

$$P_{avg} = \rho \Omega_S^2 \frac{\theta^2 a_S^2 (3\pi + 8 a_S \varepsilon \theta_n)^2}{18\pi^2 (1 + \rho^2 \Omega_S^2)}.$$

For the case of linear harvester, the averaged power is

$$P_{avgl} = \frac{\rho \Omega_S^2 \theta^2 a_S^2}{2(1+\rho^2 \Omega_S^2)}. \tag{6.104}$$

Comparing the averaged powers for the nonlinear and linear harvesters it is

$$\frac{P_{avg}}{P_{avgl}} = \left(1 + \frac{8\varepsilon\theta_n a_S}{3\pi}\right)^2. \tag{6.105}$$

It is obvious that the ratio of powers depend on the steady state amplitude of vibrations and on the nonlinear coefficient of the piezoelectric element. However, the steady state amplitude of vibration is the function of properties of vibrating systems, angular velocity of rotation and the motor torque properties, but also of the linear and nonlinear piezoelectric constants. The higher the value of the steady-state amplitude of vibration, the higher is the value of the harvested power of the nonlinear piezoelectric element in comparison to the linear one. The relation is quadratic.

### 6.3.5 Comparison of the Analytical and Numerical Solutions

Introducing the new variables

$$x_1 = z, \quad x_2 = z', \quad x_3 = \varphi, \quad x_4 = \varphi', \quad x_5 = q, \tag{6.106}$$

the Eqs. (6.73)–(6.75) in the state space are

$$\begin{aligned}
x_1' &= x_2, \\
x_2' &= -x_1 - \varepsilon k_z x_1^3 - \varepsilon\zeta_1 x_2 + \varepsilon\theta(1+\theta_n |x_1|)x_5 + \varepsilon\mu_1 \left(x_4' \cos x_3 - x_4^2 \sin x_3\right) \\
x_3' &= x_4, \\
x_4' &= \varepsilon\, \mathcal{M}(x_4) + \varepsilon\eta_1 x_2' \cos x_3, \\
x_5' &= \frac{1}{\rho}\left(\theta(1+\theta_n |x_1|)x_1 - x_5\right).
\end{aligned} \tag{6.107}$$

This form is appropriate for numerical simulation of the problem. To analyze the nonlinear dynamic of non-ideal system, we used the following values for the parameters

$$\begin{gathered}
\varepsilon = 0.10, \quad \zeta_1 = 1, \quad k_z = 0.025, \quad \eta_1 = 1, \quad \mathcal{M}_0 = 2, \\
\mu_1 = 1, \quad \rho = 1, \quad \theta = 0.5, \quad \theta_n = 0.1, \quad \omega = 1,
\end{gathered} \tag{6.108}$$

and the initial conditions

$$x_1(0) = 0, \quad x_2(0) = 0, \quad x_3(0) = 0, \quad x_4(0) = 0, \quad x_5(0) = 0. \tag{6.109}$$

Based on (6.98) and (6.100) for the numerical data (6.101) the steady state amplitude $a_S$ and frequency $\Omega_S$ as functions of $\Omega_0$ are plotted (Fig. 6.12).

For $\Omega_0$ whose values are between 2a and 2b there are three solutions for steady state amplitude and frequency, $a_S$ and $\Omega_S$, respectively. In Fig. 6.13 the $a_S - \Omega_0$ diagrams obtained numerically by solving the Eq. (6.121) are plotted. In Fig. 6.13a the linear piezoelement with $\theta = 0.5$ and $\theta_n = 0$ is considered, while in Fig. 6.13b the nonlinear piezoelement with $\theta = 0.5$ and $\theta_n = 0.5$ is treated. Two regimes are investigated: the case when $\Omega_0$ is increasing and when $\Omega_0$ is decreasing.

It can be seen that there is an excellent agreement between analytical and numerical results not only for the linear but also nonlinear piezoelectric element. Comparing Fig. 6.12 with Fig. 6.13 it is evident that only two of the three solutions in the region bounded with 2a and 2b lines are stable. One of them is unstable.

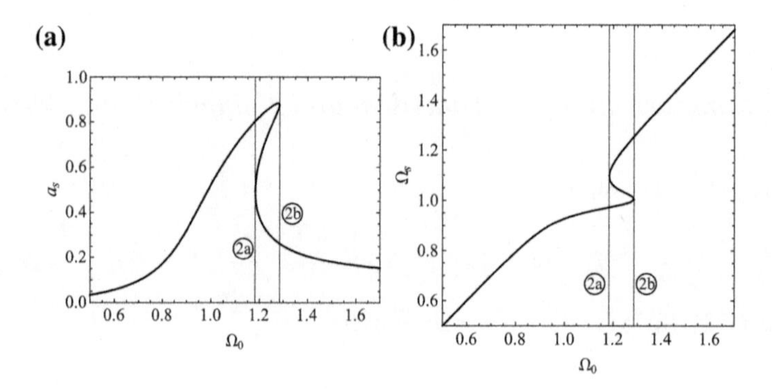


**Fig. 6.12** **a** $a_S - \Omega_0$ curve with boundary values in 2a and 2b; **b** $\Omega_S - \Omega_0$ curve with boundary values in 2a and 2b

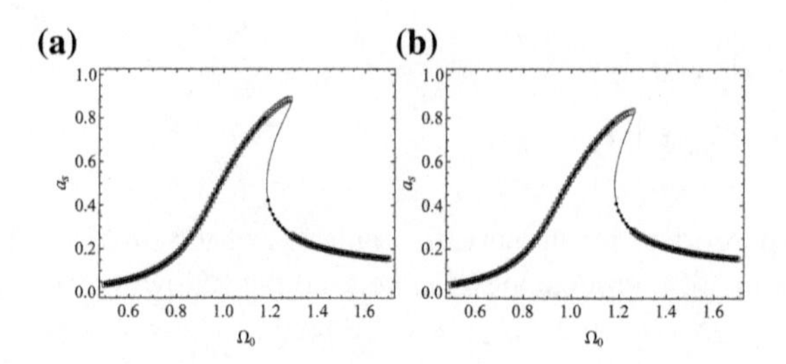


**Fig. 6.13** $a_S - \Omega_0$ curve for: **a** $\theta = 0.5$ and $\theta_n = 0$; **b** $\theta = 0.5$ and $\theta_n = 0.5$. *Thinline* is the analytical solution, *square* is the numerical solution for increasing of $\Omega_0$ and *circle* is the numerical solution for decreasing of $\Omega_0$

Comparing the curves in Fig. 6.13a, b it can be concluded that the maximal amplitude is higher for the linear piezoelement than for the nonlinear one. The nonlinearity decreases the maximal amplitude.

We separate the analysis of energy harvesting in two cases: linear energy harvesting and energy nonlinear harvesting. We seek the maximal amplitude of vibration in the graph of the resonance curve for different values of the control parameter $\Omega_0$. The power harvested was obtained by Eq. (6.101).

### 6.3.6 Linear Energy Harvester

In Fig. 6.14 the properties of the linear energy harvester as the function of $\Omega_0$ are plotted. The linear coefficient $\theta$ in linear harvesting has various values, while the coefficient of nonlinearity is zero $\theta_n = 0$.

In Fig. 6.14 the amplitude $D$ of electric charge and the averaged harvested energy as functions of control parameter $\Omega_0$ are plotted. The resonance curves of the amplitude were obtained as follows: for each value of the control parameter the maximal amplitudes of oscillations are captured. For the linear energy harvester it is evident that the maximal amplitude of electric charge and also the maximal averaged harvested energy are higher for higher values of parameter $\theta$. The system reaches the maximum amplitude D and the maximum power harvested for the same value of $\Omega_0$. For the control parameter the response of the system is stable in the region of pre-resonance with a power harvested slightly less than the power harvested in the resonance region. In the region of post-resonance the amplitude $D$ is smaller and a considerable reduction in power is captured.

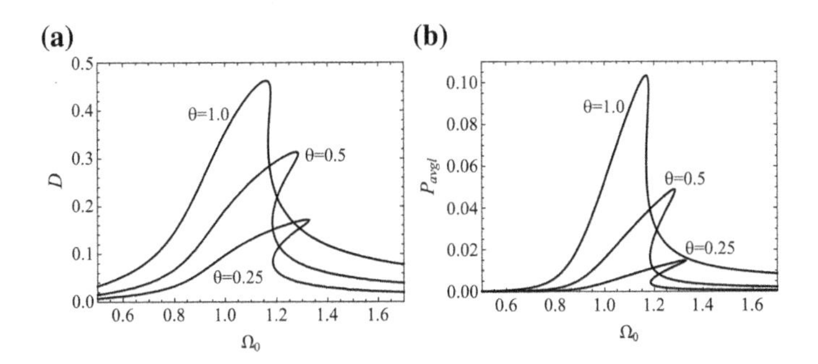


**Fig. 6.14** Linear energy harvester: **a** $D - \Omega_0$ curves; **b** $P_{avgl} - \Omega_0$ curves, for $\theta = 0.25, 0.5, 1$ and $\theta_n = 0$.

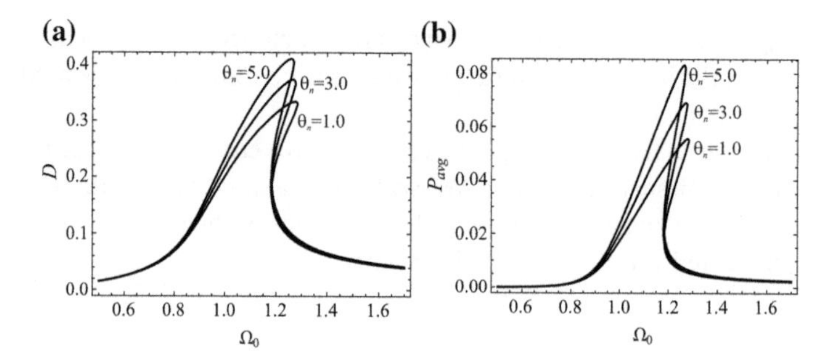


**Fig. 6.15** Nonlinear energy harvester: **a** $D - \Omega_0$ curves; **b** $P_{avg} - \Omega_0$ curves, for $\theta_n = 0.25, 0.5, 1$ and $\theta = 0.5$

## 6.3.7 Nonlinear Energy Harvesting

Let us consider the nonlinear harvesting where the nonlinear piezoelectric coupling parameter $\theta_n$ has various values ($\theta_n = 1$, $\theta_n = 3$ and $\theta_n = 5$) and the linear coefficient is $\theta = 0.5$. In Fig. 6.15 the amplitude of electric charge $D$ and the averaged harvester energy $P_{avg}$ for a nonlinear system as function of control parameter $\Omega_0$ is plotted. From figure it can be seen that for higher values of the nonlinear piezoelectric parameter, the value of the maximal amplitude of electric charge and of the averaged harvested energy is higher. Namely, by increasing of the nonlinear coupling parameter, the maximum power harvested increases. Furthermore, the behavior of the system remains stable in the region of pre-resonance and in the region of resonance. The position of maximal values move to higher values of $\Omega_0$ as the coefficient $\theta_n$ decrease.

After passage through the resonance region the significant decrease of amplitude $D$ occurs for all values of $\theta_n$. Consequently, the harvested power is also reduced. In the post-resonance region the amplitude and energy harvested is small.

## 6.3.8 Conclusion

It can be concluded:

1. Through obtained results it is concluded that the limited energy source interacts with system with piezoelectric coupling. Increasing the voltage in DC motor led the system produce a good power response.
2. Mathematical model of the non-ideal energy harvester shows that a good energy harvesting is achieved in the resonance region due to inclusion of nonlinearity property of the piezoelectric material. The value of maximal energy harvester in the resonant regime depends on the values of the linear and coupling piezoelectric parameters.

It is concluded that for higher values of both piezoelectric parameters the harvested energy is higher.

3. In the post resonance regime when the amplitude of vibration is reduced, even if stable, energy harvested is also decreased.

4. In the system the Sommerfeld effect occurs. Due to this phenomena beside stable solutions in the system also instable solutions may exist. To reduce the region of unstable solutions it is suggested to use piezoelectric element with extremely high nonlinear parameter. Then, the region of unstable solutions is extremely narrow.

## 6.4 Harvester with Exponential Type Non-ideal Energy Source

In this section the piezoelectric harvester presented in the previous section is considered. The physical properties of the system are as already described and the model is given in Fig. 6.10. However, the model differs due to the type of the energy source. Namely, the motor torque is assumed as an exponential function of the angular velocity of the motor

$$\mathcal{M}(\dot{\varphi}) = V_1 \exp(-V_2\dot{\varphi}), \tag{6.110}$$

where $V_1$ relates to voltage applied across the armature of the DC motor and $V_2$ is a constant for each model of considered DC motor. Parameter $V_1$ is suitable to be the control parameter of the system. The electric charge developed in the coupled circuit is given by $q$ and the term $\frac{\theta(x)}{C}q$ represents the piezoelectric coupling to the mechanical component, where $\theta(x)$ is the strain-dependent coupling coefficient. The voltage across the piezoelectric material is

$$V = -\frac{\theta(x)}{C}x + \frac{q}{C}, \tag{6.111}$$

and a linear function of electric intensity $i = \dot{q}$

$$V = -R\dot{q}, \tag{6.112}$$

where $C$ is the piezoelectric capacitance. The governing equations of motion are

$$\begin{aligned}
(m_1 + m_2)\ddot{x} + c\dot{x} + k_1x + k_3x^3 &= m_2d(\ddot{\varphi}\cos\varphi - \dot{\varphi}^2\sin\varphi) + \frac{\theta(x)}{C}q, \\
(J + m_2d^2)\ddot{\varphi} &= m_2d\ddot{x}\cos\varphi + V_1\exp(-V_2\dot{\varphi}), \\
R\dot{q} - \frac{\theta(x)}{C}x + \frac{q}{C} &= 0.
\end{aligned} \tag{6.113}$$

Let us introduce the following non-dimensional parameters

$$\zeta_1 = \frac{c}{\sqrt{k_1 (m_1 + m_2)}}, \quad \varepsilon = \frac{m_2}{m_1 + m_2},$$
$$\upsilon = \frac{q}{q_0}, \quad \mu_1 = \frac{V_1 (m_1 + m_2)}{k_1 (J + m_2 d^2)}, \quad \mu_2 = \frac{V_1 (m_1 + m_2)}{k_1 (J + m_2 d^2)},$$
$$\mathcal{M}(\varphi') = \mu_1 \exp(-\mu_2 \varphi'), \quad \beta = \frac{k_1}{M\omega^2} = 1, \quad \beta_1 = \frac{k_2 d^2}{k_1},$$
$$\delta_1 = \frac{m_2}{m_1 + m_2}, \quad \delta_2 = \frac{m_2 k_1}{(m_1 + m_2)^2},$$
$$\eta_1 = \frac{m_2 d^2}{J + m_2 d^2}, \quad \rho = RC\sqrt{\frac{k_1}{m_1 + m_2}}, \tag{6.114}$$

and the dimensionless length and the time variable

$$y = \frac{x}{d}, \tag{6.115}$$

$$\tau = \omega t, \tag{6.116}$$

where

$$\omega = \sqrt{\frac{k_1}{m_1 + m_2}}. \tag{6.117}$$

Based on the result of Crawley and Anderson (1990), the piezocoupling coefficient is

$$d(x) = \theta_{lin} (1 + \Theta_{nel} \left| x \right|),$$

where $\theta_{lin}$ is the linear and $\Theta_{nel} \left| x \right|$ the nonlinear part. The dimensionless piezoelectric coupling coefficient, suggested by Triplett and Quinn (2009) is

$$\hat{d}(y) = \theta (1 + \Theta \left| y \right|), \tag{6.118}$$

where the piezoelectric coefficient is represented by a linear part $\theta$ and a nonlinear part defined by $\Theta$. The governing equations of motion reduce to

$$y'' + y + \varepsilon \beta_1 y^3 - \varepsilon \theta (1 + \Theta \left| y \right|) \upsilon = -\varepsilon \zeta_1 y' + \varepsilon \left( \delta_1 \varphi'' \cos\varphi - \delta_2 \left( \varphi' \right)^2 \sin\varphi \right),$$
$$\varphi'' = \varepsilon \eta_1 y'' \cos\varphi + \varepsilon \mathcal{M} \left( \varphi' \right),$$
$$\rho \upsilon' - \theta (1 + \Theta \left| y \right|) y + \upsilon = 0. \tag{6.119}$$

Introducing the new variables

$$x_1 = y, \quad x_2 = y', \quad x_3 = \varphi, \quad x_4 = \varphi', \quad x_5 = \upsilon, \tag{6.120}$$

the Eq. (6.119) are rewritten in state space representation

$$\begin{aligned}
x_1' &= x_2, \\
x_2' &= -x_1 - \varepsilon\beta_1 x_1^3 - \varepsilon\zeta_1 x_2 + \varepsilon\theta(1 + \Theta\,|x_1|)x_5 + \varepsilon\left(\delta_1 x_4' \cos x_3 - \delta_2 x_4^2 \sin x_3\right) \\
x_3' &= x_4, \\
x_4' &= \varepsilon\,\mathcal{M}\,(x_4) + \varepsilon\eta_1 x_2' \cos x_3, \\
x_5' &= (\theta(1 + \Theta\,|x_1|)x_1 - x_5)\,/\rho.
\end{aligned} \tag{6.121}$$

The equations are numerically solved by Iliuk et al. (2013a). The system response, the beam displacement-time relation, the time variation of the angular velocity and the harvested power, calculated from the mechanical component $V^2/R$ (see Triplett and Quinn 2009) and according to the relation for the non-dimensional electric power harvested

$$P = \rho\upsilon'^2, \tag{6.122}$$

are presented in following figures.

### 6.4.1 Numerical Simulation Results

Parameter values, used to analyze the nonlinear dynamics of non-ideal system, are:

$$\begin{aligned}
\varepsilon = 0.10, \quad \varepsilon\zeta_1 = 0.10, \quad \varepsilon\beta_1 = 0.25, \quad \varepsilon\eta_1 = 0.60, \\
\varepsilon\delta_1 = \varepsilon\delta_2 = 0.40, \quad \rho = 1, \quad \mu_2 = 1.50,
\end{aligned} \tag{6.123}$$

and the initial conditions

$$x_1(0) = 0, \quad x_2(0) = 0, \quad x_3(0) = 0, \quad x_4(0) = 0, \quad x_5(0) = 0. \tag{6.124}$$

The resonance curves of the amplitude were obtained as follows: for each value of the control parameter the maximal amplitudes of oscillations are captured. Thus, for $\mu_1 = 0.690$ the region of resonance is reached and the amplitude of vibration of the system has maximal value.

Figure 6.16 illustrates the response of the system without energy harvesting, where the both piezoelectric coupling parameters are zero: linear parameter $\theta = 0.00$ and nonlinear $\Theta = 0.00$.

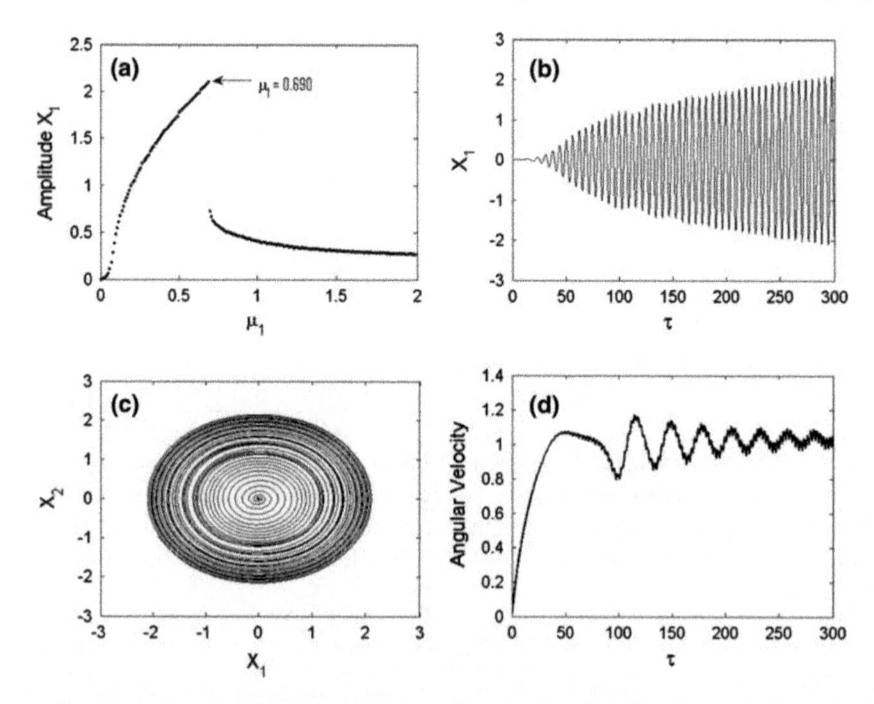


**Fig. 6.16** System response without piezoelectric coupling $\theta = 0.00$ and $\Theta = 0.00$: **a** resonance curve, **b** displacement of the beam, **c** maximum power harvested, **d** angular velocity (Iliuk et al. 2013a)

We separate the analysis of energy harvesting in two cases: linear energy harvesting and energy nonlinear harvesting. We seek the maximal amplitude of vibration in the graph of the resonance curve for different values of the control parameter $\mu_1$. The power harvested was obtained by Eq. (6.122).

### 6.4.2 *Linear Energy Harvesting*

In Fig. 6.17 the system response, when the piezoelectric coupling is linear with coupling parameters $\theta = 0.50$ and $\Theta = 0.00$, is considered. For the value of the control parameter $\mu_1 = 0.640$ the resonance region is reached (see Fig. 6.17).

Maximum amplitude of displacement and the maximum power harvested are determined. It is seen that for the control parameter $\mu_1 = 0.512$ the response of the system in the region of pre-resonance is stable but the power harvested is slightly less than the power harvested in the resonance region. For the control parameter $\mu_1 = 0.768$ the system response in the region of post-resonance has a small amplitude of vibration and a considerable reduction in power is captured, demonstrating that for this region the system begins to lose the stability.

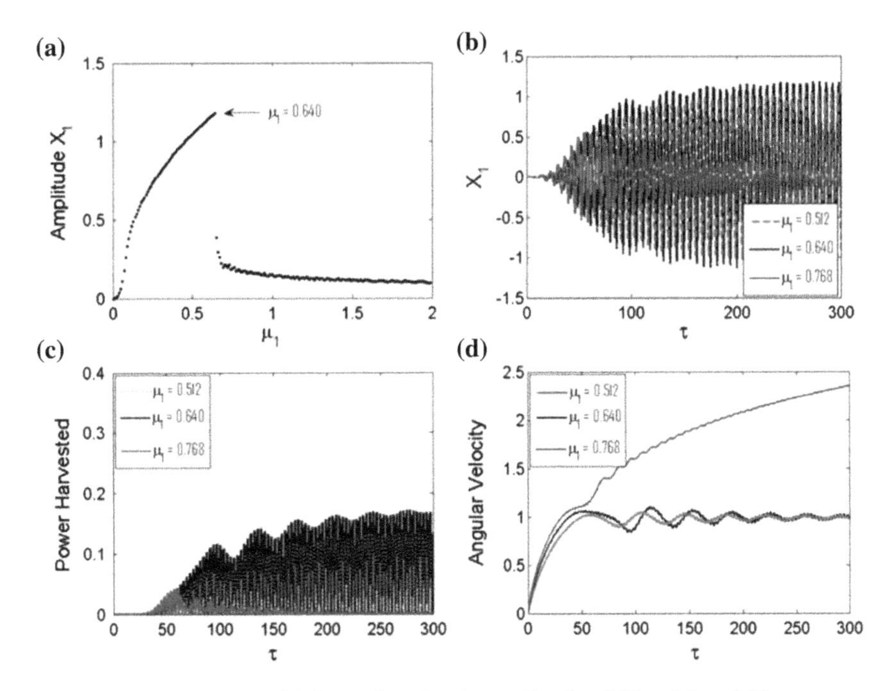


**Fig. 6.17** System response with linear piezoelectric coupling $\theta = 0.50$ and $\Theta = 0.00$: **a** resonance curve, **b** displacement of the beam, **c** maximum power harvested, **d** angular velocity (Iliuk et al. 2013a)

### 6.4.3 *Nonlinear Energy Harvesting*

In Figs. 6.18, 6.19 and 6.20 the system response for linear piezoelectric parameter $\theta = 0.50$ and various values of nonlinear piezoelectric coupling parameter ($\Theta = 0.50$, $\Theta = 1.00$ and $\Theta = 1.50$, respectively), is plotted.

Analyzing the obtained results it can be concluded that by increasing of the nonlinear coupling parameter, the maximum power harvested increases. Furthermore, the behavior of the system remains stable in the both regions: in the region of pre-resonance and in the region of resonance. The maximum power harvested is obtained for $\Theta = 1.50$ when the resonance region is reached for a lower value to the control parameter $\mu_1$(see Figs. 6.18c, 6.19c and 6.20c).

After passage through the resonance region the significant decrease of vibration amplitude occurs for all values of $\Theta$. Consequently, the harvested power is also reduced (Figs. 6.18a, 6.19a and 6.20a). In Figs. 6.18b, 6.19b and 6.20b the maximal amplitudes of the beam are plotted. In the post-resonance region the small amplitude is blue. In Figs. 6.18d, 6.19d and 6.20d, the angular velocity of the DC motor for various values of the control parameter is shown.

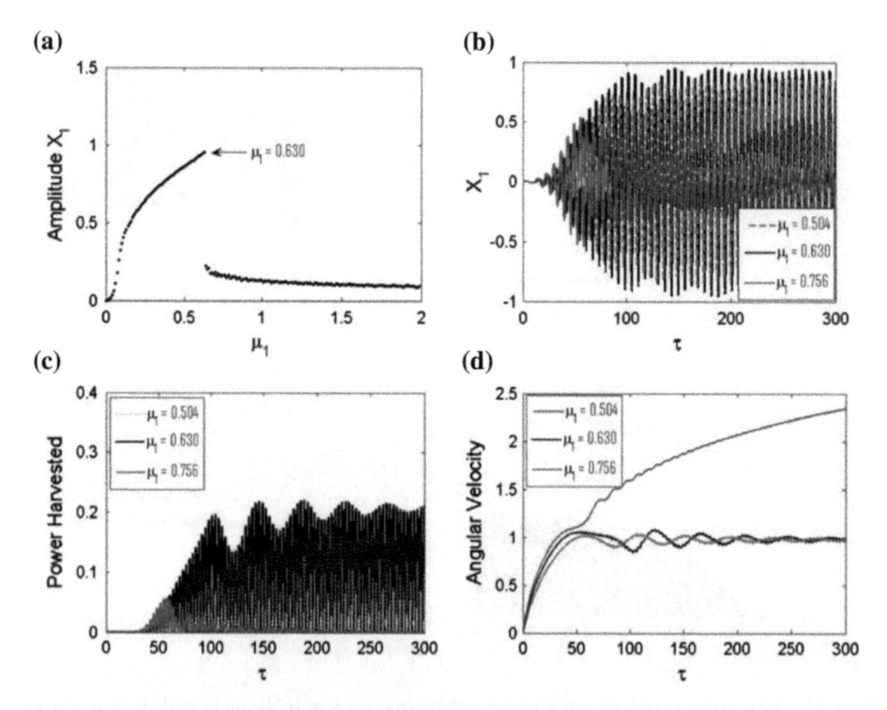


**Fig. 6.18** System response with nonlinear piezoelectric coupling $\theta = 0.50$ and $\Theta = 0.50$: **a** resonance curve, **b** displacement of the beam, **c** maximum power harvested, **d** angular velocity (Iliuk et al. 2013a)

If the control parameter is increased for 20% to the nominal value, the system passes the resonance region, reaches higher angular velocity but vibration amplitude is diminished.

### *6.4.4 Chaos in the System*

For the linear piezoelectric coupling parameter $\theta = 0.10$ and nonlinear piezoelectric coupling parameter $\Theta = 0.50$, chaotic behavior occurs (Fig. 6.21a, d). The simulation is done in 50,000 points.

In Fig. 6.21b the Sommerfeld effect and the region of instability after the resonant regime is detected. In Fig. 6.21c it is shown that for the mentioned parameter values positive Lyapunov exponent exists. It confirms the chaotic behavior.

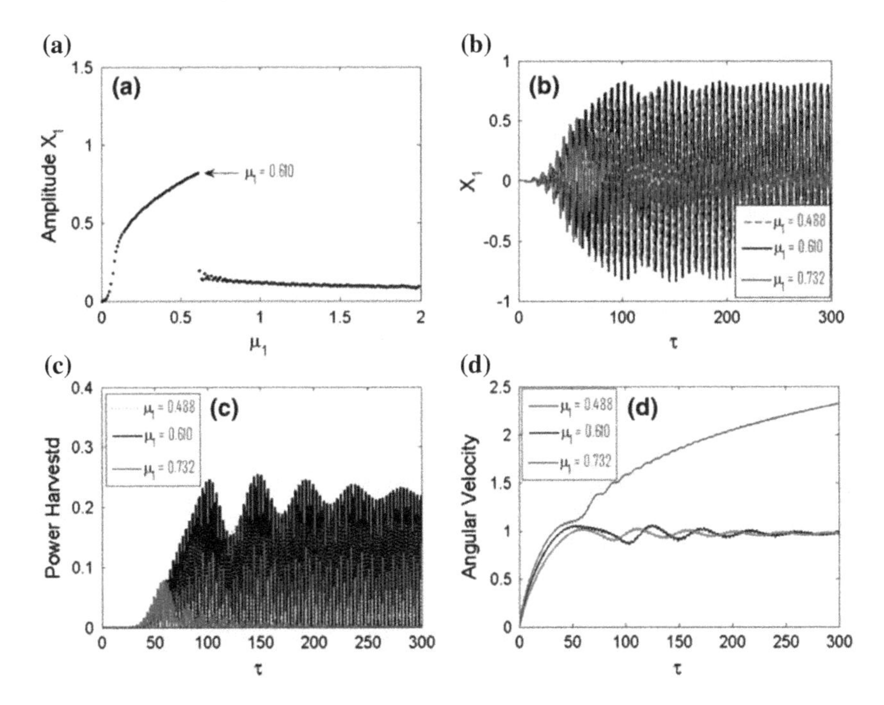


**Fig. 6.19** System response with nonlinear piezoelectric coupling $\theta = 0.50$ and $\Theta = 1.00$: **a** resonance curve, **b** displacement of the beam, **c** maximum power harvested, **d** angular velocity (Iliuk et al. 2013a)

## 6.4.5 *Control of the System*

A control technique has to be developed to transform the chaotic motion into the asymptotically stable periodic orbit motion. The control $u$ is introduced in the system (6.121):

$$
\begin{aligned}
x_1' &= x_2, \\
x_2' &= -x_1 - \varepsilon\beta_1 x_1^3 - \varepsilon\zeta_1 x_2 + \varepsilon\theta(1+\Theta\,|x_1|)x_5 \\
&\quad +\varepsilon\left(\delta_1 x_4' \cos x_3 - \delta_2 x_4^2 \sin x_3\right) + u, \\
x_3' &= x_4, \\
x_4' &= \varepsilon\,\mathcal{M}\,(x_4) + \varepsilon\eta_1 x_2' \cos x_3, \\
x_5' &= \left(\theta(1+\Theta\,|x_1|)x_1 - x_5\right)/\rho.
\end{aligned} \tag{6.125}
$$

The control function has the form

$$
u = -\kappa \tanh(\eta_1 x'), \tag{6.126}
$$

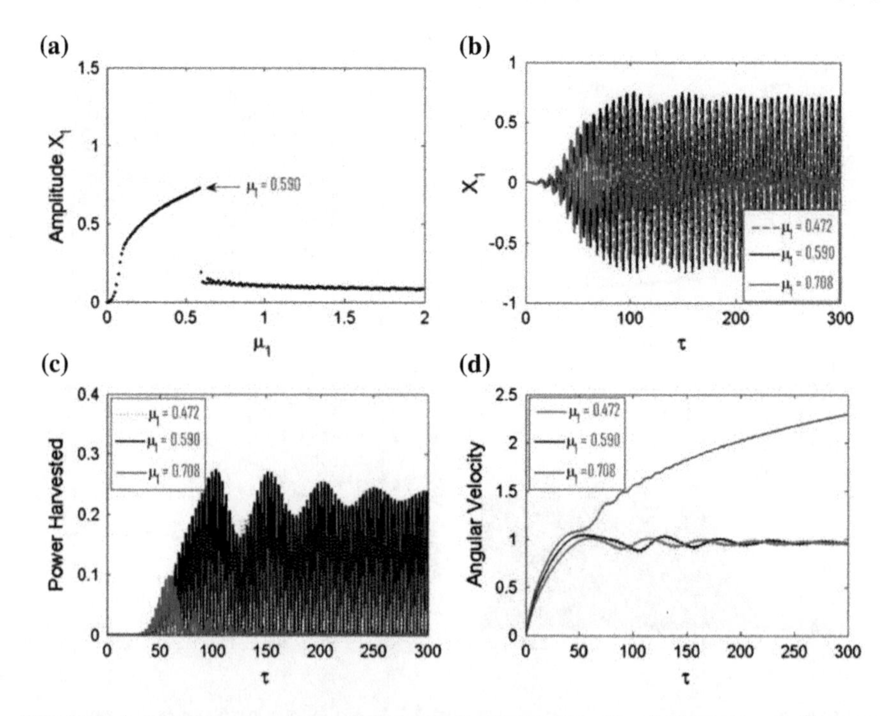


**Fig. 6.20** System response with nonlinear piezoelectric coupling $\theta = 0.50$ and $\Theta = 1.50$: **a** resonance curve, **b** displacement of the beam, **c** maximum power harvested, **d** angular velocity (Iliuk et al. 2013a)

where $\kappa$ and $\eta_1$ are positive parameters. After applying the proposed control technique, the dynamic response of the system is shown in Fig. 6.22.

The control keeps the system stable with one periodic solution. To compare the controlled and the uncontrolled systems, according to (6.122) the average power is calculated. After elimination of the transient of the system response the averaged harvested power is

$$P_{avg} = \frac{1}{T} \int_0^T P(\tau) d\tau. \tag{6.127}$$

Results plotted in Fig. 6.23 show the efficiency of the proposed control to bring the system which is vibrating along a chaotic trajectory to a stable periodic orbit.

The control responses are significant but the stabilized orbits are small. The average power of the controlled system is approximately 10% of the average power of the controlled system.

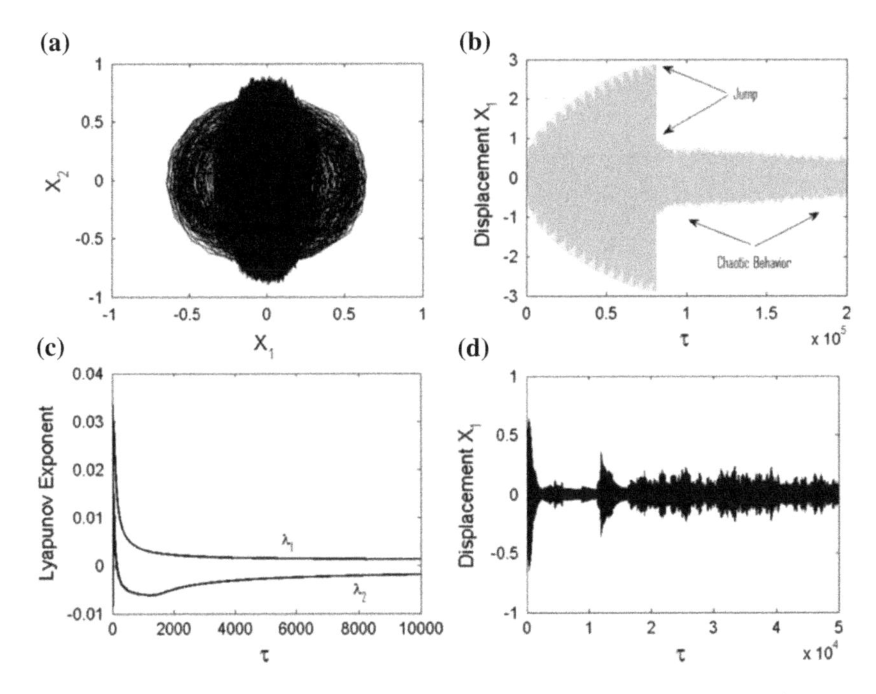


**Fig. 6.21** System response with nonlinear piezoelectric coupling $\theta = 0.10$ and $\Theta = 0.50$: **a** phase portrait, **b** displacement of the beam, **c** Lyapunov exponent ($\lambda_1 = 0.001214$, $\lambda_2 = -0.001945$), **d** displacement of the beam (Iliuk et al. 2013a)

## 6.4.6 Conclusion

Mathematical model of the non-ideal energy harvester shows that a good harvesting is achieved in the resonance region becoming stable due to inclusion of nonlinearity of the piezoelectric material. Based on numerical results it is concluded that the limited energy source interacts with system with piezoelectric coupling. Increasing the voltage in DC motor led the system to produce a good power response, especially in high-energy orbits in the resonance region. The Sommerfeld effect occurs in the system and a chaotic behavior was found in the post-resonance region. The power harvested along the time decreases because it causes loss of energy due to interaction between energy source and structure. Keeping the energy harvested constant over time is essential to make possible the use of energy harvesting systems in real applications. However, in the system chaotic motion is also possible. To achieve the requirement of constant energy harvested a control technique to stabilize the chaotic system in a periodic stable orbit is developed. The advantage of the presented control technique is that it does not require to find the analytical solution of the problem initially for obtaining of stable energy harvesting. However, due to reduced amplitude of vibration, even if stable, energy harvested is also decreased. The main advantage of

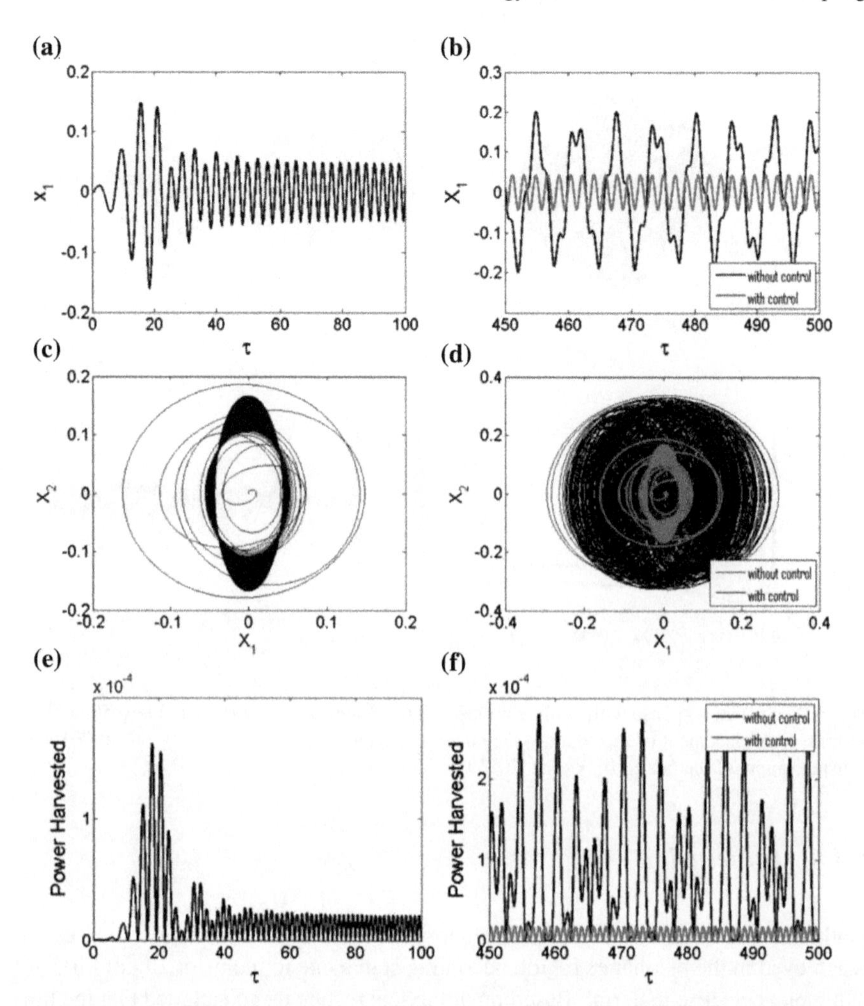


**Fig. 6.22** Dynamic response of the system after applying the control technique: **a** displacement of the beam after control, **b** displacement with and without control, **c** phase portrait after control, **d** phase portrait before and after control, **e** maximum power harvested after control, **f** maximum power harvested before and after control (Iliuk et al. 2013a)

using a control technique to stabilize the system is in reducing the need for complex filters rectifiers in the output of energy harvesting system (Iliuk et al. 2013a).

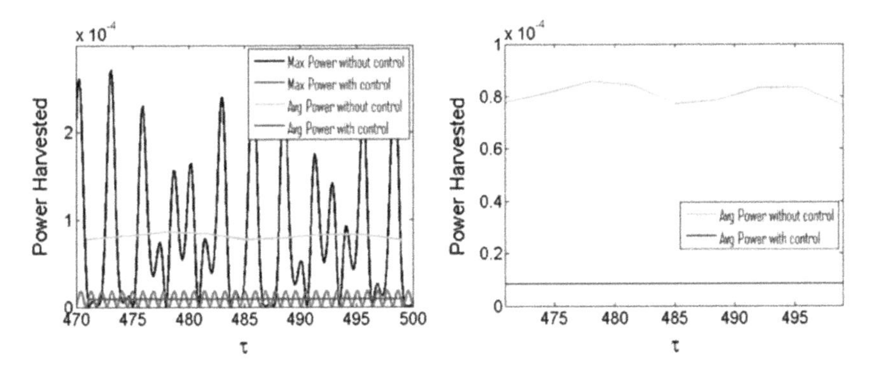


**Fig. 6.23** Comparison between average power harvested controlled and uncontrolled system. *Left* The maximum and average power harvested; *Right* Average power harvested (Iliuk et al. 2013a)

## 6.5 Non-ideal Portal Frame Energy Harvester Controlled with a Pendulum

A simple portal frame structure can be used as energy harvester. The procedure is as follows: A DC motor with limited power supply is mounted on a nonlinear structure with piezoelectric coupling. For some values of parameters the system has chaotic behavior. Varying the certain parameters of system the averaged power changes and transforms the chaotic motion into periodic orbit. It gives the energy harvesting. The control is passive and is realized with a pendulum.

The energy harvester contains a simple portal frame structure represented by the non-ideal bistable Duffing oscillator (Fig. 6.24) coupled with piezoelectric material.

The nonlinear piezoelectric coupling, assumed according to Triplett and Quinn (2009), gives the mechanical strain in the axial direction, while the voltage is in perpendicular direction. A pendulum is connected with the structure and represents a passive control which regulate the energy harvesting.

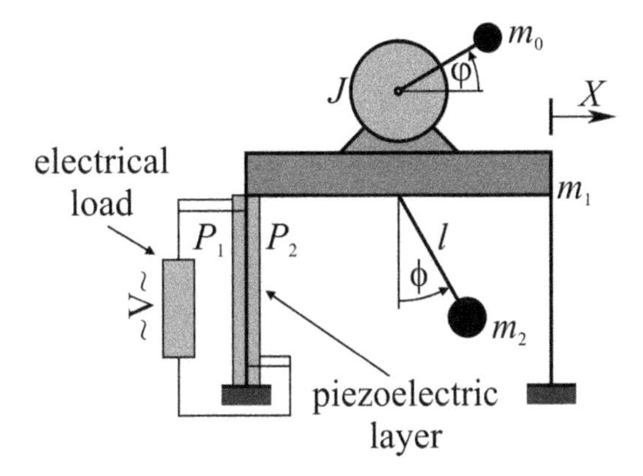


**Fig. 6.24** Portal frame for passive control (Iliuk et al. 2013b)

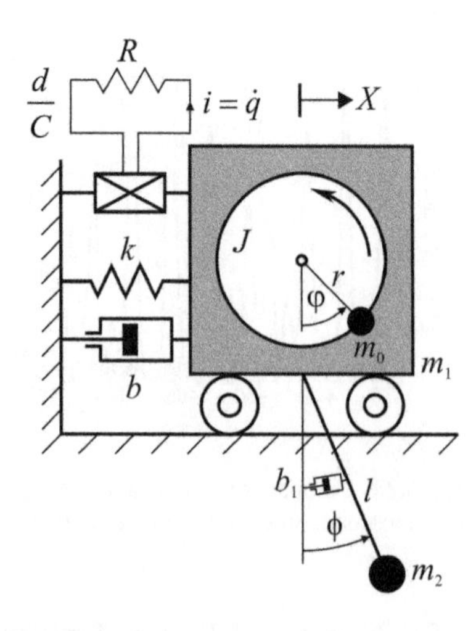


**Fig. 6.25** Model of the oscillator with passive control (Iliuk et al. 2013b)

Portal frame, non-ideal system coupled with piezoelectric material and a pendulum for passive control is shown in Fig. 6.24. Physical model of the controlled energy harvester is a nonlinear mass-spring-dashpot oscillator with piezoelectric coupling and a mathematical pendulum (Fig. 6.25). Mass of the system is $M = m_1+m_2+m_3$, where $m_1$ is the mass of the frame with motor, $m_2$ is the unbalance mass of the pendulum and $m_3$ is the unbalance mass of the DC motor. Eccentricity of the unbalanced mass is $d$. Length of the pendulum is $l$. Moment of inertia of the motor's rotor is $I = J + m_3 d^2$. The system has three-degrees-of-freedom. The generalized coordinates are the axial position coordinate $x$, the rotation angle of the DC motor $\varphi$ and the angle position of the pendulum $\phi$. The elastic force is assumed to be of bistable Duffing type: $-k_1 x + k_3 x^3$, where $k_1$ and $k_3$ are coefficients of linear and cubic rigidity. The torque of the motor is assumed to be of linear form

$$\mathcal{M}(\dot{\varphi}) = V_1 - V_2\dot{\varphi},$$

where $V_1$ relates to voltage applied across the armature of the DC motor and $V_2$ is a constant which depends on the model of DC motor. According to Triplett and Quinn (2009) the piezoelectric coupling to the mechanical system is $\frac{\theta(x)}{C}q$ where q is the electric charge, $C$ is the piezoelectric capacitance and $\theta(x)$ is a strain-dependent coupling coefficient. The voltage is represented with

$$V = -\frac{\theta(x)}{C}q + \frac{q}{C}, \tag{6.128}$$

i.e., in rewritten form

$$V = -R\dot{q}. \tag{6.129}$$

Mathematical model of the system shown in Fig. 6.25 is

$$(m_1 + m_2 + m_3)\ddot{x} + c\dot{x} - k_1 x + k_3 x^3 = m_2 d(\ddot{\varphi}\sin\varphi + \dot{\varphi}^2\cos\varphi) + \frac{\theta(x)}{C}q$$
$$+ m_3 l\dot{\phi}^2 \sin\phi - m_3 l\ddot{\phi}\cos\phi, \quad (6.130)$$
$$(J + m_2 d^2)\ddot{\varphi} = m_2 d\ddot{x}\sin\varphi + V_1 - V_2\dot{\varphi},$$
$$m_3 l^2\ddot{\phi} + m_3\ddot{x}l\cos\phi + m_3 gl\sin\phi = -b\dot{\phi}, \quad (6.131)$$
$$R\dot{q} - \frac{\theta(x)}{C}x + \frac{q}{C} = 0. \quad (6.132)$$

Let us introduce the following non-dimensional parameters

$$\zeta_1 = \frac{c}{\sqrt{k_1(m_1 + m_2 + m_3)}}, \quad \varepsilon = \frac{m_2}{m_1 + m_2},$$
$$\upsilon = \frac{q}{q_0}, \quad \mu_1 = \frac{V_1(m_1 + m_2)}{k_1(J + m_2 e^2)}, \quad \mu_2 = \frac{V_1(m_1 + m_2)}{k_1(J + m_2 d^2)},$$
$$\mathcal{M}(\varphi') = \mu_1\exp(-\mu_2\varphi'), \quad \beta = \frac{k_1}{M\omega^2} = 1, \quad \beta_1 = \frac{k_2 d^2}{M\omega^6},$$
$$\delta_1 = \frac{m_2\omega^2}{M}, \quad \delta_2 = \frac{m_2 l k_1}{(m_1 + m_2 + m_3)^2},$$
$$\rho_2 = \frac{V_1}{(J + m_2 d^2)\omega^2}, \quad \rho_3 = \frac{V_2}{(J + m_2 d^2)\omega},$$
$$\eta_1 = \frac{m_2 d^2}{(J + m_2 d^2)\omega^2}, \quad \rho = RC\sqrt{\frac{k_1}{m_1 + m_2}},$$
$$\rho_4 = \frac{m_2\omega^2}{d}, \quad \rho_5 = \frac{m_2 g}{\omega^2}, \quad \rho_6 = \frac{b_1}{l\omega}, \quad \varepsilon = \frac{1}{m_2 l}. \quad (6.133)$$

and the normalized coordinate and the time variable are

$$y = \frac{dx}{\omega^2}, \quad (6.134)$$

$$\tau = \omega t, \quad (6.135)$$

where

$$\omega = \sqrt{\frac{k_1}{m_1 + m_2 + m_3}}, \quad (6.136)$$

and $\varepsilon$ is the control variable which directly depends on mass and the length of the pendulum. Based on the result of Crawley and Anderson (1990), the piezocoupling coefficient is

$$\theta(x) = \theta_{lin}(1 + \Theta_{nel}\,|x|),$$

where $\theta_{lin}$ is the linear and $\Theta_{nel}\,|x|$ the nonlinear part. The dimensionless piezoelectric coupling coefficient, suggested by Triplett and Quinn (2009) is

$$\hat{\theta}(y) = \theta(1 + \Theta\,|y|), \tag{6.137}$$

where the piezoelectric coefficient is represented by a linear part $\theta$ and a nonlinear part defined by $\Theta$. The governing equations of motion reduce to

$$\begin{aligned} y'' - y + \beta_1 y^3 - \theta(1 + \Theta\,|y|)\upsilon &= -\zeta_1 y' + \delta_1\varphi'' \sin\varphi - \delta_1\varphi'^2 \cos\varphi \\ &\quad -\delta_2\varphi'' \cos\varphi + \delta_2\phi'^2 \sin\phi, \\ \varphi'' &= \eta_1 y'' \sin\varphi + \rho_2 - \rho_3\varphi', \\ \phi'' &= -\varepsilon\rho_4 y'' \cos\phi - \varepsilon\rho_5 \sin\phi - \varepsilon\rho_6\phi', \\ \rho\upsilon' &= \theta(1 + \Theta\,|y|)y - \upsilon. \end{aligned} \tag{6.138}$$

Introducing the new variables

$$x_1 = y, \quad x_2 = y', \quad x_3 = \varphi, \quad x_4 = \varphi', \quad x_5 = \upsilon, \tag{6.139}$$

the equations are rewritten in state space representation

$$\begin{aligned} x_1' &= x_2, \\ x_2' &= x_1 - \beta x_1^3 - \zeta x_2 + \varepsilon\theta(1 + \Theta\,|x_1|)x_7 + \delta_1 x_4' \sin x_3 - \delta_1 x_4^2 \cos x_3 \\ &\quad + \delta_2 x_6^2 \sin x_5 - \delta_2 x_6' \cos x_5 \end{aligned} \tag{6.140}$$

$$\begin{aligned} x_3' &= x_4, \\ x_4' &= \eta x_2' \sin x_3 + \rho_2 - \rho_3 x_4, \end{aligned} \tag{6.141}$$

$$\begin{aligned} x_5' &= x_6, \\ x_6' &= -\varepsilon\eta_1 x_2' \cos x_5 - \varepsilon\rho_5 \sin x_5 - \varepsilon\rho_6 x_6, \end{aligned} \tag{6.142}$$

$$x_7' = (\theta(1 + \Theta\,|x_1|)x_1 - x_7)\,/\rho. \tag{6.143}$$

The power harvested from the mechanical component is calculated according to (6.122) and eliminating the transient of the system response, the averaged power is calculated as (6.127).

### 6.5.1 Numerical Simulation

To analyze the nonlinear dynamic of non-ideal system, in the paper of Iliuk et al. (2013b) the following values for the parameters

$$\begin{aligned} &\varepsilon = 0.10, \quad \zeta_1 = 0.10, \quad \beta_1 = 0.20, \quad \eta_1 = 0.05, \\ &\delta_1 = \delta_2 = 8.373, \quad \rho = 1, \quad \mu_2 = 1.50, \quad \rho_2 = 100, \\ &\rho_3 = 200, \quad \rho_4 = 2.254, \quad \rho_5 = 0.23, \quad \rho_6 = 0.4, \\ &\theta = 0.20, \quad \Theta = 0.60, \quad 0 \le \varepsilon \le 1, \end{aligned} \tag{6.144}$$

and initial conditions

$$x_1(0) = 0, \quad x_2(0) = 0, \quad x_3(0) = 0, \quad x_4(0) = 0, \quad x_5(0) = 0, \tag{6.145}$$

are used.

The numerical simulation is carried out in the dimensionless time range $0 \le \tau \le 500$.

In Fig. 6.26 the phase portrait of the chaotic strange attractor, the time history diagram of the voltage, the Poincare map of the strange attractor and the maximum dimensionless power harvest for the case without control and piezoelement is plotted.

In Fig. 6.27 the bifurcation diagram and the average harvested power for the control parameter $\varepsilon$ in the interval [0,1] are plotted. For $\varepsilon = 0$ the system is without passive control. From the bifurcation diagram, given in Fig. 6.27a, the periodic and chaotic behavior are evident. Using the results presented in Fig. 6.27, the control strategy is defined. From Fig. 6.27b it is evident that the maximum average power is

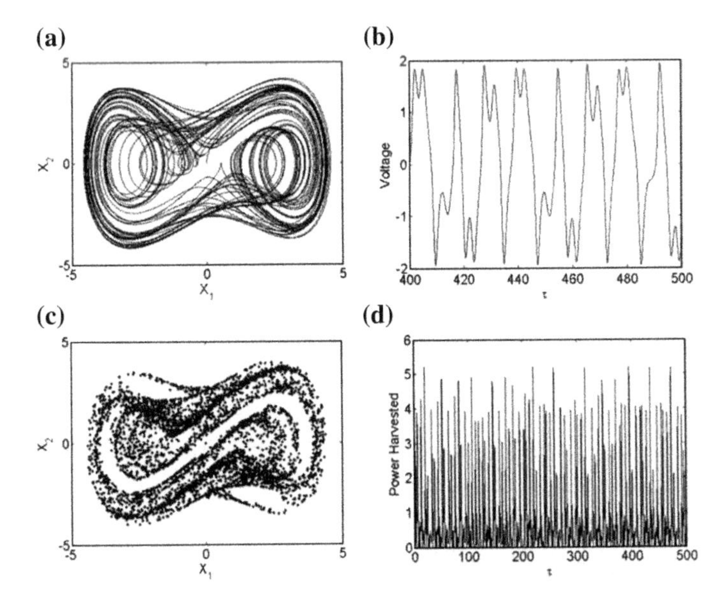


**Fig. 6.26** Portal frame without pendulum and piezoelectric element: **a** Phase portrait, **b** Theoretical voltage, **c** Poincare map, **d** Maximum power harvester (Iliuk et al. 2013b)

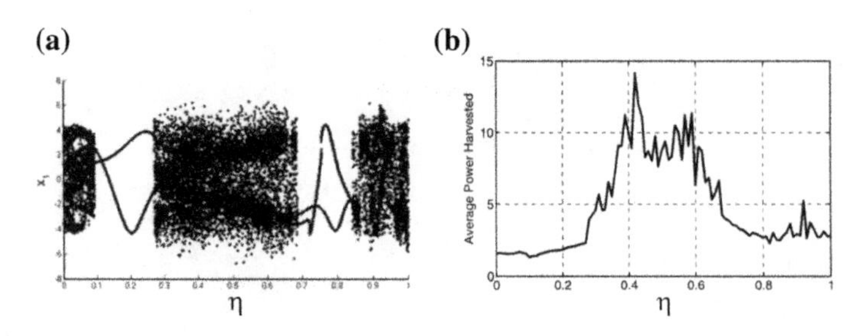


**Fig. 6.27** **a** Bifurcation diagram versus control parameter, **b** Average power versus control parameter (Iliuk et al. 2013b)

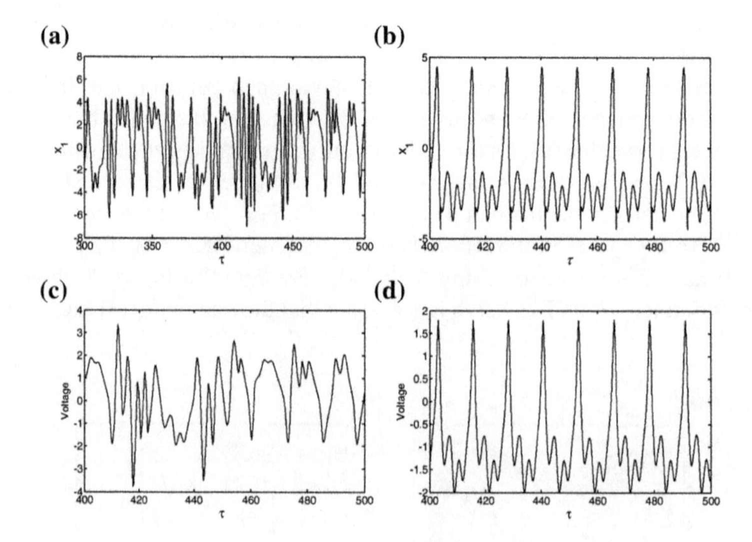


**Fig. 6.28** **a** Displacement - time diagram for $\varepsilon = 0.42$, **b** Displacement - time diagram for $\varepsilon = 0.70$, **c** Voltage - time diagram for $\varepsilon = 0.42$, **d** Voltage - time diagram for $\varepsilon = 0.70$ (Iliuk et al. 2013b)

obtained for $\varepsilon = 0.42$, while for $\varepsilon = 0.7$ the motion is periodical and the average power is without reduction.

In Fig. 6.28 the result of passive control is shown: (a) and (c) displacement and voltage diagrams for $\varepsilon = 0.42$, (b) and (d) displacement and voltage diagrams for $\varepsilon = 0.70$. Finally, in Fig. 6.29 the harvested power distribution for uncontrolled ($\varepsilon = 0$) and controlled systems (for $\varepsilon = 0.42$ and $\varepsilon = 0.7$) are compared. The calculated average power values are: $P(\varepsilon = 0) = 1.547$, $P(\varepsilon = 0.42) = 15.7$ and $P(\varepsilon = 0.70) = 3.662$.

The instantaneous energy exchange between the non-ideal system and the pendulum is defined as

$$E^*_{NIS_i} = \frac{E_{NIS}}{E_{NIS} + E_p},$$

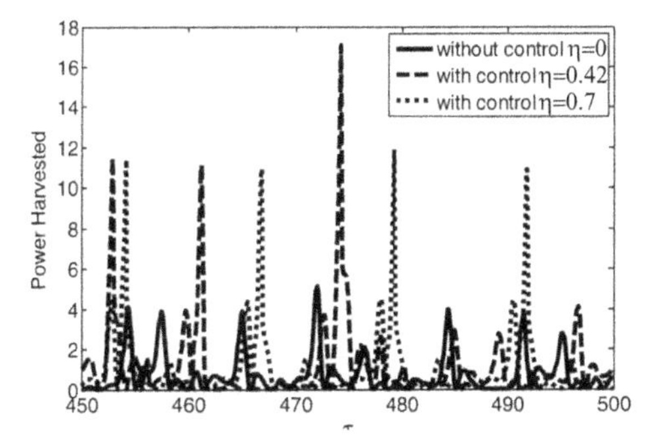


**Fig. 6.29** Power harvester for uncontrolled and controlled system (Iliuk et al. 2013b)

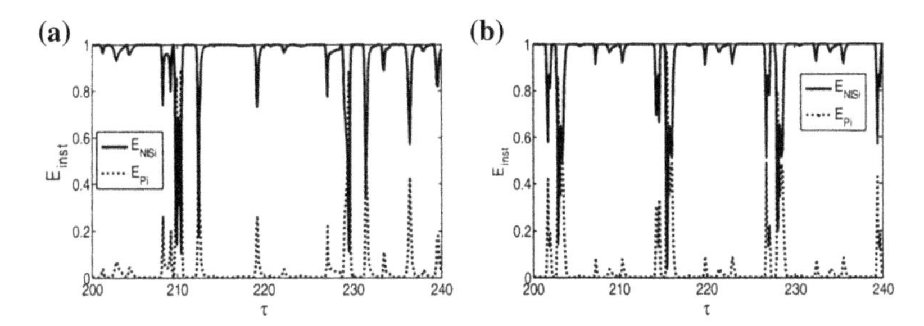


**Fig. 6.30** Time history diagrams for instantaneous energy for: **a** $\varepsilon = 0.42$, **b** $\varepsilon = 0.70$ (Iliuk et al. 2013b)

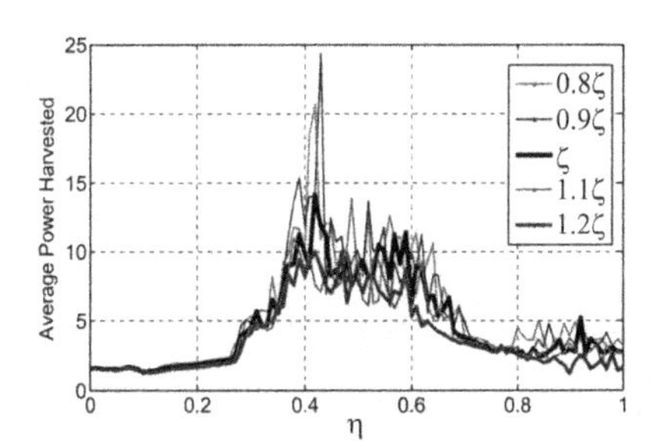


**Fig. 6.31** Average power harvester for various values of $\zeta$ (Iliuk et al. 2013a)

and

$$E^*_{p_i} = \frac{E_p}{E_{NIS} + E_p},$$

where

$$E_{NIS} = \frac{x_2^2 + x_4^2}{2}, \quad E_p = \frac{x_6^2}{2}.$$

The calculated values are plotted in Fig. 6.30. It is of special interest to investigate whether the damping and nonlinear stiffness coefficients have an influence on vibrational energy harvesting.

In Fig. 6.31 the influence of the damping coefficient $\zeta$ on the average power is plotted. It can be seen that the averaged power is sensitive on the variation of damping, but the maximal value is achieved for the control parameter $\varepsilon = 0.42$.

### 6.5.2 Conclusion

From the previous text it is obvious:

1. The energy harvesting may be improved by introducing of the passive control. Namely, with proper choice of the value of the control variable $\varepsilon$ the energy harvesting is optimized.
2. The main advantage of the technique of passive control is that it does not need any electronic component to control the system. The attached mechanical element, i.e., the pendulum, acts as the control device.
3. The suggested control system available to suppress the chaotic motion and to transform it into a periodical orbital motion.
4. The suggested control provides a way to regulate the energy captured to a desired operating frequency.

## References

Bendame, M., Abdel-Rahman, E., & Soliman, M. (2015). Electromagnetic impact vibration energy harvesters. In M. Belhaque (Ed.), *Structural Nonlinear Dynamics and Diagnosis* (Vol. 168, pp. 29–62), Springer Proceedings in Physics. Berlin: Springer.

Crawley, E. F., & Anderson, E. H. (1990). Detailed models of piezoceramic actuation of beams. *Journal of Intelligent Material System and Structure*, *1*, 4–25.

Cveticanin, L. (2010). Dynamics of the non-ideal mechanical systems: A review. *Journal of the Serbian Society for Computational Mechanics*, *4*(2), 75–86.

Cveticanin, L., & Zukovic, M. (2015a). Motion of a motor-structure non-ideal system. *European Journal of Mechanics A/Solids*, *53*, 229–240.

Cveticanin, L., & Zukovic, M. (2015b). Non-ideal mechanical system with an oscillator with rational nonlinearity. *Journal of Vibration and Control*, *21*(11), 2149–2164.

Cveticanin, L., Zukovic, M., & Cveticanin, D. (2017). Non-ideal source and energy harvesting. *Acta Mechanica*. Accepted for publication in 2017.

Daqaq, M. F., Masana, R., Erturk, A., & Quinn, D. D. (2014). On the role of nonlinearities in vibratory energy harvesting: a critical review and discussion. *Applied Mechanics Reviews*, *66*(4), 040801.

du Toit, N. E., & Wardle, B. L. (2007). Experimental verification of models for microfabricated piezoelectric vibration energy harvesters. *AIAA Journal*, *45*, 1126–1137.

Erturk, A., Hoffmann, J., & Inman, D. J. (2009). A piezomagnetoelastic structure for broadband vibration energy harvesting. *Applied Physical Letter*, *94*(25), 254102 209.

Iliuk, I., Balthazar, J. M., Tusset, A. M., Felix, J. L. P., & de Pontes, B. R, Jr. (2013a). On non-ideal and chaotic energy harvester behavior. *Differential Equations in Dynamical Systems*, *21*(1–2), 93–104.

Iliuk, I., Balthazar, J. M., Tusset, A. M., Piqueira, J. R. C., de Pontes, B. R., Felix, J. L. P., et al. (2013b). A non-ideal portal frame energy harvester controlled using a pendulum. *The European Physical Journal, Special Topics*, *222*, 1575–1586.

Iliuk, I., Balthazar, J. M., Tusset, A. M., Piqueira, J. R., de Pontes, B. R., Felix, J. L., et al. (2013c). A non-ideal portal frame energy harvester controlled using a pendulum. *The European Physical Journal, Special topics*, *222*(7), 1575–1586.

Iliuk, I., Brasil, R. M. L. R. F., Balthazar, J. M., Tusset, A. M., Piccirillo, V., & Piqueira, J. R. C. (2014a). Potential application in energy harvesting of intermodal energy exchange in a frame: FEM analysis. *International Journal oof Structural Stability and Dynamics*, *14*(8), 1440027.

Iliuk, I., Balthazar, J. M., Tusset, A. M., Piqueira, J. R., de Pontes, B. R., Felix, J. L., et al. (2014b). Application of passive control to energy harvester efficiency using a nonideal portal frame structural support system. *Journal of Intelligent Material Systems and Structures*, *25*(4), 417–429.

Litak, G., Friswell, M. I., Kwuimy, C. A. K., Adhikari, S., & Borowiec, M. (2012). Energy harvesting by two magnetopiezoelastic oscillators with mistuning. *Theoretical and Applied Mechanical Letter*, *2*(4), 043009.

Litak, G., Friswell, M. I., & Adhikari, S. (2016). Regular and chaotic vibration in a piezoelectric energy harvester. *Meccanica*, *51*(5), 1017–1025.

Rocha, R. T., Balthazar, J. M., Tusset, A. M., Piccirillo, V., & Felix, J. L. P. (2016). Comments on energy harvesting on a 2:1 internal resonance portal frame support structure using a nonlinear energy sink as a passive controller. *International Review of Mechanical Engineering (IREME)*, *10*(3), 147–156.

Stephen, N. G. (2006). On energy harvesting from ambient vibration. *Journal of Sound and Vibration*, *293*(1), 409–425.

Tereshko, V., Chacon, R., & Preciado, V. (2004). Controlling chaotic oscillators by altering their energy. *Physica Letters A*, *320*, 408–416.

Triplett, A., & Quinn, D. D. (2009). The effect of non-linear piezoelectric coupling on vibration-based energy harvesting. *Journal of Intellignet Material System and Structure*, *20*(16), 1959–1967.

# Chapter 7
# Instead Conclusions: Emergent Problems in Nowadays and Future Investigation

Nowadays the majority of engineering systems have, at least, one electromechanical sub-system in their composition. These systems fall into three groups: the conventional electromechanical systems, the micro electromechanical systems and the nano electromechanical systems. Note that micro and nano electromechanical system technologies are still in their infancies, with global research and development actively under way. Often many practical electromechanical devices are discussed in the context of simple lumped mechanical masses, electric and magnetic circuits. However, the interest is to discuss the existence of a full interaction between mechanical and electrical field quantities and to investigate the relevant dynamics, in order to predict the structural response due to the excitations, in macro and micro scales. In general, all of the studies, involving electromechanical vibrating systems, were based on assumptions that the external excitations are produced by an ideal source of power, with prescribed time history, magnitude, course and frequency, or in random problems with prescribed characteristics. However, the fact is that the excitation sources are non-ideal: they have always limited power supply, limited inertia and their frequencies vary according to the instantaneous state of vibrating (oscillating) system. The behavior of the vibrating systems departs from the ideal case, as power supply becomes more limited. It is noted that, in non-ideal systems, near resonance, an increase of power will usually be accompanied by an increase of amplitude of oscillations, without significant increase in frequency. Only after the amplitude of oscillation reaches the maximum there will be a significant change in frequency. The dynamic process is called Sommerfeld effect. There is a problem of passage through resonance of unbalanced equipment: the need of power may not be able to be supplied with an operational speed higher than the lower frequencies of vibration of the supporting structure. Hence, for this property, a large part of its energy is used in shaking the structure and not in accelerating the rotation of the shaft. Thus, the introduction of real torque–speed curves for non-ideal motors yields the system as nonlinear and capable of multiple steady-state periodic motions whose stability must be assessed: that is the vibrations provide an "energy sink". In summary, it may be

© Springer International Publishing AG 2018

L. Cveticanin et al., *Dynamics of Mechanical Systems with Non-Ideal Excitation*, Mathematical Engineering, DOI 10.1007/978-3-319-54169-3_7

said that jump phenomena and the increase of power required by a source operating near resonance are manifestations of a non-ideal energy source. This referenced phenomenon suggests that the vibrational responses provide an energy sink, and thus vibrating the structure rather than to operate the machinery. One of problems often faced by designers is how to drive a system through resonance and avoid this kind of energy sink.

In this book, main properties of non-ideal systems have been reviewed, such as the Sommerfeld effect, i.e., jump phenomena and the increase in power required by a source operating near resonance; the possibility of occurrence of saturation phenomenon, i.e., a transfer of energy to low frequency high amplitude mode; and the existence of regular (periodic) motion and irregular (chaotic) behavior, depending on the value of control parameters (voltage of a DC motor). The purpose of the book is twofold: in one hand to give the explanation for motion in non-ideal systems mainly concentrated in non-ideal vibrations and in the other hand to provide an overview of the main engineering applications, analyzing both physical phenomena involved and the adequate methodologies to deal with them.

In this chapter the novelties in considering non-ideal vibrating systems are presented.

Nowadays, instead of the DC electric motors which affect the base-excitation, the motion is obtained by electro-mechanical shakers with limited power supply. It is concluded that the electromechanical shaker has the same dynamic properties as the DC motor, i.e., non-ideal energy source. Already a flexible portal frame excited by an electrodynamic shaker was considered. The electro-mechanical device consists of an electric system magnetically coupled to a mechanical structure. In the system the occurrence of the dynamic jump is verified, which is an important characteristic of nonlinear dynamic system. Besides, an important point associated with the proposed system is the energy transfer between the shaker and the vibrating structure. Using this property of the system, the electro-mechanical shaker may be included into new type of micro electro-mechanical systems (MEMS) such as, micro-gyroscopes, electrostatic transducers and into some energy harvesters.

One of devices with electro-mechanical shaker is the tuning fork gyroscope. In the tuning fork micro-gyroscope energy recovery or capture occurs. The gyroscope is assumed to consists of two vertical posts embedded on a mass of suspension excited by an electrodynamic shaker driven by a sinusoidal voltage. The tuning fork beam is modeled by two inverted pendulums. If the beams oscillate in phase, the excitation can lead to vibration. Due to internal resonance between the vibration modes the energy transfer occurs from vertical toward horizontal direction. Future investigation of the tuning fork gyroscope properties is necessary.

As it is presented in Chap. 6, in the last years, it has been seen a need of an energy source smaller than the usual and more efficient, for design of vibrating systems based on new technologies. With that, the research about energy harvesting has increasing substantially. To build energy harvesting devices, many researchers have concentrated their efforts on finding the best configuration for these systems and to optimize its power output. In the process of energy harvesting, the electrical energy is obtained through the conversion of mechanical energy created by an ambient vibration source by a type of transduction, for example, as a piezoceramic thin film. Based on the results given in Chap. 4 for two degrees-of-freedom structure it is concluded that the energy harvesting is possible to be achieved applying the saturation phenomena which is evident for the two-to-one internal resonant case. When the excitation frequency reaches near resonance conditions with the natural frequency 2, the amplitude of this mode grows up to a certain level and then it saturates, transferring the surplus of vibrating energy, as it "spilled over", to the other mode, which experiences a sudden increase in its amplitude. Namely, the energy pumped into the system through one of the modes is partially transferred to the other mode. Energy is transformed from low amplitude high frequency motion into high amplitude low frequency motion. This is the so-called saturation phenomenon. The most recent investigation show that the phenomenon of saturation is suitable for harvesting. One of the most promising device to harvest energy is the piezoelectric material. If in the two degree-of-freedom portal frame support structure, considering a nonlinear piezoceramic coupled to a column, the vertical motion of the mid-span of the beam possesses its natural frequency twice of the column natural frequency, and if is based-excited in the vertical direction with same frequency of the beam, saturation phenomenon occurs, transferring the surplus energy of the beam vibration to the column, acquiring higher amplitude than in the beam. Further investigation are suggested.

A constantly sustained energy harvesting is essential for using harvester devices in real applications; for this, a control strategy is required. In the non-ideal systems usually the control is introduced to suppress the vibrations and to reduce the transient responses.

The saturation phenomena may be applied in the nonlinear control technique in non-ideal systems, as it is suitable to suppress the high amplitude of the first-mode vibration. A two-to-one internal resonance condition is maintained between the plant and the controller. In this case, energy is transferred completely from one part of the combined system to the other. Thus, when the plant is forced at resonance, this energy transfer mechanism limits the response of the plant.

Usually the passive control specified through energy pumping is applied. The passive control is obtained by means of a nonlinear substructure with properties of nonlinear energy-sink. It is known that passive control leads the non-ideal oscillatory system to a stable periodic orbit, allowing a more efficient energy harvesting due to the higher peak-to-peak amplitude of oscillation mean value. Besides, it eliminates the need of an active micro-controller to stabilize the system in a periodic orbit, improving the energy budget (harvested vs. expended). The application of passive control is successful through the suppression of the chaotic motion, leading the system to a periodic orbit with stable amplitude of vibration, without damaging the structure.

Considering nonlinear energy sink as vibration absorber, it is possible to reduce drastically the amplitudes of oscillations of the system and reduce the Sommerfeld effect in the passage resonance region. Furthermore, the proposed absorber of nonlinear type is quite effective to reduce amplitude, Sommerfeld effect and jump phenomenon of non-ideal system while the linear absorber is not. Namely, based on an extension of the ideal system and taking into account that in a vibrating system with small low viscous dissipation the energy initially imparted from the primary sub-system can be transferred to the nonlinear energy-sink, reducing the amplitudes of vibrations of the system and eliminating or reducing the occurrence of the Sommerfeld effect, inside and outside resonance regions, respectively, it is suggested to develop a strong nonlinear vibration absorber. Vibration absorption in resonance, nonlinear and chaotic motions have to be analyzed for a simple portal frame excited by a non-ideal power source connected to a device, which renders descriptions that are close to engineering situations encountered in practice.

Recently, a wide group of new materials are developed. Functionally graded materials and other composites, shape memory alloys, magneto rheological materials, dielectric elastomers, polymers and thin-films have shown great potential for applications in all engineering fields. They may form intelligent and adaptive material systems and structures whose fundamental characteristic is their ability of adapting to environmental conditions. One of the new class of materials with promising applications in structural and mechanical systems is the shape memory alloy. Shape memory alloy consist of a group of metallic materials that demonstrate the ability to return to some previously defined shape or size when subjected to the appropriate thermal procedure. A shape memory alloy may represent the connection between the mass and the rigid support in the non-ideal system. It is believed that the shape memory alloy part may turn around a fixed point in structures, i.e., to rotate around an axis. This kind of motion is called slewing motion. At the moment, slewing motions are present in many applications in aerospace and industrial robots. In this kind of systems it is evident that the lighter the structure, the better is the efficiency of the system. This is the case in satellite appendages and aerospace robotic manipulators, for example. It is expected that the shape memory alloy may be incorporated into an actuator for active control of vibrations of the flexible structural beam driven by a DC motor, i.e., in the non-ideal system.

Recently, the tendency is the various types of active control systems developed for ideal systems to be adopted for non-ideal ones. The active control with tuned liquid column dampers is suggested for practical application. Dampers are U-tubes filled with some liquid, acting as an active vibration damper in structures of engineering interest like buildings and bridges. The damper may be mounted on the structural portal frame which contains a horizontal beam and two identical columns with an unbalanced DC motor of limited power supply (non-ideal source). It is necessary to analyze the control of vibration with the tuned liquid column damper.

Another type of active controller is the so called magneto-rheological damper which is suitable to bring the system into the desired orbit. It is possible to control the force of the magneto-rheological damper by controlling the applied electric current in the coils of the damper. Namely, the control force of the damper is a function of the electric current applied in the coil of the damper, which is based on the force given by the controller and on the velocity of the damper piston displacement. The attenuation of the jump phenomena, associated with the Sommerfeld effect introduced by the nonlinearity of a magnetic rheological damper, is expected in a non-ideal vibrational system, excited by a DC motor modeled as limited power source.

In spite of positive results obtained in passive and active control separately, it is suggested to use a combination of passive control (energy pumping) and active control. It is predicted that the strategy of combining the two control techniques seems to be more effective in suppressing oscillations of the main system than using each of the strategies separately. It remains to be proved.

In order to improve the design of previously mentioned structures and devices and to predict the structural response due to the excitations further investigation in dynamics of non-ideal systems is necessary. A lot of vibrating phenomena of real systems are not explained yet. Using the linear theory a significant number of real system properties is neglected and omitted. It is important to introduce nonlinear characteristics into the mathematical models of these systems. For example, many different electromechanical coupling mechanisms have been developed for harvesting devices and because of the constitutive laws of piezoelectric materials, the role of nonlinearity, in the electromechanical coupling of the design of energy harvesting system must be taken into account. Despite of great advances of the vibrating (oscillatory) theory, some kind of vibrations that are still met cannot be well explained by the current vibration theory.

Finally, the mentioned emergent and actual problems in non-ideal systems have to be studied and solved in near future. New phenomena and emergent areas addressed to structures supporting unbalanced machines capable of a limited power output and systems with electromechanical shakers need to be investigated. Research on the full interaction between non-ideal sources and their supports is eligible. Based on these results the direction of future investigation in the non-ideal vibration systems has to be defined.

# Index

**A**
Active control, 224
Amplitude-frequency function, 26, 39
Analytical solving procedure, 34
Asymptotic solution
  first approximation, 54
Asynhronous AC motor, 23
Ateb function
  cosine, 115
    Fourier expansion, 78
  period, 76, 115
  properties, 116
  sine, 114
    Fourier expansion, 78
Averaging procedure, 14, 26, 56, 73

**B**
Beta function, 76
  complete, 113
  incomplete, 113, 114
Bifurcation diagram, 42, 111, 112, 215

**C**
Chaos
  control
    in harvester, 207
Chaos control, 44
Chaotic motion, 43, 102, 112
  control, 103
  nonlinear harvester, 206
Characteristic point, 58, 66, 80, 98
Control
  active, 5
  passive, 4, 213, 216
Cutting force, 148, 152, 163
Cutting mechanism, 141
  structure, 143
  with elastic support, 155
  with ideal forcing, 161
  with non-ideal forcing, 163
  with rigid support, 147

**D**
DC series wound motor, 52
Discontinual elastic force, 31
Dissipation function, 61
Dissipative function, 92, 157, 188
  for two degree-of-freedom, 123
  of cutting mechanism, 147

**E**
Electro-mechanical force, 175
Electro-mechanical shaker, 222
Electro-mechanical system
  conventional, 221
  micro, 221, 222
  nano, 221
Energy exchange, 216
Energy harvester
  linear, 174, 197, 204
  nonlinear, 174, 197, 205
  on vibration, 218
Energy harvesting, 173
  device, 222, 223
Energy sink, 3, 18, 221, 223

**G**
Generalized force, 12, 24, 147, 157
  of motor torque, 190

© Springer International Publishing AG 2018

L. Cveticanin et al., *Dynamics of Mechanical Systems with Non-Ideal Excitation*, Mathematical Engineering, DOI 10.1007/978-3-319-54169-3

Generalized reactive force, 24

**H**
Harmonic balance method, 107
Harmonic excitation, 176
Harvested energy
  averaged, 183
Harvested power, 178, 196, 205
  averaged, 178, 196
Harvesting system
  model, 187
History diagram, 215

**J**
Jump phenomena, 3

**K**
Kinetic energy, 61
  Duffing oscillator
    linear motor torque, 92
  for two degree-of-freedom system, 123
  mass variable system, 24
  of cutting mechanism, 147
    on elastic support, 157
  of harvesting system, 189
  of one degree-of-freedom system, 11

**L**
Lagrange equation, 92
Lagrange's equation of motion, 189
Linear oscillator
  linear non-ideal source, 9
Linstedt–Poincare method, 179
Lyapunov exponent, 43, 102, 111
Lyapunov spectrum, 111

**M**
Magneto-rheological damper, 224
Motor torque, 190
  exponential, 201
  linear, 15, 72, 106, 212
  nonlinear, 51
  of cubic type, 68
  virtual work, 72
Motor-oscillator system, 11

**N**
Non-ideal
  continual system, 4
  in three degrees-of-freedom system, 4
  one degree-of-freedom system, 2
  two degrees-of-freedom systems, 3
Non-ideal source, 1
Non-ideal system, 1
Non-ideal vibrating system, 1
Non-stationary motion, 56
  in cutting mechanism, 168

**O**
Order of nonlinearity, 60, 71
Oscillator
  bistable Duffing, 105
  Duffing, 91
  pure nonlinear, 60, 71
  with clearance, 30
  with strong nonlinearity, 51
Oscillator with variable mass, 22

**P**
Passive control, 223
Period doubling bifurcation, 102
Phase portrait, 215
Phase space, 100
Physical analogy, 18
Piezoelectric element, 174
Piezoelectric material
  constitutive equation, 175
  non-dimensional model, 177
Piezoelectricity
  linear, 182
  nonlinear, 185
Piezoelement
  linear, 198
  nonlinear, 198
Poincare map, 100, 215
Portal frame
  in energy harvester, 211
Potential energy
  Duffing oscillator
    linear motor torque, 92
  for two degree-of-freedom, 123
  of cutting mechanism
    on elastic support, 157
  of nonlinear system, 188
  of one degree-of-freedom system, 11
  of pure nonlinear oscillator, 61

**Q**
Quenching of amplitude, 110

**R**
Resonance
  primary, 3
  principle parametric, 107
  subharmonic, 3
Resonance frequency
  in two orthogonal direction, 134
    equal, 130
Resonant case, 54

**S**
Saturation phenomenon, 223
Shape memory alloy, 224
Slewing motion, 224
Slider-crank mechanism
  double, 142, 145
  eccentric, 145
  simple, 141, 145
Solution
  non-trivial, 110
  semi-trivial, 109
Solution in the form of Ateb function, 74
Sommerfeld effect, 2, 17, 41, 67, 73, 85, 196
  criteria, 60
  pure nonlinear oscillator
    linear motor torque, 81
    motor torque of cubic type, 70
  suppression, 59, 68, 70, 82, 99
  two degrees-of-freedom system, 3
Stability analysis, 57
Stability of motion, 21
Steady state
  amplitude, 65, 128, 182, 195
  equation, 79
  phase, 66, 128, 182, 195
Steady state motion, 127, 193
  in cutting mechanism, 165
  stability
    two degree-of-freedom, 129
Steady state solution, 14, 21, 57, 65, 95
  stability, 96
  system with clearance, 38
Strange attractor, 215

**T**
Transient motion, 33, 214
  first approximation, 55
  in resonant case, 79
Tuning fork gyroscope, 222

**U**
U-tube damper, 224
UnitStep function, 148

CPSIA information can be obtained
at www.ICGtesting.com
Printed in the USA
LVHW081004171119
637602LV00002B/68/P